DICTIONARY OF DRYING

DICTIONARY OF DRYING

CARL W. HALL

College of Engineering
Washington State University
Pullman, Washington

MARCEL DEKKER, INC.   New York and Basel

Library of Congress Cataloging in Publication Data

Hall, Carl W
Dictionary of drying.

Bibliography: p.
1. Drying--Dictionaries. I. Title.
TP363.H274 660.2'8426 79-14783
ISBN 0-8247-6652-0

MARCEL DEKKER, INC.

270 Madison Avenue, New York, New York 10016

Current printing (last digit):

10 9 8 7 6 5 4 3 2 1

PRINTED IN THE UNITED STATES OF AMERICA

PREFACE

Drying is the oldest method of preservation--primarily using solar energy. The use of solar energy is still probably the most important method on the basis of energy used. Moisture removal from products has gone from a very primitive art to very sophisticated operations using both art and science. Research and practice have evolved a vocabulary on drying which, along with some special words and phrases, draws heavily on many other subject matter fields. Fields of food, crops, lumber, livestock, fuel, coal, heat and mass transfer, energy, industrial processes, and many more have words which have become common in drying.

The *Dictionary of Drying* is based on the literature in the field and the experience of the author. An effort has been made to include those words encountered in reading the literature. Also included is a summary of recommendations on types of dryers and procedures for drying. A listing of bibliographies on drying and book references in the field is a unique feature. The processes described are not intended to provide all of the information necessary for design of dryers or specific products. The design of a dryer for a specific product is a complicated and involved subject requiring investigation of past and present practices, literature in the field, and objectives to be accomplished with the design. The material, however, should assist a designer in developing a new dryer as well as improving old ones. In addition to the 4,000 entries on drying, there are considerable cross entries to assist the reader in tying together a more complete picture than might be received from reading one entry only.

Special acknowledgment is given to the following individuals who reviewed the book and made valuable suggestions. Dr. Fred W. Bakker-Arkema, Michigan State University, East Lansing, MI; Mr. Wayne Bennett, Manager, Lumber Research and Development, Weyerhauser Company, Tacoma, WA, (sections on wood and lumber); Professor Donald Brooker, University of Missouri, Columbia, MO; Mr. John F. Collins,

P.E., Aeroglide Corporation, Raleigh, NC; Dr. Richard Crain, Jr., Professor and Chairman, Mechanical Engineering, Washington State University, Pullman, WA; Dr. Glenn E. Hall, Andersons, Maumee, OH; and Mr. James F. Kelly, President, Aeroglide Corporation, Raleigh, NC.

Thanks to the following persons who contributed materials for special sections. George T. Austin, A. P. Cobra, H. Göhlick, W. Hammer, Hironichi Hayashi, Gad Hetsroni, Arne Hilmerson, Ernest H. Kidder, Roger Martinot, N. N. Mohsenin, E. H. Oksanen, Duvilio A. Ometto, Stanislaw Pabis, Russell L. Perry, Sverker Person, T. Tougaard Pederson, Jorge Rodriguez-Arias, E. Paul Taiganides, and Jaw-Kai Wang.

Several organizations and companies furnished illustrations and tables which are identified in the book.

Carl W, Hall

DICTIONARY OF DRYING

A

A A prefix meaning not, such as asymmetric, which means not symmetric; aperiodic, not periodic; acyclic, not cyclic.

Abalone Drying The abalone is an ear-shaped shell marine animal containing edible meat commonly harvested on the Pacific Coast, which is dried in the sun after removing the shell, washing, and boiling.

Ablimation The process of condensation of vapor directly to ice without going through the liquid phase. Ablimation is a process which occurs in a direction opposite to sublimation, as commonly defined. Also see *Sublimation*.

Abscission The separation of a body part, such as seeds, fruits, and leaves, from the plant either naturally or artificially. A special layer of cells weakens and dies as a result of natural aging, or through the use of a chemical to induce aging or death. Plants may be treated to cause abscission previous to drying so as to control growth of plant, timing of harvest, or moisture content of the product.

Absolute Density The mass per unit volume of a substance, m/v, which has dimensions of M/L^3, and for which the value is for the actual substance involved, not including free space which might be between particles.

Absolute Filter See *Filter*.

Absolute Humidity The weight of water vapor, usually in pounds or grains per pound of dry air, or kilograms per cubic meter or kilograms per kilogram of dry air for particular temperature and pressure conditions.

Absolute Pressure Controller An instrument used in the operation of a vacuum pan or evaporator to control the vacuum on the product. The absolute pressure is independent of the atmospheric pressure. Absolute pressure measuring elements use a bellows or U-tube and do not have to be compensated for atmospheric conditions because these units measure from zero absolute pressure. These units are preferred for evaporator control because the boiling temperature depends on the absolute pressure. Also see *Absolute Pressure Gage; Air Loading; Evaporator; Vacuum Pan.*

Absolute Pressure Gage An instrument for measuring the pressure, usually in a confined space, based on zero pressure or a perfect vacuum. Two major elements are used: bellows and U-tube (with closed end)(Fig. A-1). At sea level the standard absolute pressure is 14.7 psi or 760 mm of Hg greater in numerical value than the gage pressure.

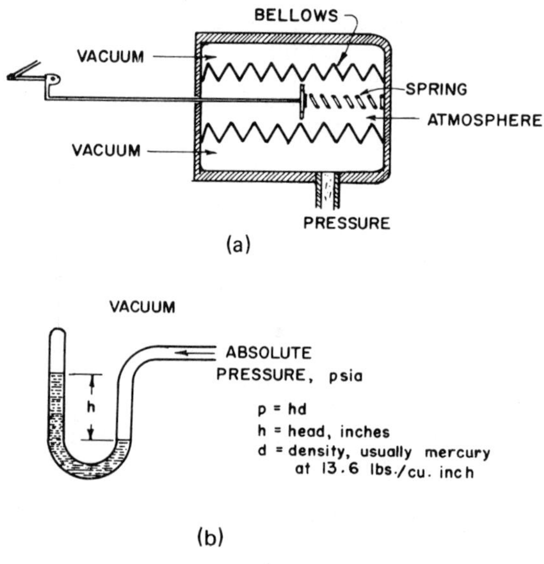

Figure A-1. Absolute pressure gage. *From Hall, C. W., and Hedrick, T. I. Drying of Milk and Milk Products (1971). Used with permission of Avi Publishing Co., Westport, CT.*

Absolute Temperature A scale of temperature based on a zero value at absolute zero. Absolute temperature values are required for problems involving multiplication or division by temperature (T) or temperature ratios. The common absolute temperature scales are the Kelvin scale (°K = °C + 273) and the Rankine scale (°R = °F + 460).

Absolute Units Various absolute systems of physical units are generally used including primarily the British absolute system, the metric absolute system, and system international units (SI). The absolute systems of units are based on fundamental units of length, mass, and time; for the British absolute system, foot-pound-second (fps); for the metric absolute system, centimeter-gram-second (cgs); and for the SI units, meter-kilogram-seconds. Variations from these are also used.

Absolute Viscosity The resistance to flow (flowability or internal fluid friction) of fluid expressed in the absolute system of units, such as the cgs system, and is not affected by density (or mass). Also known as the dynamic viscosity. The unit of absolute viscosity in the cgs system is the poise (or centipoise) which is 1 dyne sec/cm^2 which is equal to 1 g/sec cm and in the fps system is the Reyn which is lb mass/ft sec or poundal sec/sq ft; in SI units is in kg/m or Ns/m^2 or Pas. Also see *Kinematics Viscosity; Viscosity*.

Absolute Zero The lowest value of temperature as energy is removed from a material. At absolute zero the material would be at its lowest possible thermal energy state.

Absorbent A substance that absorbs or is capable of absorbing a liquid or a gas, usually through physical, chemical, or molecular means, such as $CaCl_2$ or LiCl. The absorbate penetrates the absorbent with a possible resultant loss of identity of the original material.

Absorption A unit operation in which a soluble component of a mixture of gases is dissolved in a solid or liquid. The process of the taking up of one substance by another substance, generally by solution. Also see *Absorbent; Adsorption; Sorption*.

Absorption Bands See *Infrared Dryer*.

Absorption Gage An instrument for determining the rate of absorption of fabric in which the specimen impregnated with potassium bichromate is partially immersed in water and the velocity of migration of the solution determined.

Absorptivity A measure of the portion of radiant energy falling on a body which is converted to heat, with the remainder of the energy being reflected or transmitted. The absorptivity depends on the wavelength of the energy and the roughness and color of the absorbing surface (Table A-1).

Table A-1
Absorptivity and Emissivity of Surfaces for Solar Energy

Material	Absorptivity	Emissivity
Copper	0.26	0.03
Copper plate, oxide layer		0.78
Iron, 38°C (100°F)	0.45	0.30
Iron, oxidized layer, 100°C (212°F)	0.74	0.74
Stainless steel, 18-8	0.51	0.44
Tungsten filament, 3300°C (6000°F)		0.35
Red brick building, 21°C (70°F)	0.75	0.93
Aluminum foil, 100°C (212°F)		0.087
Oil paints, all colors, 100°C (212°F)	0.75 black 0.15 white	0.93
Aluminum plate, polished, 260°C (500°F)	0.26	0.04
Water, 0 to 100°C (32° to 212°F)	0.93	0.96
Ice, 0°C (32°F)	0.95	0.96
Asphalt pavement	0.85	
Glass		0.93
Snow	0.10-0.20	

Accelerated Freeze-Drying In freeze-drying heat is added to a product, usually food or pharmaceutical, such that the moisture is removed as a vapor, during which the structure of the solid material is not severely damaged. Heat is supplied to the product by conduction from a warm platen or from a radiation or microwave source. The vapor removed from the product condenses on a cold coil. In the AFD method open metal mesh is placed on the two sides of the product to permit paths for moisture flow (Fig. A-2). The AFD method is about twice as rapid as the conventional freeze-drying methods. Also see *Freeze-Drying; Vacuum Contact Dehydration (VCD)*.

Accessory Material Drying A procedure in which an auxiliary or accessory material is added to the material being dried. The material is used to hasten the speed of heating and/or to absorb moisture to increase the rate of drying. An accessory material such as filter paper when used with sand increases the drying rate, whereas filter paper with a hygroscopic material, such as potatoes or carrots, increases the first part of the drying rate then decreases the remainder of the drying rate. The decrease in drying rate is caused by a decrease in heat and mass transfer. Accessory materials can also be used for heating a material to be dried, such as the use of heated sand, which has been used for drying cereal grains. For drying with accessory materials, conduction heating predominates in contrast to convection heating for many drying operations.

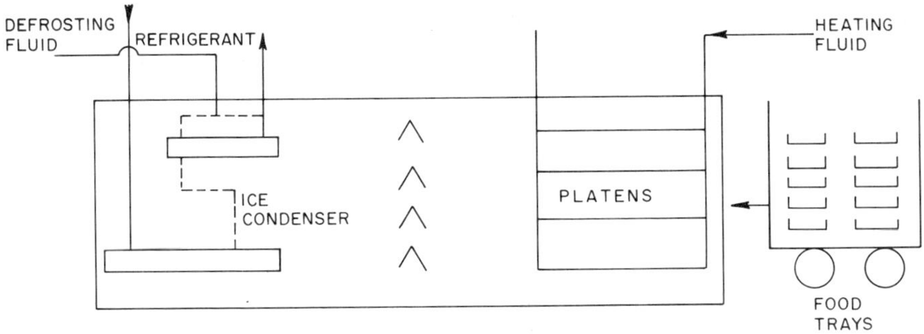

Figure A-2. Accelerated freeze drying. *From Keey, R. B. Drying Principles and Practices (1972). Used with permission of Pergamon Press, Ltd., Oxford, England.*

Accuracy The correctness of a measurement, calibration, or a value. Certain commercial practices or standards may be followed and various statistical techniques may be used to represent the accuracy of a value or an instrument (Table A-2). Precision refers to the reproducibility of data by an instrument or procedure.

Acetic Acid See *Organic Acid.*

Acid A chemical compound or material containing hydrogen which will ionize and a part or all of the hydrogen ions may be exchanged for metal or base ions. Acids have a relatively high affinity for water molecules and thus can be considered as a moisture absorber for some applications. The strength of an acid is represented by the concentration of the hydrogen ions, a neutral solution having a pH of 7, with a lower number representing a stronger acid. Also see *Alkali; Organic Acid; pH.*

Acid Number A measure of the free fatty acid (FFA) in a fat or oil, or in a product containing fat and/or oil. One method of designation gives the acid number by determining the milligrams of potassium hydroxide required to neutralize the free fatty acids in one gram of substance. The determination is usually performed by titrating the fat with 0.1 or 0.5 normal alkali using phenolphthalein as an indicator. Fatty acids are organic acids in the form of glycerides, R•COOH, where R is hydrogen or carbon and hydrogen atoms. Also called acid value. Also see *Saponification Number.*

Acid Salt See *Salt.*

Acid Value See *Acid Number.*

Table A-2
Checklist on Accuracy of Measurements

Category	Items to be Considered
Instrument selection	Range, sensitivity, scale length, scale graduation, precision or accuracy, resolution, damping, response or time constant.
Instrument condition	Leveling, zero setting, friction pen resistance, pointer drag, hysteresis, wear and backlash, capillarity.
Instrument calibration	Calibration before use, accuracy at calibration points, amplification and linearity, repeatability, calibration after use, zero drift, damping.
Environment	Temperature, heat transfer, humidity, air motion, barometric pressure, vibration, local gravity, leakage, grounding or short-circuiting, "noise," accessibility, convenience of setup.
Observational	Parallax, accidental scale-reading errors, inaccurate estimates of average reading, timing of readings, inaccurate interpolation, inaccurate conversion of units, mistakes, extrapolation, variety of users.
Test planning	Inadequate control of variables, nonrepresentative samples, poor scheduling, nontypical conditions, timely maintenance.

From Hall, Carl W., Errors in Experimentation (1978). Used with permission of Matrix Publishers, Inc., Portland, OR.

Acidulous Water Water which gives an acidic reaction due to relatively large quantities of carbon dioxide and/or sulphate in the water.

Aconitase An enzyme found in many plants and animals which catalyzes dehydration reactions in aerobic breakdown of carbohydrates in certain reactions.

Acoustic Drying *See Resonating Drying; Ultrasonic Drying; Vibration Drying.*

Acre A unit of area consisting of 43,560 ft^2, 160 sq rods, or 4,047 m^2. One acre is 0.4047 hectare.

Actinometer An instrument used for meteorological measurements of solar radiation, such as a pyrheliometer. The intensity of photochemically active radiation is measured by determining the extent of a chemical decomposition reaction by the incident radiation. Also see *Pyrheliometer*.

Activated Alumina Aluminum oxide which has been treated to give a large surface area to provide a strong ability to absorb water vapor;

used for dehydrating gases. Approximately two seconds contact are required for the moist gas to be adsorbed by the alumina.

Activated Carbon A solid chemical compound made up of 90-95% carbon used for adsorbing vapors and gases, particularly those affecting odor and taste. The material has 500-1000 square meters/gram of surface area. Organic vapors displace water present on the carbon. The carbon can be stripped of adsorbents with steam. Vapors are adsorbed approximately inversely to the ratio of the molecular volumes, inversely with temperature, and directly with pressure. Activated carbon may adsorb a weight equal to the dry weight. Activated carbon is also called activated charcoal, active carbon, adsorbent carbon, decolorizing carbon, and vegetable carbon.

Activated Charcoal See *Charcoal*.

Activation Energy The energy required to activate one mole of a substance to cause a reaction. Represented by the constant A in the Arrhenius equation and determined empirically. Also known as the energy of activation. Also see *Arrhenius Equation*.

Active Evaporation The length of time that constant rate drying or constant rate of moisture removal occurs from a droplet in a spray dryer. For a typical spray dryer, the active evaporation for a product of 20-30% solids is 4-5 sec.

Activity The ratio of change from one to another condition or with time, such as activity coefficient, biological activity, chemical activity (electromotive series), radioactivity, and water activity.

Activity Coefficient The ratio of effective and actual contribution of a substance to a phenomenon under consideration. For a gas the effective pressure is represented by the fugacity (escaping tendency) f, and the actual pressure, p. The activity coefficient is $\gamma = f/p$. For a gas, one means of determining f, and thus γ, is on the basis of the deviation of the volume of a real gas from an ideal gas. (Raoult's law and Henry's law.) The principle is also applied to a component in a solution, in which case a measure of nonideal behavior is represented. Also see *Fugacity*.

AD A standard lumber abbreviation meaning air dried, in common use in contracts and other documents used in transactions of purchase and sale of lumber. Also see *KD*.

Adhesive Hardening The process of hardening of an adhesive is variously called curing, drying, hardening, and setting. Curing is generally used to represent the process when chemical polymerization is involved. Drying is generally used with starch adhesives. Setting is the term commonly used for emulsion and gelatin glues.

Adiabatic A thermodynamic process which takes place without a transfer of heat to or from an external source.

Adjutage A tube inserted into an orifice to obtain pressure values.

ADMI American Dry Milk Institute which has headquarters at 130 N. Franklin, Chicago, IL 60606.

Adsorbate The substance which is adsorbed on a surface during an adsorption process, which when drying is water.

Adsorption The process in which gases, liquids, or dissolved substances are held on the surfaces of solids. Also see *Absorption; Deliquescent.*

Adsorption Isotherms A graphical or mathematical representation of the moisture adsorbed by a hygroscopic material in various humidities at a given temperature. Several equations may be used to represent adsorption characteristics of hygroscopic materials. One adsorption isotherm developed on an empirical basis is given by Freundlich:

$$y = k\ p^{1/n}$$

where

y = weight or volume of gas absorbed per unit area or unit mass
p = equilibrium pressure
n and k are constants depending on the nature of the gas and solid and the temperature.

Also see *B-E-T; Freundlich; Harkins-Jura; Henderson; Sigmoid.*

Adsorptionally Bound Water Primarily represented by a monomolecular layer of water on the internal and external surfaces of a capillary-porous body, represented by equilibrium moisture content curves or similar relationships.

Advection The natural movement of air resulting in horizontal motion. The horizontal motion of air is due to local pressure differences.

Aeration The movement of air at a low flow rate through a product to maintain or improve product quality by maintaining a uniform, usually low, temperature, for removing high humidity air, or for circulating chemicals. Air flow rates usually do not exceed 0.0167 m^3/m^3s (1 cfm/ft^3 or 1 cfm/bu) through a dry product. Also see *Airflow.*

Aeration System Arrangements Several systems can be utilized to accomplish aeration (Fig. A-3). Also see *Aeration.*

AFD See *Accelerated Freeze-Drying; Freeze-Drying.*

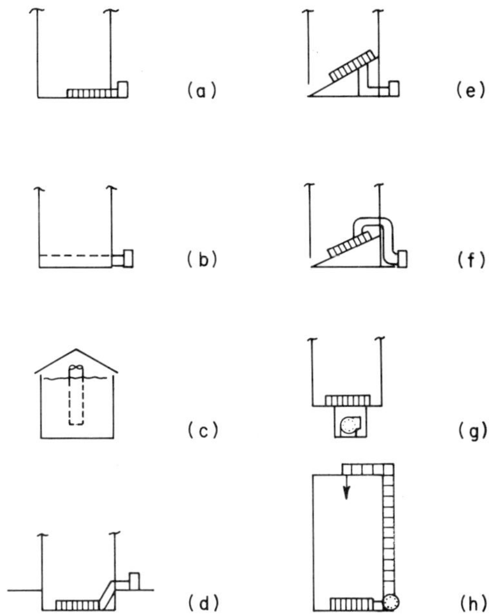

Figure A-3. Aeration system arrangements. (a) flat bottom, no basement. (b) flat bottom, perforated floor. (c) vertical tube. (d) flat bottom, high grade level. (e) hoppered bottom, low grade level. (f) hoppered bottom, high or low grade level. (g) flat bottom, no basement. (h) flat bottom, circulation of fumigant.

AGA American Gas Association, 605 Third Ave., N. Y., NY 10016. An organization which establishes standards of design and performance of systems utilizing gas-fired equipment and control devices, such as used on dryers.

Agglomeration The process of gathering fine particles together into larger particles to form coarser particles (Fig. A-4). The process is used to join small particles into agglomerates of food particles to form instantized dry products. Agglomerated products are more readily soluble in water when reconstituted than the fine particles.

Aging A process of holding or storing materials, such as whiskey, wine, cheese, tobacco, dough, etc., under certain conditions, such as temperature, moisture, acidity, etc., for a period of time until the desired changes have occurred. Also see *Air Curing; Curing*.

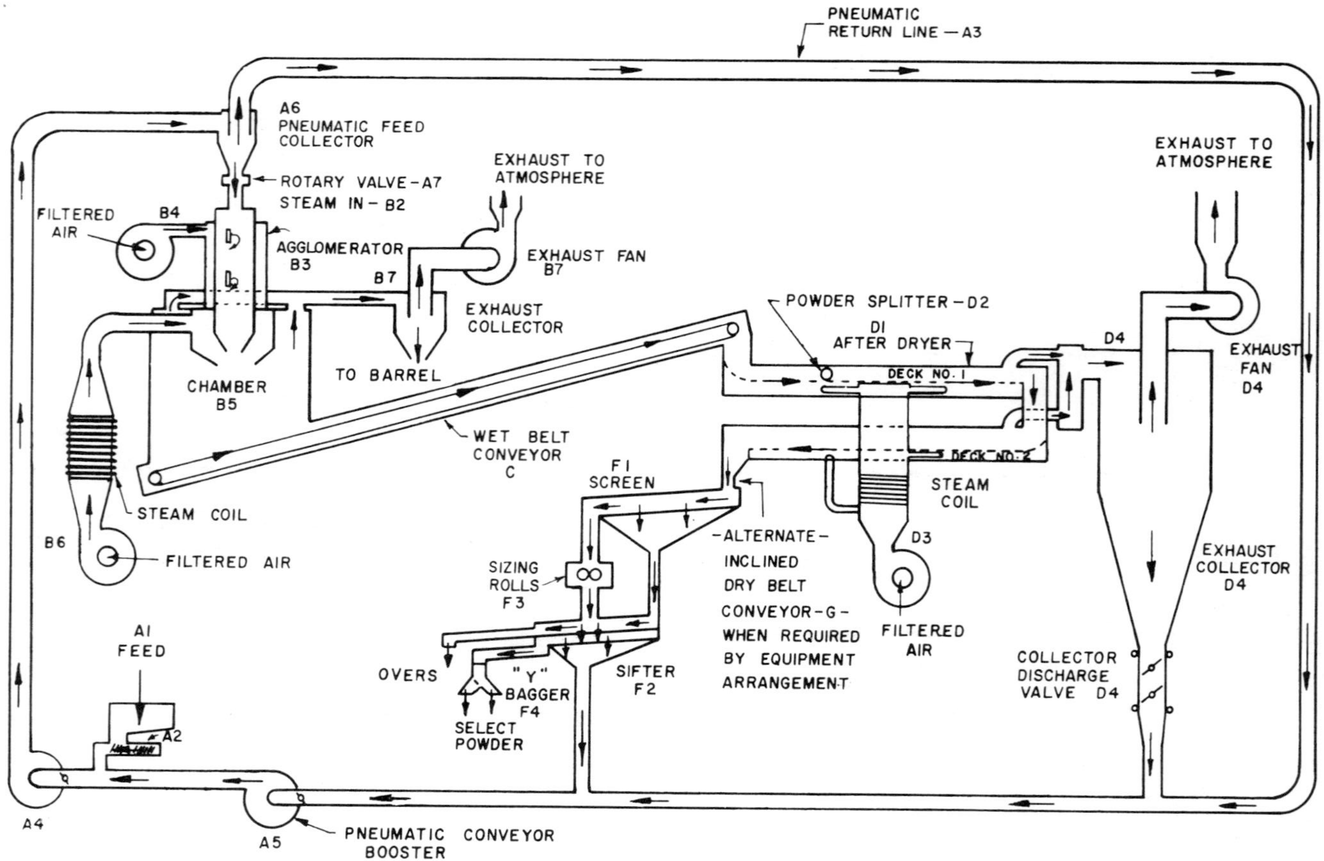

Figure A-4. Agglomeration-flow diagram of instantizing process.
From Hall, C. W., and Hedrick, T. I. Drying Milk and Milk Products (1971). Used with permission of Avi Publishing Co., Westport, CT.

Agitated Batch Dryer   A device for removing moisture from a wet product in a batch holding bin in which the product is stirred with a mechanical agitator (Fig. A-5). Also see *Horizontal Agitated Batch Dryer; Vertical Batch Dryer.*

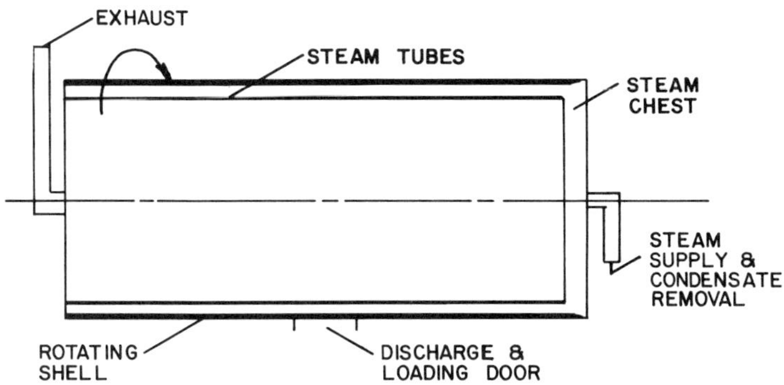

Figure A-5. Agitated batch rotary dryer.

AIChE   American Institute of Chemical Engineers, 345 E. 47th St., N. Y., NY 10017.

Air   A mixture of gases and vapors that surrounds the earth. Dry air is a colorless and odorless mixture which at sea level consists of 78% nitrogen and 21% oxygen by volume (76.8% nitrogen and 23.2% oxygen by weight) plus several minor elements and usually some water vapor. Air is compressible and at constant temperature the density increases in proportion to the change in the absolute pressure ratio (Table A-3). The properties of air are summarized in Table A-4. Also see *Atmosphere.*

Air Agitated   See *Fluidized Bed; Pneumatic Dryer.*

Air Brush   A method by which a dried product is separated from the air in a spray dryer. Air from outside of the dryer, either at room temperature or conditioned to a lower temperature, is used to direct a jet of air to move the product from the bottom of the dryer.

Air Conditioning   The control of atmospheric conditions of an enclosed space for human comfort at temperatures from 18.3 to 24°C (65 to 75°F) and relative humidities 25 to 75%. Also controlled at other temperatures and humidities depending on the purpose. Other factors such as air movement and cleanliness are often considered but are secondary to temperature and humidity control for most applications.

Table A-3
Components of the Atmosphere, not Including Water Vapor, at Sea Level

Component	Content by Volume, %	Content by Weight, %
Nitrogen, N_2	78.03	75.5
Oxygen, O_2	20.99	23.2
Argon, A	0.94	1.33
Carbon dioxide, CO_2	0.03	0.045
Hydrogen, H_2	0.01	
Neon, Ne	0.00123	
Helium, He	0.0004	

Source: International Critical Tables, vol. I, 1926. McGraw-Hill, N.Y., N.Y.

Table A-4
Air Properties

Properties	English Units	SI Units
Density	lb/ft^3	kg/m^3
0°C (32°F)	0.08065	1.2929
20°C (68°F)	0.07516	1.2050
Specific heat (heat capacity)	Btu/lb °F	J/kgK
Constant pressure, STP	0.241	1008.34
Constant pressure, 1 atm, 100°C	0.2404	1005.83
Constant volume, STP	0.17	711.28
Standard pressure	in. Hg	kN/m^2 (kPa)
Sea level (760 mm Hg)	29.92	101.325
10,000 ft (3048 m)	20.58	69.695
Thermal conductivity	Btu/hr ft^2°F/ft	W/mK
0°C (32°F)	0.014	0.0242
38°C (100°F)	0.0154	0.0266
93°C (200°F)	0.0174	0.0301
Vapor pressure, saturated	in. Hg	kN/m^2 (kPa)
-18°C (0°F)	0.03764	0.12746
0°C (32°F)	0.18035	0.61073
10°C (50°F)	0.36240	1.22723
25°C (77°F)	0.93523	3.16482

AIR CORE The central portion of the core, primarily air, produced in a pressure or atomizing nozzle of a spray dryer. Also see *Atomizer; Spray Dryer*.

Air Curing The curing of products such as lumber, meat, hides, and tobacco in normal atmospheric air. The term air cured is most often applied to curing of tobacco and lumber. Air curing of tobacco is carried out in a barn, usually with the tobacco leaves hung from the ceiling to the floor. The roof and sides of the barn are designed to be opened to provide ventilation by natural air flow. The tobacco leaves hang in the barn for about three months until the leaves become brown. Air-cured tobacco is produced in many countries of the world, but the best known areas are Kentucky and Tennessee in the USA and Malawi, Mexico, and Brazil. Also see *Air Drying; Curing.* For tobacco see *Bright-Leaf; Burley; Shade Grown;* also see *Lumber.*

Air Dehumidification The process of removing a water vapor from a mixture of air and water vapor in which the humidity is reduced. Dehumidification is carried out by condensing the moisture on a surface at a temperature below the dew-point, by spray dehumidifiers, with a chemical such as an acid in an absorption tower, or with chemical dehydrating agents such as calcium chloride. Also see *Calcium Chloride.*

Air-Dry A material in a condition in which it will yield no further moisture loss by exposure to atmosphere or air.

Air Drying The removal of moisture from a product by naturally occurring evaporation to air.

Air Dry Lumber Lumber dried by exposure to air outdoors or under an unheated shed for any length of time. The range of the final moisture content is from 6% (in the southwest) to 24% (in winter in the northwest), but usually in the range from 12-15% moisture content.

Air Flow; Airflow The flow of air, heated or unheated, through fans, ducts, floor, or product in a drying installation. Refers to a rate of flow through a cross-sectional area, m^3/m^2s (cfs/ft^2); through a volume of product, m^3/m^3s (cfm/ft^3 or cfm/bu). The air flow rate for drying is usually greater for heated air than unheated air. An airflow rate of 1 m^3/m^3s (70 cfm/ft^3) is common for drying a 0.5 m (1.5 ft) layer with heated air. An airflow rate of 0.03 to 0.25 m^3/m^3s (2 to 15 cfm/ft^3) is common for unheated air drying. For these calculations, 1 bu = 1.25 cu ft. Also see *Aeration; Barn Drying; Cereal Grain Drying; Orifice; Velocity.*

Air Horsepower The work done by a fan on air in ft-lb per min divided by 33,000. Also refers to the power required to move air through a system of ducts or other resistances to air flow. Also called fluid horsepower. Also see *Fluid Horsepower.*

Air Humidification The process of evaporating a liquid such as water into air and the diffusion and convection of the moisture throughout the air. The humidity of the air is increased as a result of humidification. Humidification may be done to increase the moisture content of a product or to condition a product for subsequent processing.

Air-Lift Dryer The name may be applied generally to a fluidized bed or pneumatic dryer. A more precise use of the term is for a dryer which was used to partially dry potato granules during World War II. The dryer consisted of a vertical feed chamber into which the product was fed by a vibrator into a hot air stream at a venturi section. The product was carried vertically while being dried in the hot air stream. The vertical chamber had an enlarged cross-section as the exit was approached, thus decreasing the velocity and therefore the carrying capacity of the air. The product hit a deflector and was dropped to a collector and removed. The moist air exited vertically after passing around the deflector. Reference: von Loesecke, H. W., 1955, Reinhold Publishing Co. Also see *Fluidized Bed Dryer; Pneumatic Dryer.*

AIR LOCK A chamber with valves located between two other chambers at different pressures or between a chamber at a vacuum or pressure and the outside air. The air lock is operated in a procedure similar to the lock in a canal to move a ship from one level to another. The purpose of the air lock is to permit moving a product from one pressure chamber to another without losing the pressure (or vacuum) of the chamber, such as with a vacuum freeze-dryer and for moving grain into and out of a dryer.

Air Moving Device (AMD) A revolving, wheel-type, mechanical device, driven by a power source, used to move air for drying or aeration. For the ASAE standard, an AMD includes the wheel or blade assembly, mounting structure and casing, and may or may not include a power source. Also see *Axial Flow Fan; AMCA; Centrifugal Fan; Disc Fan; Vaneaxial Fan.*

Air Oven A nearly closed container, with or without air circulation, usually heated with electricity, having a carefully controlled temperature, with or without vacuum, used to dry weighed samples of product to determine the moisture content. The time and temperature of dehydration have been standardized for most commercial products. Also see *Moisture Content Determination.*

Air Separator A mechanical device for separating liquid and/or solid contents from air. The most common principle used is centrifugal force. The device is also known as a dust collector, centrifugal separator, cyclone collector, and cyclone separator. Also see *Cyclone Collector.*

Air-Straightening-Vanes Mechanical channels, guides, etc., placed in airflow ducts to assure a more uniform distribution of the air at bends or elbows and throughout pipes, tubes, or ducts.

Air-Suspension Dryer See *Fluidized Bed Dryer.*

Air Trunking The ducts, tunnels, or channels which direct air to storages, silos, or drying chambers. The velocity of air flow normally should not exceed 7.6-10 m/s (1500 to 2000 ft/min) to prevent excessive pressure drop. Trunking is constructed from sheet metal, wood, plywood, chipboard, etc. It is a general term for an air distribution system. The term is used more widely in England than in the USA.

Air Velocity The velocity of the air through a duct, pipe, or space, m/s (ft/sec, ft/min). The air flow is equal to the cross sectional area times the average velocity of flow. The average velocity of flow is determined by dividing the cross sectional area into a number of annular rings of equal area, with the velocity measurement taken at the point in the annular ring dividing the annular ring in half. For a circular duct or pipe of radius, r, the recommended locations from the center for reading the velocity are at 0.1316 r, 0.548 r, 0.707 r, 0.837 r, and 0.949 r. A location at 0.8 r from the center is sometimes used to obtain one reading to represent the average velocity. Also see *Superficial Air Velocity*.

Albedo The ratio of the reflected radiant energy to the incident (incoming) radiant energy for a surface generally applied to the earth's surface, where the approximate albedo values are: dark forest, 0.04; bright sand dunes, 0.28; ocean, 0.05-0.06; fresh snow, 0.80-0.90; land surfaces, 0.12; wood, pine, 0.40; and earth, wet, 0.08.

Alfalfa Drying See *Hay Drying*.

Alimentary Paste Products such as macaroni, spaghetti, alphabets, noodles, and sea shells, each having a characteristic shape and dried doughs from flour of semolina (semola), farina, wheat or durum flour, or a mixture of these or similar products. The dough is partially dried rapidly in hot air, then dried more slowly to produce the final product.

Alumina A granular, porous absorbent of Al_2O_3 used to absorb gases and vapors from gaseous mixtures for dehydration or drying of air, gases, and liquids. When saturated, the moisture content is 20-25%, dry basis. Alumina is used primarily for drying air and non-reactive gases, and for drying gaseous fuels. Under this subject is included absorbents or drying agents identified as activated alumina, alumina gel dessicant, aluminum oxide, driocel, and florite dessicant. Also see *Alumina Gel Dessicant; Oven Method of Moisture Measurement*.

Alumina Gel Dessicant A dehydrating agent with a surface area of 400 m^2/g, available as microspheres, pellets, and granules. Also see *Alumina*.

Aluminum Oxide A drying agent. Also see *Alumina*.

Ambient Temperature The temperature of the surrounding atmospheric air. Ambient air temperature changes could affect the operation of equipment, instruments, or controls. Compensating devices may be used to adjust for ambient air temperature changes to provide greater accuracy of measurement and control.

AMCA Air Moving and Conditioning Association, 30 W. University Drive, Arlington Heights, IL 60004.

AMD Air Moving Device. Also see *Fan; Fan Rating.*

American Gas Association See *AGA.*

Amino Acids See *Protein.*

Ancillary Equipment Equipment added to a primary piece of equipment to assist in meeting the functional requirements of the system. The operation of a dryer is influenced by ancillary equipment, used preceding, following, surrounding, or associated with the dryer. Ancillary equipment for dryers include: mechanical dewaterers, grinders, choppers, lacerators, pulverizers, mixers, pelleters, materials handling devices, heating equipment, dust collectors, exhaust systems, and ventilation (external) equipment.

Anemometer An instrument used for measuring the velocity of air or wind. The anemometer utilizes cups, vanes, a pressure plate, or electrically heated wire as the active element.

Aneroid A type of instrument used for measuring barometric or atmospheric pressure which uses a metal bellows as a sensing device. Also see *Aneroid Manometer; Barometer.*

Aneroid Manometer A device for measuring the level of fluid for evaporator control (Fig. A-6). The aneroid is an "open type" manometer (open to atmosphere) which has connecting sensing lines between the level taps and the instrument.

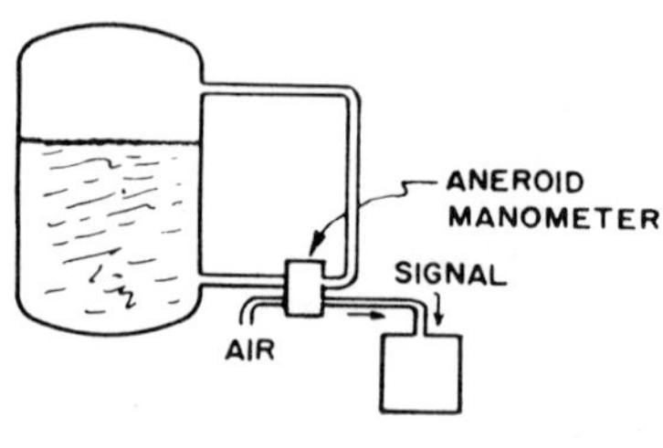

Figure A-6. Aneroid manometer.
From Hall, C. W., and Hedrick, T. I. Drying of Milk and Milk Products (1971). Used with permission of Avi Publishing Co., Westport, CT.

Angle of Repose The maximum angle from the horizontal which a bulk material assumes when piled, generally 25-35° for cereal grains (Table A-5). The angle of repose is an internal angle of friction. Also see *Friction.*

Table A-5
Angle of Repose, ϕ, Approximate

Product	Angle from horizontal, degree
Anthracite coal (hard)	27
Ashes	40
Beans	32
Bituminous coal (soft)	35
Cement	40
Coke	30-45
Corn, shelled, dry	27
Limestone, crushed and screened	37
Liquid	0
Malt	22
Rice	36
Sand, dry	25-35
Wheat	25

Anhydrate To dehydrate. The process of removing water from a substance.

Anhydrator A device for dehydrating, or more specifically, a device to remove primarily the water of hydration from a material. Also see *Water of Hydration.*

Anhydride An oxide which produces an acid or a base when water is added.

Anhydrone A trade name for a dehydrating agent consisting of anhydrous magnesium perchlorate, $Mg(ClO_4)_2$.

Anhydrous A term applied to a solid substance that contains no water.

API American Petroleum Institute, 1271 Avenue of the Americas, N.Y., NY 10020.

Apparent Density The weight per unit volume of a material, including voids, which is a measure of the bulk of the material. Apparent density refers to the weight per unit volume whereas the apparent specific gravity denotes the specific gravity of a bulk volume, not including voids. Also see *Apparent Specific Gravity.*

Apparent Specific Gravity The apparent specific gravity denotes the specific gravity of a porous solid where the volume, V, excludes the permeable voids. The bulk specific gravity refers to measurements in which the volume, V, of the solid includes permeable and impermeable voids. Thus, bulk specific gravity is less than apparent specific gravity. Also see *Specific Gravity.*

Apparent Velocity The calculated speed of air flow through a cross section where drying is occurring assuming that no product is present. The apparent velocity is determined by dividing the quantity of air flow by the cross section of the drying chamber, as if empty. The actual or average velocity of air flow will be greater than the apparent velocity, depending on the amount of void space.

Apple Drying Apples are peeled and usually sliced, ringed, or diced before drying. Drying or dehydrating with heated air is usually done in a tunnel, kiln, or belt dryer. In the tunnel dryer, sectioned apples at 85% moisture content are loaded on trays, 9.76 kg/m^2 (2 lb/ft^2), and moved through the tunnels where the temperature is about 74°C (165°F). Using countercurrent flow, about 8 hr is required to dry apple slices to 20%. With two stage drying, temperatures of 85°C (185°F) in the first stage and 74°C (165°F) in the second stage are used. Dry apples are 24% moisture and dehydrated apples are 2.5% or less for US standards. Sulfuring of the peeled fruit is done to reduce discoloration, either before drying or after the first stage of drying. One hundred pounds of apples produces approximately 10 lb of dried apple pieces. Also see *Apple Powder; Dehydrofrozen Apple Slices*.

Apple Powder Drum-dried apple powder is a product used in cake mixes and in baked pastries. The dried product rehydrates readily and may reasonably be termed "instant apple sauce." Considerable amounts of drum-dried apple powder are now being produced by processors in the Pacific Northwest. It may be prepared by grinding low-moisture dried apples, or preferably by drum-drying apple purée. Apples are peeled, cored, cooked in clean steam, and dried on a double drum dryer of the type once used for drying milk. The conventional equipment is usually modified by plating the steel rolls with chromium and by installing air ducts to cool the dried product as it leaves the drum. The purée is applied to the drum by a travelling spray feeder which distributes the product uniformly on the upper surface of the drum. Reference: Powers, M. J. 1973. CSIRO Food Res. Q 33(2):37.

Apricot Drying Apricots are dried on trays using sun or heated air. Sun drying for 3-5 hr may precede mechanical drying. Apricots are sliced and placed on trays at a loading of 9.8 kg/m^2 (2 lb/ft^2). The moisture content is reduced to 18-20% moisture content after drying 8 hr at 65.6°C (150°F). Another method of drying is a two-step process involving drying, blanching, and continued drying (known as DBD process). Sulfuring is often done by burning sulfur to get 1500-3000 ppm of SO_2 absorbed previous to drying. Sulfuring may also be accomplished by dipping the apricots in a bisulfite solution before drying.

Aqua A prefix often used in combination with another word to represent water, such as aqua ammonia.

Aqueous Applies to solutions with water as the solvent.

ARCS A commercial system of agglomerating fine particles of dry products which includes the wetting, agglomerating, drying, and cooling. A trade name for equipment of Cherry-Burrell Co. Also see *Agglomeration*.

Arrhenius Equation; Arrhenius Plot A relationship in which a variable, such as energy of activation, vapor pressure, speed of change, etc., provides a linear relationship when the log of the variable is plotted versus the reciprocal of the absolute temperature. The empirical Arrhenius equation is

$$\frac{d \ln k}{dT} = \frac{A}{RT^2}$$

where

k = specific reaction velocity
T = absolute temperature
A = energy characteristic of the reaction
R = gas constant

Artificial Drying The removal of water or moisture from solids or semi-solids under other than naturally occurring air conditions.

ASAE American Society of Agricultural Engineers, St. Joseph, MI 49085. See *ASAE Standards: ASAE S248.2 for Construction and Rating of Equipment for Drying Farm Crops*.

Ascorbic Acid Vitamin C, ($C_6H_8O_6$), essential for human health which is present in nuts, fresh vegetables and fruits, which is destroyed during sun drying and heated air drying. Treatment of some products with sulfur is done to prevent complete loss of vitamin C. For example, for a reasonable retention of vitamin C, apricots need about 450-500 ppm of sulfur dioxide. Also see *Vitamins*.

Aseptic The condition of a surface, substance, or air in which microorganisms are either not present or are killed and not able to grow. Most often aseptic refers to absence of pathogenic organisms in contrast to sterile which refers to absence of all organisms. Substances used to provide aseptic conditions are antiseptics.

ASHRAE American Society of Heating, Refrigerating and Air Conditioning Engineers, Inc., 345 East 47th St., N.Y., NY 10017. Preceded in name by the American Society of Refrigeration Engineers and the American Society of Heating and Ventilating Engineers.

ASME American Society of Mechanical Engineers, 345 East 47th Street, N.Y., NY 10017.

Aspiration Psychrometer An instrument used to measure the relative humidity of air, in which the air is moved over a thermometer sensing element (bulb) to evaporate moisture using an aspirator to propel the air.

Aspirator A device for moving air by creating a slight vacuum. An aspirator can be used to separate light from heavy material as, for example, before or after drying.

ASTM American Society for Testing and Materials, 1916 Race Street, Philadelphia, PA 19103.

Atmosphere A unit of pressure in which one atmosphere (or a standard atmosphere) is the normal sea level atmospheric pressure of 760 mm Hg, 29.92 in Hg, 14.69 psi, 101.325 kN/m^2, 101.325 kPa, or 760 torr. The atmosphere is the body of gases, primarily air, surrounding the earth, traceable for 833 km (500 mi) into space but is usually considered as the layer of air surrounding the earth, approximately 100 km (60 mi) or less thick, consisting of nitrogen, oxygen, hydrogen, argon, carbon dioxide, carbon monoxide, moisture, and traces of other components. The portion of the atmosphere which will support life is the biosphere. The atmosphere may be considered as made up of a lower area of troposphere, middle layer or stratosphere, and outer layer or ionosphere which extends beyond the atmosphere. Also see *Air*.

Atmospheric Double Drum Dryer Two hollow drums into which steam is fed for heating, with the two drums rotating towards each other, with the product placed above and between the two rolls, evaporation taking place at atmospheric pressure, and the dried product removed from the rotating drying drums. The dried product is usually removed after 3/4-7/8 revolution of the drum. Devices are available to hold the knife off the drum for more than one revolution. The product is in contact with the drum for drying about 3 sec or less. The spacing between the drums when cold is 0.05-0.10 cm (0.02-0.04 in). The speed of the drum may be varied. Condensed product, or products at 30-70% moisture, are usually placed on the dryer and dried to 2-8% moisture content. Also see *Drum Dryer; Nip*.

Atmospheric Dryer A device for removing moisture from products using atmospheric air forced through or over the product at atmospheric pressure.

Atmospheric Evaporator or Atmospheric Concentrator A steam jacketed cylinder inside of which a moving blade moves the liquid product over the heating surface, with evaporation at atmospheric pressure and at a temperature only slightly above 100°C (212°F). Vapor is removed and discharged into the atmosphere.

Atmospheric Pressure The force per unit area or height of a mercury column which is supported by the weight of the atmosphere. In many cases the local atmospheric pressure found using a barometer must be used. Also see *Atmosphere*.

Atmospheric Tray Dryer Trays containing materials to be dried are placed in one or more compartments at atmospheric pressure. These dryers are operated as batch or continuous units. Hot air is blown over the trays, around the trays, or through the product and trays, some having screened or perforated bottoms (Fig. A-7). Because of the high labor cost these units are usually expensive to operate. The atmospheric tray dryer can be used for drying paste-like as well as other conventional materials.

Atomization The process of breaking or dividing a liquid into very small drops or droplets. Atomizers are used for fuels and for drying or dehydration of liquid products in spray dryers. The total surface area of a unit volume available for heat and mass transfer is greatly increased as the size of drop is decreased (Fig. A-8). During spray drying the temperature of the atomized particles approaches the wet-bulb temperature of the drying air. Also see *Spalding Number; Spray Dryer.*

Atomizer A device for producing droplets from a liquid, such as for atomization of fuels for a burner and for atomization of liquid products for spray drying. The atomized fuel is burned in a combustion chamber. The atomized product for drying is introduced in the drying chamber where heated air evaporates moisture from the atomized product and carries the moisture out of the dryer. Atomization of products for drying is accomplished by a high pressure nozzle, a two-fluid (air or steam) atomizing nozzle, or a centrifugal disc (disk) (Fig. A-9). Also see *Spray Dryers.*

Atomizing Burner Devices for producing droplets from fuel oil for combustion. Various types of burners for fuel oil are high pressure air atomizing, low pressure air atomizing, spray nozzle, rotary mechanical atomizing, and steam atomizing.

Attrition Drying See *Mill Drying.*

Automation of Dry Kiln Automation for Lumber The automation of kiln operation which consists of (1) the automatic determination of the moisture content and (2) the automatic changing of drying conditions to meet needs of the process. The moisture content is determined by automatic continuous weighing of a part of the kiln load from which the moisture content is determined (Fig. A-10). Drying times can be reduced about 20% with an automated system.

Auxiliary Material Drying See *Accessory Material Drying.*

Available Energy A general term applied to the quantity of energy available for the process in question. Generally, the available energy is calculated as that energy above the lowest temperature for discarding energy. The available energy is represented in a thermodynamic system by $(h - T_0 s)$ where h is the enthalpy and $T_0 s$ is the unavailable energy, and T_0 is the lowest temperature.

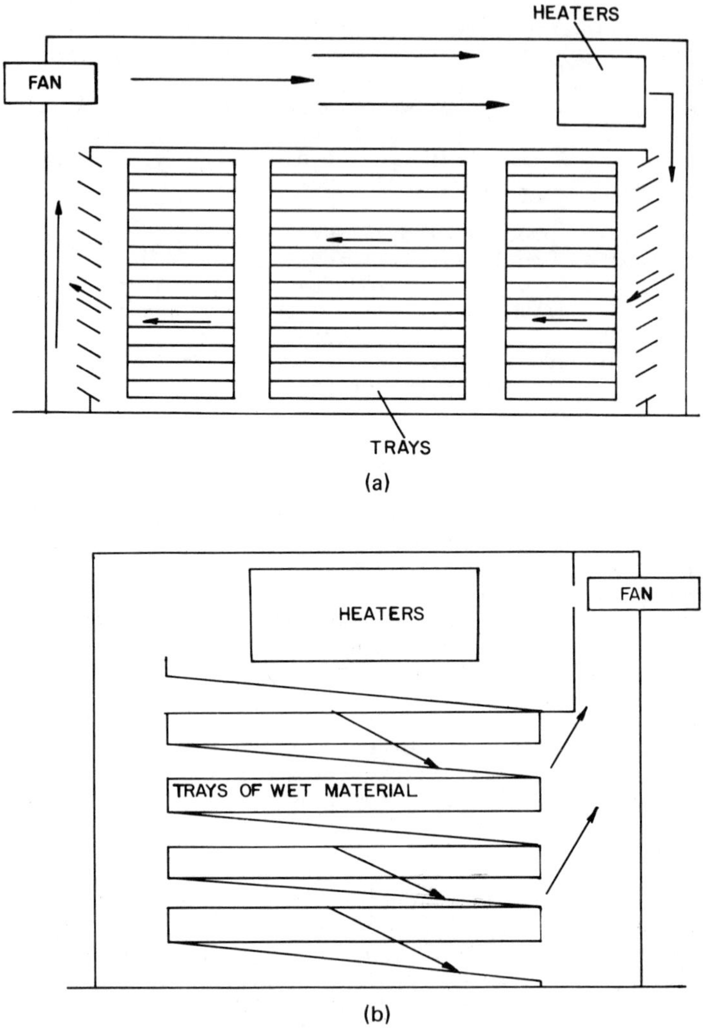

Figure A-7. Atmospheric tray dryers. (a) tray or compartment. (b) through-circulation.
From A. Williams-Gardner Industrial Drying (1971). Used with permission of George Godwin, Ltd., London.

Average Velocity The rate of air flow through product void space. It is determined by dividing the apparent velocity by the product void space expressed as a decimal. The average velocity is always greater than the apparent velocity. Also see *Air Velocity; Apparent Velocity; Air Flow.*

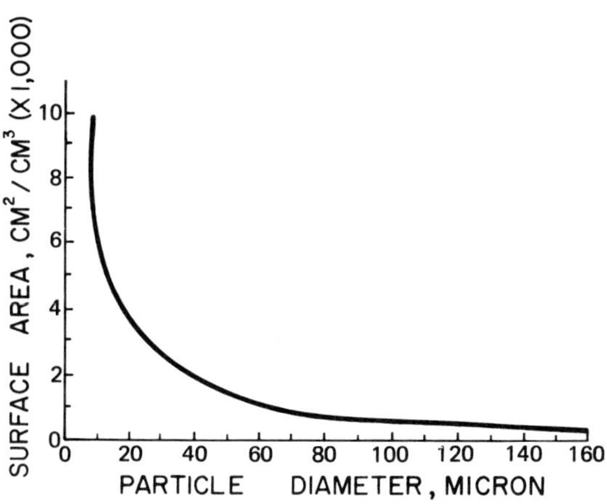

Figure A-8. Atomization-surface area by particle diameter.

Axial-Flow Fan A mechanical device consisting of a power driven blade or propeller which moves air or gas parallel to the axis of rotation of the fan blade. Axial flow fans are characterized by operating at high air flows and low static pressures. Axial flow fans consist of propeller fans or other axial flow fans with an impeller or propeller fans with an enlarged hub, known as a disc fan. The other major classification of fans is the radial flow fan. Also see *Centrifugal Fan; Disc Fan; Radial Flow Fan; Vane-axial-flow Fan.*

Azeotropic Drying The removal of moisture from a product (such as wood) in a heated chemical mixture which has a constant boiling point. The process may be carried out in a vacuum. An azeotrope is a given mixture of liquids which has a constant boiling point, for which the composition of the liquid does not change as it is converted to vapor.

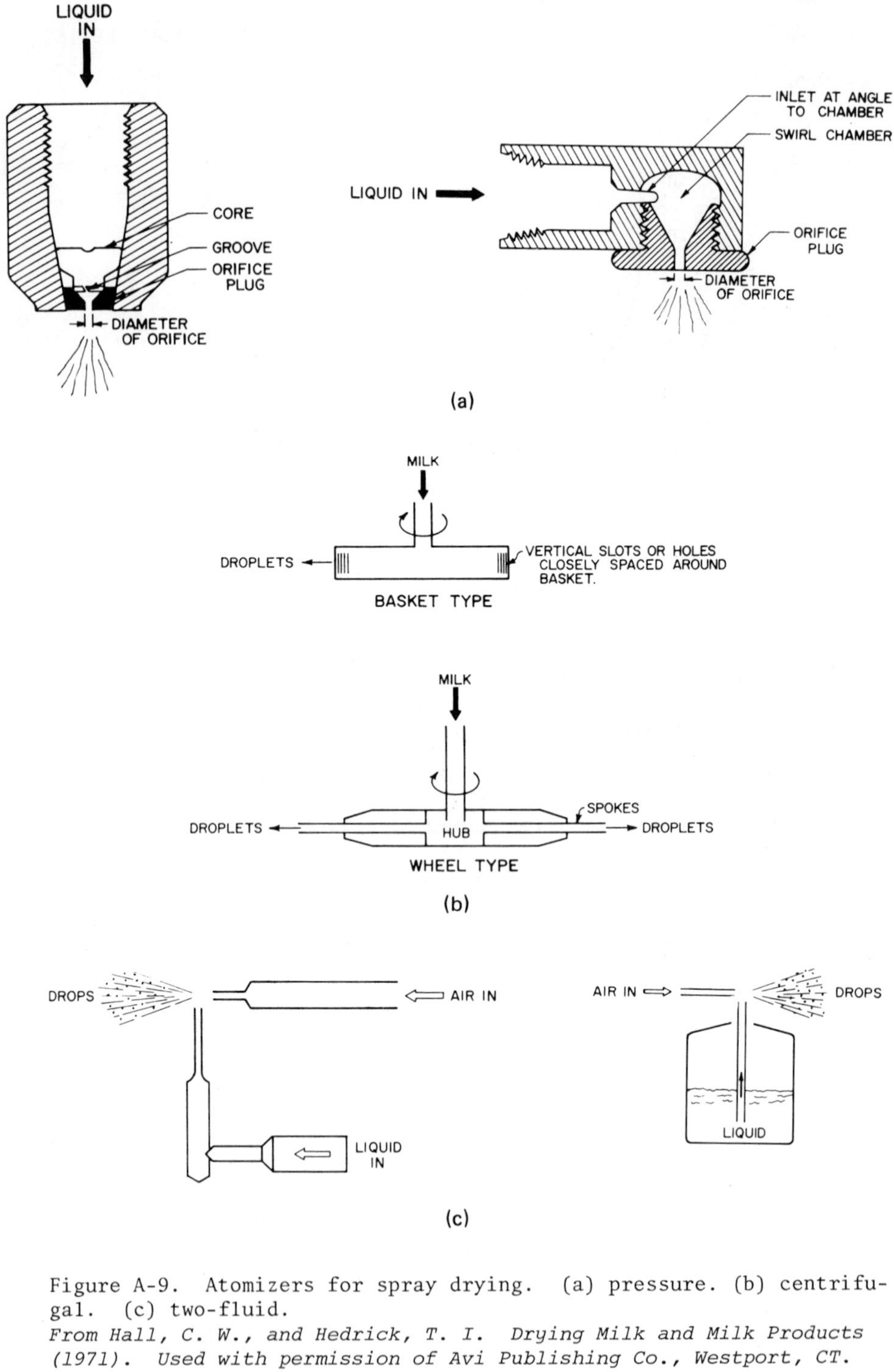

Figure A-9. Atomizers for spray drying. (a) pressure. (b) centrifugal. (c) two-fluid.
From Hall, C. W., and Hedrick, T. I. Drying Milk and Milk Products (1971). Used with permission of Avi Publishing Co., Westport, CT.

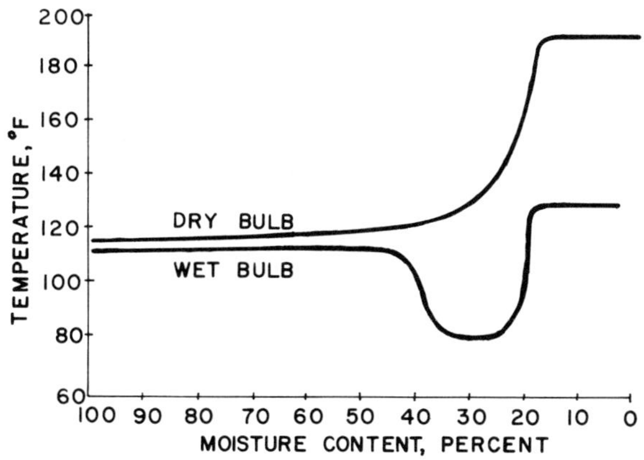

Figure A-10. Automation of kiln, change of moisture and temperature.

B

Backward-Curved Centrifugal Fan A power-driven mechanical device of the radial flow and centrifugal type used to move air through ventilation and drying systems. A wheel containing from ten to over fifty blades around the periphery rotates within a spiral housing. The leading edge of the curved blades is on the inner edge of the outer periphery. As compared to axial flow fans, the centrifugal fan has a low noise level. This fan is often used where a large variation in volumes of air flow may occur. Also see *Centrifugal Fan; Straight-Bladed Fan; Forward-Curved Fan; Radial Flow Fan.*

Back Radiation The radiation loss from a body to another body, or to the environment, such as from the earth's surface to outer space. When back radiation is larger than incoming absorbed radiation, the temperature of the body goes to a value below the immediate surrounding temperature. The phenomenon is also called negative radiation. Also see *Emissivity.*

Bagasse The residue which remains after mechanical pressing of sugar cane which has a moisture content in the range of 40-50% wet basis. Used for boiler fuel having a heating value of 18,500-21,000 kJ/kg (8000-9000 Btu/lb) of dry matter. Also see *Solid Fuel Burner; Sugar Drying.*

Bag Dryer A unit especially designed to remove moisture from seed or grain in bags or sacks by forced air. A unit consists of a fan and air distribution system upon which sacks of seeds or grain are placed over 25 cm x 60 cm (10 in x 24 in) openings with each opening provided with a door. Also see *Platform Dryer.*

Baghouse A chamber for supporting and housing fabric filters which remove fine particles from air and dried material from the air leaving a dryer. The filtered air is discharged to the atmosphere and the collected product is delivered to a hopper. The bags have a rating based on air flow of 0.001 to 0.005 m^3/m^2s (1/5 to 1 cfm/ft^2) of fabric area. The bags may be vibrated periodically to remove the material collected on the surface of the fabric. Fabric bags are made of natural fibers, such as cotton, and glass fibers.

Banana Drying Bananas may be dried in many forms using one of several different methods. Sun-dried bananas are used for flour or powder. Diced or sliced peeled bananas are dried on trays in a heated air chamber or tunnel. Air heated to not more than 93°C (200°F) initially and 65.5°C (150°F) finally is used to dry bananas to 20% moisture content. The bananas can be pulped in a disintegrator or homogenizer and dried on a double drum atmospheric dryer. Flakes are produced on a drum dryer. The flakes may be pulverized to make a powder. The pulped material can be diluted with water and dried in a spray dryer. Discolorization can be reduced by sulfuring after peeling, before processing.

Band Through-Circulation Dryer A device for removing moisture in which the wet material is in particulate or granular form of nearly uniform size, such as pellets, placed on a perforated band or conveyor, and moved through a forced air stream in a drying chamber (Fig. B-1). Also see *Convection Solid Band Dryer*.

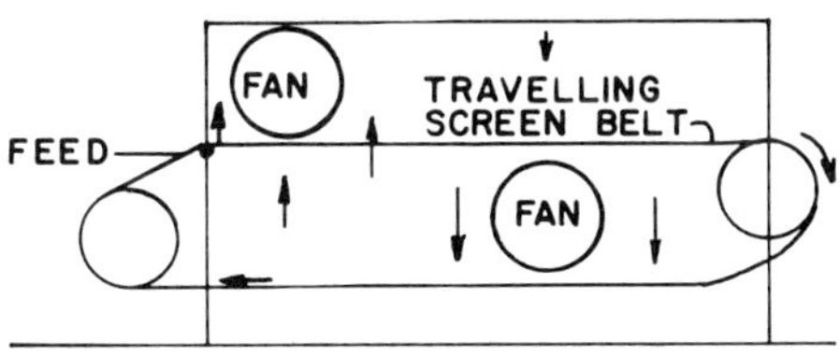

Figure B-1. Band continuous through-circulation dryer.
From Nonhebel, G., and Moss, A. A. H., Drying of Solids in the Chemical Industry (1971). Used with permission of Butterworth Publishers, London.

Bar A unit of pressure equal to 1 million $dynes/cm^2$ which is 0.9868 atmosphere or 10^5 Pascal. A microbar is 1 $dyne/cm^2$.

Bark The refuse of wet-wood from a paper, lumber, and plywood operation includes a variety of wet bark along with sawdust, butt ends, and culled pieces. These materials can be used for fuel but a certain amount of pre-processing is desired to provide consistent burning. Bark from the barking devices may contain as much as 80% moisture, wet basis. For wet bark there may not be enough heat available to support combustion. Dry bark has a heating value of about 21,000 kJ/kg (9000 Btu/lb). Mechanical presses can be used to reduce the moisture content from 80 to 65%, wet basis. Stack or flue gases can be used for drying the pressed wood in bins. Large operations with uniform supply are needed to economically justify rotary dryers or other commercial drying units. Wood waste for fuel is also called hogged fuel

Barley Drying   See *Cereal Grain Drying*.

Barn Drying; Barn Hay Drying   The process of completing the drying of hay, grass, or other forage in the mow of the barn using forced air or forced heated air after wilting the forage in the field, drying a maximum of 50 to 15% moisture content, wet basis. Duct systems are provided for distributing air through the storage or mow. A center main duct, a main duct with laterals, or a slatted floor is used to distribute the air through the forage. For heated air systems, temperatures up to 82.2°C (180°F) are commonly used. For unheated air, airflows of 0.076 to 0.10 m^3/m^2s (15 to 20 cfm/ft^2) are used, and for heated forced air, airflows of 0.10 to 0.15 m^3/m^2s (20 to 30 cfm/ft^2) are used. Airflows are also designated on the basis of 0.16×10^{-3} to 0.26×10^{-3} m^3/kgs (300 to 500 cfm/ton), with higher airflows for heated air systems. These systems will have a static pressure of 0.19 to 0.38 kPa or 19 to 38 kg/m^2 (0.75 to 1.5 in H_2O) (Fig. B-2).

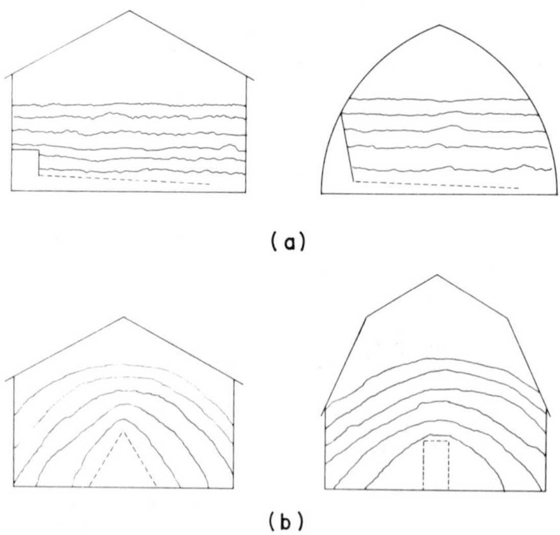

Figure B-2. Barn drying system duct arrangement.

Batch A quantity of a harvested crop, chemical, fuel, or other substance placed into a bin or container on a repetitive basis specifically for treatment or processing, such as for drying.

Batch Dryer A device in which the product to be dried is placed in batches for drying, the complete drying or drying and cooling operations performed, and then removed (Fig. B-3). Usually the unit is self-contained comprising of a drying compartment, either horizontal or vertical, through which heated air is forced or drawn from a central portion usually called a plenum chamber. A large range of heated air temperatures is used. Cooling is usually carried out in the same manner. Integral conveying equipment provides for loading and unloading the dryer. Also included as part of the dryer are the air moving device (AMD), burner, and control system. Batch dryers are often portable, but may be stationary. Included in the category of batch dryers are the following: tray dryer, agitated batch dryer, and rotary dryer, all of which may be for either atmospheric or vacuum operation. Also see specific types and applications of dryers.

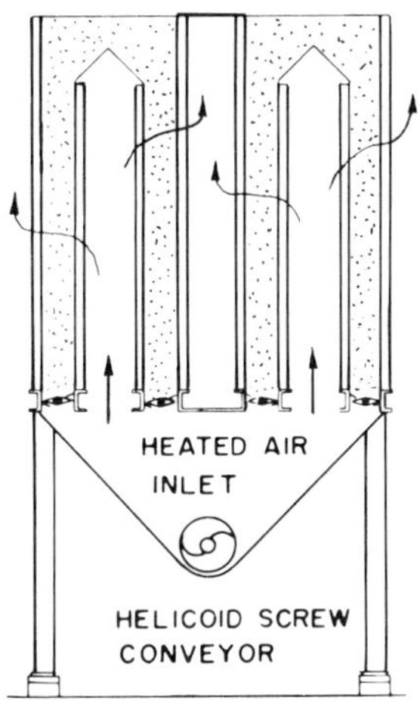

Figure B-3. Batch farm drying chamber, column type.

Batch-In-Bin Dryer A unit in which wet grain in shallow depths, usually 1.25 m (4 ft) or less, is placed in a bin with a false floor and is dried with forced heated air, 50 to 70°C (120 to 160°F), at flow rates of 0.1 to 0.25 m^3/m^3s (8 to 15 cfm/bu). The system for batch-in-bin drying is characterized by (1) all of the floor on which grain is placed is exposed to the air plenum; (2) improved equipment has made loading and unloading the grain, and in-bin materials handling, workable; (3) possibility of a large variation in moisture throughout the bed; and (4) cooling of grain with fan but without the heater.

Bean Drying Edible beans, such as pea (navy), yelloweye, kidney, pinto, great northern, and lima beans or soybeans, are often harvested when too wet (above 18% moisture) to store, so must be dried. Drying is done with forced air or heated air. If dried too slowly, spoilage may occur. If dried too rapidly, the seed coat and bean may crack and/or split, reducing the market quality. Continuous heated air drying can be done at 36°C (95°F) or less. Cracking is prevalent when more than 3 to 4% moisture is removed rapidly. Following a 4 to 6 hr holding period, drying may be continued. Beans handled by conventional means when cold, below 1.7°C (35°F), or too dry (less than 12% moisture) may incur intensive damage. Also see *Green Bean Drying*.

Beef Dehydration See *Freeze Drying; Nutritive Value*.

Beet (Red) Drying Red beets are used for coloring, flavoring, and food. Beets are sliced, diced, or cut into strips prior to drying. Beets are dried to 5% moisture content, or below, often in two steps. Blanched diced beets are dried at 93.3°C (200°F) on conveyor or in tunnel dryers with final drying which may take place in a bin dryer. Also see *Red Beet Drying; Sugar Beet Drying*.

Beet Pulp Drying Beet pulp is the wet material leaving the diffusers following the removal of sugar from sugar beets. The beet pulp is dewatered in a press. It may be fed, stored, or processed. One process of preparing an animal feed consists of mixing the pulp with molasses followed by drying.

Bell Peppers See *Pepper Drying*.

Belt Dryer A unit for continuously drying granular, aggregated, lump, pelleted, and similar products on a belt passing through a moving heated air stream. The product usually does not exceed 4 in depth. The belt may be an open mesh type, such that air will pass through while retaining the product. The unit may be provided with a heating section, primarily for drying, and a cooling section in which the temperature of the product is reduced and additional drying occurs. The unit may operate under vacuum or atmospheric conditions (Fig. B-4). Examples of products dried on belt dryers are fruits and vegetables, particularly apples, beets, carrots, onions, and potatoes. Also see *Continuous Band Dryer; Conveyor Dryer*.

Belt-Trough Dryer A type of continuous dryer in which the granular material being dried flows through the unit on a slowly overturning bed in the path of heated air forced through the bed.

Bernoulli Theorem The total head in steady flow of a fluid system at any section is equal to the total head at any further point in the

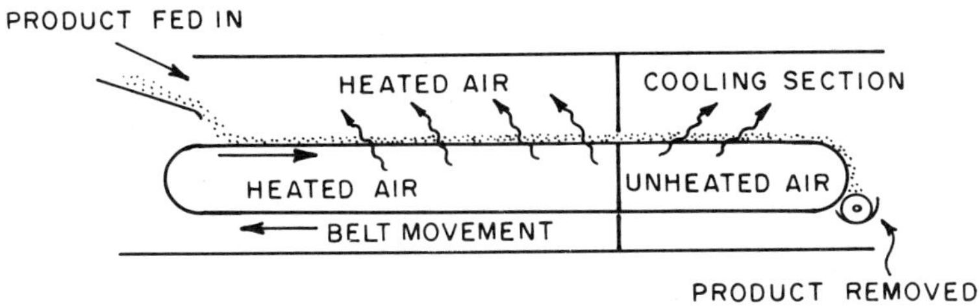

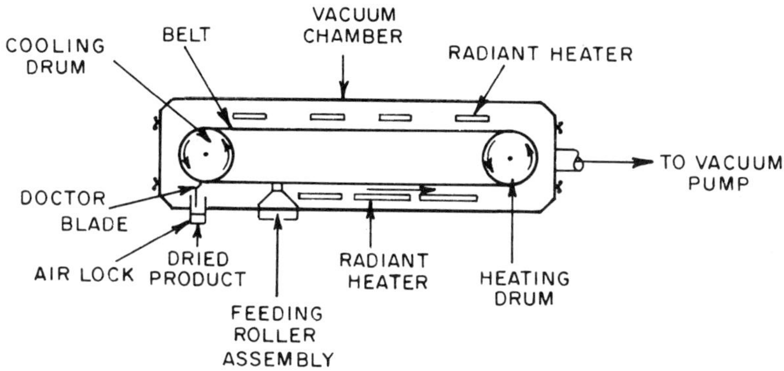

Figure B-4. Belt dryer, continuous: (a) atmospheric. (b) vacuum.

system plus the loss of head from friction between these two points. The total head is the sum of the pressure head, the potential head, and the velocity head.

$$\frac{p_1}{w} + z_1 + \frac{v_1^2}{2g} = \frac{p_2}{w} + z_2 + \frac{v_2^2}{2g} + \text{losses}$$

The theorem is applicable only for incompressible fluids with no work and no heat transfer.

B-E-T Equation Represents the Brunauer, Emmett, and Teller theory of multilayer adsorption which is based on the assumption that the same forces which produce condensation of moisture are also responsible for the binding energy of multilayer adsorption. The two constant B-E-T equation is useful in predicting the equilibrium moisture contents of cereal grains in 5 to 50% relative humidity environments. Also see *Adsorption; Equilibrium Isotherm; Smith Equation*.

Beverage Drying See *Fruit and Vegetable Juice; Milk Drying; Tomato Juice Drying.*

Bibliographies on Drying The following major bibliographies in English are devoted exclusively to drying:

Anon. 1945. Bibliography on Dehydration of Foods, 1938-43. Bibliographical Bulletin No. 6 USDA Library, Washington, D.C. 120 p.

Anon. 1951. Bibliography of Industrial Drying. Department of Scientific and Industrial Research (3 volumes), Great Britain, London, England, Dec. 802 p.

Anon. 1969-1970. Freeze Drying--Vol. I, Vol. II. A DDC Bibliography. Defense Documentation Center, AD702 700; AD 866 830. Alexandria, VA. 105 p., 136 p.

Anon. 1975. Bibliography for the Drying of Forest Products Materials, 1971-75. Forest Products Research Society, Madison, WI. 67 p.

Beary, Eugene G. 1967. Moisture equilibrium in relation to the chemical stability of dehydrated foods: a bibliography. AD 656 927. Defense Documentation Center, Alexandria, VA. 146 p.

Bennett, E. F. 1940. 1941. A Review of Driers and Drying. Paint Technology. Chemical Publishing Co., Brooklyn, NY. 90 p.

Bourton, Kay (Kathleen). 1967. 1968. Drying Bibliography, in Chemical and Process Engineering. Unit Operations - A Bibliographical Guide. MacDonald and Co., London. IFI/Plenum, NY. 534 p.

Corridon, G. A. 1963. Freeze Drying of Foods--A List of Selected References (1954-1962). Library List 77. National Agricultural Library, USDA, Washington, D.C. 79 p.

Dietz, V. R. 1944. 1956. Bibliography of Solid Absorbents. U. S. Cane Sugar Refiners and NBS, USDC. 887 p. 1528 p.

Graf, Dorothy. 1938. Artificial Drying of Farm Crops in U.S.--A Selected Bibliography. Bureau of Agr. Engr., USDA. 46 p.

Great Britain. 1914(1). 1917(2). Patent Office. Patents for inventions. Fifty Years Subject Index, 1861-1910. Class 34, Drying, Part 1. 1914. Part 2. 1917. London.

Hall, Carl W. 1976. Bibliography of Drying--International Books and Bibliographies. Publication 2-76, ASAE, St. Joseph, MI. 26 p.

Hansen, A. J. 1972. Engineering Drying Bibliography. CA Report No. 46, TAPPI, Atlanta, GA. 87 p.

Kemler, E., and LaJoy, M. 1953. I. Survey of Scientific Literature on Drying. II. Survey of Patent Literature on Drying. PB 134801; 307 p. PB 134803, 317 p.

Weaver, J. W., and Zerfoss, G. E. 1940. Natural and Artificial Drying of Hay. Selected References. Agr. Engr. Development Div., TVA, USDC, Knoxville, TN. 45 p.

Weiner, J., and Byrne, J. 1961. Drying of Paper and Board. Bibliographic Series No. 196. Institute of Paper Chemistry, Appleton, WI. 149 p.

Billion A large number which has different values in different countries. In the United States of America and France, one billion is 1 x 10^9; in England and other Western European countries one billion is 1 x 10^{12}. These numbers are also expressed as a thousand million and a million million, respectively.

Bin Drying A procedure commonly used to complete the dehydration or drying of vegetables, nuts, grain, or forage by holding a batch in a deep bin or container with slow movement of heated or unheated air (Fig. B-5). Also see *Batch-in-Bin Drying*.

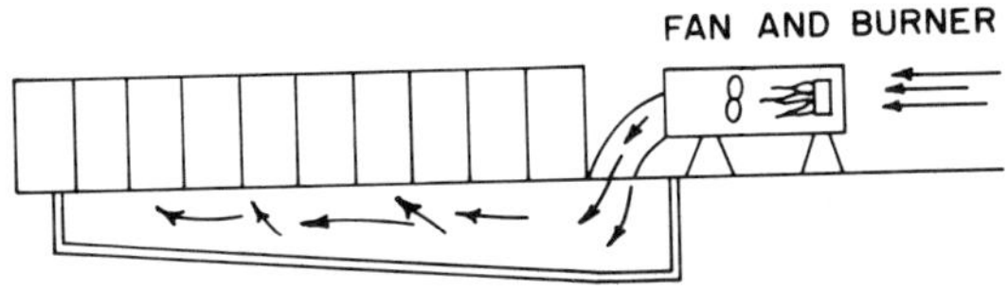

Figure B-5. Bin walnut dehydrator, pothole with forklift loading.
From Chhinnan, M. S., and Miller, M. W. (1977). ASAE Paper 77-6605.

Biological Tissue Drying Biological tissues must be dried if they are to be stored unless the water is immobilized and maintained as by freezing. Tissues may be dried for medical analysis and for surgical repair. For handling at normal temperatures dry tissues are used. Skin fails to be viable if dried below 25%. Body components containing protein are frozen to -40°C (-40°F) and moisture is removed by holding the product in a vacuum, condensing, and removing the vapor. Rapid freezing is necessary so that only small ice crystals are formed, minimizing damage to the cells. Using a refrigeration freeze-drying system may leave 1% residual moisture.

Biosphere See *Atmosphere*.

Biot Number A dimensionless group, Bi, used in unsteady-state heat transfer or mass transfer, which relates the internal resistance to the flow of heat or mass transfer to the resistance of that flow at the surface of a solid body. For heat transfer:

$$Bi_h = \frac{h\ell}{k}$$

where

h = film coefficient
ℓ = length (half thickness of plate or radius of sphere)
k = thermal conductivity of the solid

For mass transfer:

$$Bi_m = \frac{k_m L}{D}$$

where

k_m = mass transfer coefficient
L = thickness of layer
D = diffusivity at interface

BIRS The name of a process and the tower in which drying occurs in an especially designed tall tower. The product, such as a juice, is sprayed into the top of the tower, perhaps 77 m (250 ft) high. Dry air is forced vertically upward in the tower. Drying takes place at low temperatures, below 30°C (86°F), so that damage to the organoleptic properties and the flavor is avoided.

Black Body In heat transfer, a body which absorbs all radiation falling on the body. A black body has an absorptivity of one and a reflectivity of zero. The emissive power of a black body is dependent only on the temperature, being proportional to the fourth power of the absolute temperature. Also see *Stefen-Boltzmann*.

Blanching The procedure of briefly heating a fruit or vegetable, usually in hot water or steam, to inactivate enzymes and to produce other desirable product changes. The procedure is sometimes called "scalding." If vegetables such as sweet potatoes, carrots, and spinach are blanched before dehydration, there may be less loss of carotene during dehydration.

Block A product, tablet-like in appearance, made from condensed or dried milk or milk products. Agglomerated nonfat dry milk is compressed with 1300 to 4100 kPa (200 psi to 600 psi) presses to a density of 450 to 600 kg/m^3 to make milk blocks or tablets. Another procedure is to concentrate milk in an evaporator to one-third the original volume, add sugar, homogenize, dry in a scraped surface heat exchanger-dryer, and solidify in molds to form blocks or cakes. Cream tablets are made from a high fat product (30 to 55% fat) by compressing a spray-dried product.

Blood Drying Animal blood can be dried with a spray dryer using an entering drying air temperature of 166°C (330°F) and an exit temperature of 71.1°C (160°F) for an incoming product at 65% moisture, wet basis. Freeze drying is also used for moisture removal so there will be minimal thermal effect on the product while drying.

Two general approaches are used for drying blood for food and feed purposes: (1) Blood is collected from the animal and held in a cool place and clotted or coagulated. The coagulated material is placed on screen-bottom trays, and the serum is drained and dried in pans or trays in a cabinet dryer. (2) Blood is collected and treated with an anticoagulant (sodium citrate and sodium chloride solution).

The blood is screened and centrifuged. The blood is separated into 40% hemoglobin (28 to 30% solids) and 60% plasma (9% solids). The fibrin is removed from the plasma, leaving serum. The serum is condensed to 25% solids. The serum and the hemoglobin are spray-dried separately to 8% moisture. One hundred pounds of whole beef blood yields 4 lb of dry serum, 16 lb of dry hemoglobin, and 4 lb of wet fibrin. Animal blood may also be dried with a gas fired rotary dryer (kiln).

A beef animal of 1100 lb will give 33 lb of blood at 82% moisture, which upon drying gives about 6 lb of dry blood at 5% moisture and is 90% protein. In one procedure, the blood is coagulated with steam before drying. Disagreeable odors may be minimized by injecting an aerosol mist (propylene glycol base) into the stack discharge of the dryer.

Blood Meal   A by-product of the meat packing industry which is ground dried blood.

Blower   The terms fan and blowers are often used interchangeably. A blower is a mechanical device designed to deliver air with pressures up to 35 psi. For drying installations blowers are used for pressures up to 15 in water. Radial flow fans are commonly referred to as blowers and axial flow fans are commonly called fans. Also see *Fans and various types of fans*.

Board   See *Lumber*.

Boerner Sampler   A laboratory device used for separating or dividing grain samples into equal parts, often for the purpose of comparing or obtaining moisture contents of the aliquot parts.

Boiler Efficiency   The ratio of energy increase of the water in the boiler to the heat supplied by the fuel, multiplied by 100 to obtain the efficiency in percent.

$$\text{Efficiency} = \frac{G_a(h_1 - h_2)}{G_f(H)}$$

where

G_f = weight of fuel burned, kg/hr or lb/hr
H = calorific heating value of fuel, kJ/kg or Btu/lb
G_a = evaporation rate kg/hr or lb/hr
h_2, h_1 = enthalpy of steam entering and leaving the boiler, respectively, kJ/kg or Btu/lb

Boiler Horsepower (BHP)   Rating of a boiler based on the equivalent of evaporation of water at the rate of 34.5 lb/hr for 1 BHP at atmospheric pressure. In SI units one BHP = 9.8 kW.

Boiling   The process of evaporation occurring throughout the liquid. Boiling of liquids may occur in the various regimes of natural convection from a heating surface, nucleate boiling, partial film boiling, and film boiling. The temperature of boiling for a particular liquid depends on the pressure above the liquid (Table B-1). As the rate of heating is increased, the temperature of boiling does not increase above a certain value as determined by the pressure, but the rate of evaporation is increased. During boiling the pressure of the vapor in the liquid must exceed the pressure over the liquid by an infinitesimal amount. As the concentration of a solution increases during evaporation the temperature of boiling increases. One gram molecular weight of substance added to 1 kg of water increases the boiling point 0.5°C (0.9°F) at 1 atm. Also see *Raoult's Law*.

Table B-1
Boiling Temperature vs. Pressure and Latent Heat of Water

Temperature, °F	Pressure, psia	Latent Heat, Btu/lb	Temperature, °C	Pressure, kPa	Latent Heat, kJ/kg
32	0.0885	1075.8	0.01	0.61	2501.3
40	0.1217	1071.3	10	1.23	2477.7
50	0.1781	1065.6	20	2.34	2454.1
60	0.2563	1059.9	30	4.25	2430.5
68	0.3390	1055.5	40	7.38	2406.7
70	0.3631	1054.3	50	12.35	2382.7
80	0.5069	1048.6	60	19.94	2358.5
90	0.6982	1042.9	70	31.19	2333.8
100	0.9492	1037.2	80	47.39	2308.8
125	1.942	1022.9	90	70.14	2283.2
150	3.718	1008.2	100	101	2257.0
175	6.715	993.3	125	232	2188.5
200	11.526	977.9	150	476	2114.3
212	14.696	970.3	175	892	2032.4
250	29.825	945.5	200	1554	1940.7
300	67.013	910.1	250	3973	1716.2
350	134.63	870.7	300	8581	1404
400	247.31	826	350	16513	893

Boltzmann (Gas) Constant (k) A constant representing the ratio of the general gas law constant, R, and the number of molecules, N; a modification of the universal gas constant, where R is the gas constant for one gram mole;

$$k = \frac{R}{N}, \quad k = 1.381 \times 10^{-16} \text{ erg/K}$$

where R is the gas constant in $pv = RT$, $R = 8.314$ J/mole K, and N is Avogadro's number. k is the gas constant per molecule.

Boltzmann Number A dimensionless number, Bo, which is a parameter of radiation heat exchange representing the relation between the enthalpy of gases and heat flow emitted at the surface of an enclosure.

$$\text{Bo} = \frac{(\text{flow velocity})(\text{specific heat})(\text{specific weight})}{(\text{Stefan-Boltzmann constant})(\text{absolute temperature})^3}$$

$$\text{Bo} = \frac{w\ c\ \gamma}{\sigma T^3}$$

where

w = flow velocity
c = specific heat
γ = specific weight
σ = Stefan-Boltzmann constant
T = temperature

See also *Boltzmann Constant; Stefan-Boltzmann.*

Bone-Dry A general term used to identify a product which is very dry, containing only traces of moisture.

Books on Drying The following list includes only those books printed in English of at least 100 pages in length devoted principally to drying and dehydration published since 1950. Several excellent books in foreign languages are available, principally in German, French, Spanish, Russian, Polish, Hungarian, etc., which are listed in bibliographies. Many excellent discussions of drying equipment, processes, and products appear as chapters in books devoted to a broader range of subjects. Many bulletins cover drying. Numerous articles appear in technical journals. Many papers on drying have been presented at technical society meetings, principally AIChE, ASAE, ASME, etc. These are not included in the following list.

Anderson, Harriette. 1970. Kiln-Fired Glass. Chilton Book Co., Philadelphia, PA. 185 p.

Anon. 1954. 1959. 1966. Farm Grain Drying and Storage. H.M.S.O., London. 121 p.; 119 p.; 172 p.

Anon. 1958. Fundamental Aspects of the Dehydration of Foodstuffs. Society of Chemical Industry, London. 238 p.

Anon. 1960. Freeze-Dehydration of Foods. Res. and Devel. Assoc., Food and Container Institute, Chicago. 108 p.

Anon. 1961. The Accelerated Freeze-Drying of Food Preservation. H.M.S.O., London. 169 p.

Anon. 1964. Water Removal at the Presses and Dryers. Canadian Pulp and Paper Assoc., Montreal. 218 p.

Anon. 1966. Advances in Freeze-Drying. Hermann Press, Paris.

Bateson, Richard G. 1952 (3rd ed.). Timber Drying and the Behavior of Seasoned Timber in Use. C. Lockwood, London. 143 p.

Binsted, Raymond, and Dewey, J. D. 1960. Soup Manufacture: Canning, Dehydration and Quick Freezing. Food Trade Press, London. 166 p.

Bolton, Eleanor R. 1958. Dried Flowers with a Fresh Look. Van Nostrand, Toronto. 210 p.

Brooker, Donald B., Bakker-Arkema, Fred W., Hall, Carl W. 1974. Drying Cereal Grains. Avi Publishing Co., Westport, CT. 263 p.

Brown, W. H. 1965. An Introduction to the Seasoning of Lumber. MacMillan, N. Y., and Pergamon, Oxford. 178 p.

Carico, Nita (Cox), Guynn, J. C. 1962. The Dried-Flower Book. Doubleday, Garden City, N. Y. 128 p.

Copson, David A. 1962. Microwave Heating in Freeze-Drying, Electronic Ovens and Other Applications. Avi Publishing Co., Westport, CT. 433 p.

Cotson, S., and Smith, D. B. (ed.). 1963. Freeze-Drying of Foodstuffs. Columbine Press, Manchester, London. 295 p.

Crowe, Irene. 1976. Oven Drying. Sheed and Ward, Inc., Kansas City, MO. 158 p.

Devine, J. 1959. Production Kiln Drying. Moore Dry Kiln Co., Jacksonville, FL. 149 p.

Deyarburgh-Bateson, Nina. 1973. Dried Flowers: The Art of Preserving and Arranging. Charles Scribner's Sons, N. Y.

Farm Journal Editors. 1975. How to Dry Fruits and Flowers at Home. Countryside Press, Philadelphia, PA. 124 p.

Findley, W. P. K. 1962. The Preservation of Timber. Adams and Charles Black, London; Macmillan, Toronto and N. Y. 162 p.

Fisher, Frank R. 1962. Freeze-Drying of Foods. National Academy of Science-National Research Council, Washington, D.C. 237 p.

Floyd, Harriet. 1973. Plant It Now, Dry It Later. McGraw-Hill Book Co., N. Y. 231 p.

Fraser, Henry L. 1969. Kilns and Kiln Firing for the Kraft Potter. Pitman, London. 134 p.

Gavelin, N. Gunnar (ed.). 1972. Drying of Paper and Paperboard. Lockwood, London. 309 p.

Gerzhoi, A. P., and Samochetov, V. F. 1951. 1958 (3rd ed.). Grain Drying. English translation from Russian available from U.S.O.T.S. 60-21164, 304 p.

Gibb, J. A. C. 1955. Crop Drying, Barn and Storage Machinery. Temple Press, Ltd., London. 178 p.

Gillies, M. T. 1974. Dehydration of Natural and Simulated Dairy Products. Noyes Data Corp., Park Ridge, NJ. 328 p.

Ginzburg, A. S. 1969 (English ed.). Translated by A. Grochowski. Application of Infra-Red Radiation in Food Processing. Leonard Hill, London. 412 p.

Goldblith, S. A., Rey, L., and Rothmayr, W. W. 1975. Freeze Drying and Advanced Food Technology. Academic Press, London and N. Y. 730 p.

Goose, Peter G., and Binsted, Raymond. 1964. 1973 (2nd ed.). Tomato Paste, Puree, Juice and Powder. Food Trade Press, London. 151 p. (1st ed.).

Greensmith, Maurice. 1971. Practical Dehydration. Food Trade Press, London, and Avi Publishing Co., Westport, CT. 174 p.

Hall, Carl W. 1957. Drying Farm Crops. Avi Publishing Co., Westport, CT. 336 p.

Hall, Carl W., and Hedrick, T. I. 1966. 1971. Drying of Milk and Milk Products. Avi Publishing Co., Westport, CT. 338 p., 338 p.

Hanson, S. W. F. (ed.). 1961. The Accelerated Freeze-Drying (AFD) Method of Food Preservation. H.M.S.O., London. 169 p.

Harris, R. J. C. (ed.). 1951. Freezing and Drying--Report of a Symposium, held June, 1951. The Institute of Biology, London. 205 p.

Harris, R. J. C. 1951. 1958 (2nd ed.). Freezing and Drying. Symposium of the Institute of Biology, London, Hafner, N. Y. 205 p.

Harris, R. J. C. (ed.). 1954. Biological Applications of Freezing and Drying. Academic Press, N. Y. 415 p.

James, Ronald W. 1973. Desiccants and Humectants. Noyes Data Corp., Park Ridge, NJ. 313 p.

Keey, R. B. 1972. Drying Principles and Practices. Pergamon Press, N. Y., Oxford, Toronto. 358 p.

King, C. Judson and Clark, J. Peter (ed.). 1977. Water Removal Processes: Drying and Concentration of Foods and Other Materials. AIChE Symposium Series 163 (vol. 73). N. Y. 206 p.

Lal, Radhey and Agrawal, K. K. 1968. Principles and Practices of Rice Drying. IIT, Kharagapur, India. 184 p.

Lykov, A. V. 1966. Heat and Mass Transfer in Capillary-Porous Bodies. Pergamon Press, Oxford and N. Y. 528 p.

Macey, H. H. 1950. Drying in the Heavy Clay Industry. H.M.S.O., London. 189 p. (*Also*, Paper Three, Ministry of Works, National Brick Advisory Council, London.)

McIntyre, J. M., Stone, W. K., and Peterson, M. S. 1954. 1955. Dry Whole Milk. National Academy of Sciences, Washington, D.C. 194 p.

Marshall, W. R., Jr. 1954. Atomization and Spray Drying. No. 2 (Vol. 50) Chem. Engr. Prog. Monograph Series, AIChE, N. Y. 122 p.

Martin, G. 1950. Chemical Engineering and Thermodynamics Applied to the Cement Rotary Kiln. Anglobooks, Technology Press, N. Y. and London. 244 p.

Masters, K. 1972. Spray Drying. Leonard Hill, London, and Chemical Rubber Co., Cleveland, OH. 668 p.

Morrison, Winifrede. 1975. Drying and Preserving Flowers. Hippocrene Books, N. Y.

Nonhebel, Gordon, and Moss, Arnold A. H. 1971. Drying of Solids in the Chemical Industry. Butterworths, London, and Chemical Rubber Co., Cleveland, OH. 310 p.

Noyes, Robert. 1968. Freeze Drying of Foods and Biologicals. Noyes Development Corp., Park Ridge, NJ. 313 p.

Noyes, Robert. 1969. Dehydration Processes for Convenience Foods. Noyes Development Corp., Park Ridge, NJ. 367 p.

Obayashi, Yoji. 1955. Dried BCG Vaccine. World Health Organization, Geneva. 220 p.

Parkes, A. S., and Smith, A. U. (ed.). 1960. Research in Freezing and Drying. Blackwell Scientific Publ., London, Oxford, and Charles C. Thomas Publ., Springfield, IL. 320 p.

Peck, Edward C. 1961. 1962. Lumber Seasoning. MSS Copy. University of Wisconsin, Madison. 182 p.

Peray, K., and Wadell, J. J. 1972. The Rotary Cement Kiln. Chemical Publishing Co., N. Y. 200 p.

Rhodes, Daniel. 1968. 1969. Kilns; Design, Construction, and Operation. Chilton, Philadelphia, and Pitman, London. 241 p.

Roberts, E. J. 1937. Grass Drying. (Agricultural Research Council) H.M.S.O., London. 125 p.

Robson, James T. 1954. Operating the Tunnel Kiln. Industrial Publ., Chicago, IL. 157 p.

Rutner, E., Goldfinger, P., and Hirth, J. P. 1965. Condensation and Evaporation of Solids. Gordon and Breach, N. Y. 707 p.

Sharp, J. G. 1953. Dehydrated Meat. (Dept. Sci. and Ind. Res.) H.M.S.O., London. 184 p.

Spellacy, John R. 1953. Casein, Dried and Condensed Whey. Lithotype Process Co., San Francisco, CA. 523 p.

Spicer, Arnold (ed.). 1974. Advances in Preconcentration and Dehydration of Foods. Halsted Press, John Wiley and Sons, N. Y. 526 p.

Squires, Mabel. 1958. The Art of Drying Plants and Flowers. M. Barrons, N. Y. 258 p. *and* Bonanza Books (Crown), N. Y. 258 p. *and* Gramercy Publ. Co., (Crown), N. Y. 258 p.

Stevens, William C., and Pratt, G. H. 1952. 1969. Kiln Operator's Handbook. H.M.S.O., London. 185 p. 153 p.

Stevenson, Violet. 1972 (2nd ed.). Dried Flowers for Decoration. Drake Publ., N. Y. 171 p.

Torrey, M. 1974. Dehydration of Fruits and Vegetables. Noyes Data Corp., Park Ridge, NJ. 287 p.

Van Arsdel, W. B., and Copley, Michael J. 1963. Food Dehydration. Vol. I, Principles. Avi Publishing Co., Westport, CT. 185 p.

Van Arsdel, Wallace B., Copley, Michael J., and Morgan, Arthur I. Vol. 1, 1973 (2nd ed.). Food Dehydration, Drying Methods and Phenomena. Avi Publishing Co., Westport, CT. 347 p.

Van Arsdel, W. B., and Copley, M. J. 1964. Food Dehydration. Vol. II, Products and Technology. Avi Publishing Co., Westport, CT. 721 p.

Van Arsdel, W. B., Copley, M. J., and Morgan, A. I. 1973 (2nd ed.). Food Dehydration, Vol. 2, Practices and Applications. Avi Publishing Co., Westport, CT. 529 p.

Vanacek, V., Markvart, M., and Drbohlav, R. 1966. Fluidized Bed Drying. (Translated by Josef Landau from Czech to English.) Leonard Hill, London. 195 p.

Von Loesecke, Harry W. 1955. Drying and Dehydration of Foods. Reinhold Publ. Corp., N. Y. 300 p.

Whyte, R. O., and Yeo, M. L. 1952. Green Crop Drying. Faber and Faber, London. 328 p.

Williams-Gardner, A. 1971. Industrial Drying. Leonard Hill, London, and Chemical Rubber Co., Cleveland, OH. 320 p.

Boundary Layer The boundary layer represents a region of velocity change from free stream to solid boundary of the flow pattern in a fluid stream near the vicinity of a solid body. The boundary layer terminology is applied to (1) velocity boundary layer and (2) thermal boundary layer. Outside of the boundary layer there is a negligible effect of the film on the velocity and/or temperature profile. The boundary layer thickness is equal to the velocity or temperature is equal to 95-99% of the free stream velocity or temperature (Fig. B-6).

Bound Moisture The liquid held by a solid in small capillaries, in solution in cells, and physical or chemical adsorption on solid

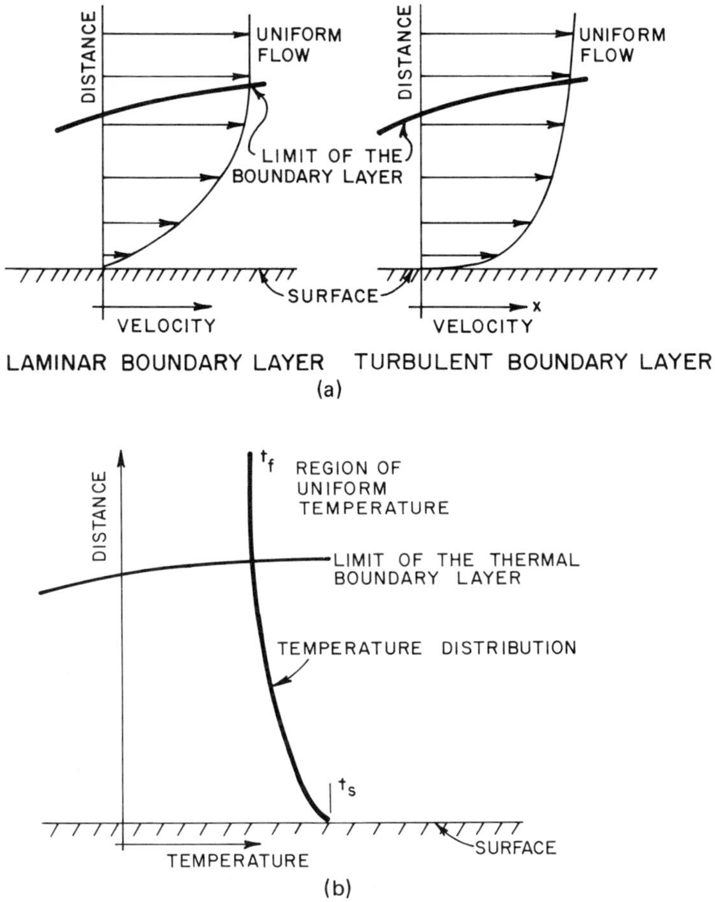

Figure B-6. Boundary layer flow. (a) velocity. (b) thermal.
Reprinted with permission of Macmillan Publishing Co., Inc. From Heat Transfer by Alan J. Chapman. Copyright 1967 by Alan J. Chapman.

surfaces, and with a vapor pressure less than that of the pure liquid at the same temperature. Unbound moisture is that moisture in excess of the equilibrium moisture content corresponding to the saturation humidity.

Bourdon Gage An instrument for measuring gas pressures or vapor pressures, such as steam, in which the operating element consists of a tube curved in a circular arc which has an elliptical cross-section. One end of the tube is fixed with the other end connected to a mechanical linkage through which the pressure is indicated on a calibrated scale (Fig. B-7).

Box Kiln See *Compartment Kiln.*

Boyle-Charles Law The combination of the Boyle's law, which relates volume, V, and absolute pressure (p), and Charles' law, which relates volume, V, and absolute temperature, T, for an ideal gas is as follows:

$$\frac{p_1V_1}{T_1} = \frac{p_2V_2}{T_2} = MR$$

where the gas constant, R, is 53.3 ft lb/lb R (0.2871 kJ/kg K) for air; 34.9 (0.1880) for carbon dioxide; 55.1 (0.2968) for carbon monoxide; 48.3 (0.2602) for oxygen; and 55.1 (0.2968) for nitrogen. M is the mass of the gas under consideration. Also see *Boyle's Law; Charles' Law; General Gas Law.*

Boyle's Law The volume, V, of a perfect gas is inversely proportional to the absolute pressure, p, of that gas at constant temperature:

$$p_1V_1 = p_2V_2$$

Brake Horsepower See *Horsepower.*

Brewer's Dried Grains The dried extracted residue of barley malt or of a mixture with other cereal grain or grain products.

Brick Drying See *Chamber Drying; Clay Drying; Hack Drying; Kiln; Tunnel Dryer.*

Bright-Leaf Tobacco Drying (Curing) The curing of bright-leaf tobacco with natural air or with a heat supply and a method of obtaining full or partial recirculation of the air in a tightly-built barn. Natural draft ventilation is often used with an adjustable ridge ventilator at the top of the storage building. The desirable temperatures and humidities at various stages of the curing are listed in Table B-2. The initial curing process consists of yellowing the tobacco during which large quantities of starch are converted to sugar and the tobacco leaf is maintained in a physical condition which permits the moisture to be removed quickly from the cells. As soon as the proper color is reached the leaf cells are killed by using a high temperature. The water must be removed from the leaves as rapidly as possible when the desired color has been obtained. The heating system for drying is larger than required for yellowing only. Drying can progress very rapidly as long as

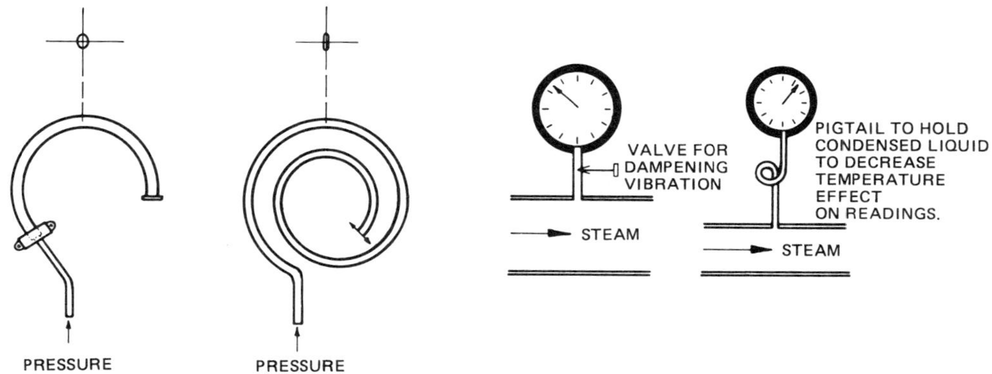

Figure B-7. Bourdon tube and spiral element.
From Hall, C. W., and Hedrick, T. I., Drying of Milk and Milk Products, 2nd edition (1971). Used with permission of Avi Publishing Co., Westport, CT.

Table B-2
Bright-Leaf Tobacco Curing Conditions

Storage	Critical Relative Humidity, %	Critical Temperature, °F
Yellowing	92-85	4-8 above outside
Color setting	75, 65, 50	105, 110, 120
Killing	below 10	170-200
Drying	25-39	140

scorching of the leaves is avoided. After the tobacco is dried it is "ordered," meaning that it absorbs moisture from the atmosphere and becomes pliable.

Brown-Duvel A method of moisture determination in which the moisture is removed from the product by heating a weighed sample, usually 0.1 kg, in oil. The moisture content is determined on the basis of the moisture removed when collected as condensed vapor or from loss of weight of the sample. It is one of the earliest standard methods for determining the moisture content of grain. The original method required about 1 hr to obtain the moisture content. The modified Brown-Duvel utilizes more rapid heating of the sample using a higher oil temperature for heating and stirring of the oil. The oil is heated to 145°C (293°F) for forages and 190°C (374°F) for grains.

Browning See *Heat Damage; Maillard Reaction.*

BTU (British Thermal Unit) A common measure of quantity of heat. One Btu will increase the temperature of one pound of water one degree Fahrenheit at 4°C (39.2°F). One Btu is equal to 0.252 kg-cal (Cal), 252 cal., 1055 Joules, 1.055 kJ, and 778.2 ft lb. Also see *B.T.U.*

BTU/Hr, British Thermal Units per Hour A measure of the rate of release or gain or flow of energy. As an example, a burner for a dryer is usually rated according to its heat output in Btu/hr. In SI units one Btu/hr is 0.293 W.

Buffalo Grass Seed Drying The removal of moisture from seed after treatment to cause germination. Only about 7-10% of new buffalo grass seed will germinate. If the seed is soaked for 24 hr in a 0.5% saltpeter solution, then chilled at 4.4°C (40°F) for 6 wk and then dried, the germination increases to about 80%. The seed is dried with heated air at 50°C (122°F).

Bulk Density The gross or apparent density of a material which is the mass per unit of gross volume including void space (Table B-3). Also see *Apparent Density.*

Bulk Fluid Temperature The temperature at a cross-section of a

Table B-3
Bulk Density of Material

	Bulk Densities, ±10%	
	lb/ft^3	kg/m^3
Cement	94-100	1505-1602
Cereals	26-48	416-769
Clay	63	1009
Coal, lump	50	801
Flour	28	449
Gravel	110	1762
Lime, gypsum	53-64	849-1025
Limestone	95	1521
Sand	90-105	1442-1682
Sawdust	12	192
Starch	96	1538

For grains and seeds, see Table B-4, Bushel Weights and Bulk Densities.

flowing medium if the medium were mixed to a uniform temperature, also called cup mixed temperature. The bulk fluid temperature is defined as the flux of the enthalpy at a cross-section divided by the product of the mass flow rate and specific heat.

Bulk Specific Gravity See *Apparent Specific Gravity.*

Bulk Temperature The average temperature of the mass of a product, usually as would occur in a bin dryer, bulk storage, or flowing mass. To determine temperatures, thermocouples may be placed at appropriate locations in the mass such that average readings will represent the bulk temperature. The bulk temperature in the free stream is the weighted mean of the temperature of a cross-section of fluid.

Bulk Ventilation Dryer A unit for removing moisture from grain, forage, or aggregates in which the air is directed by ducts or an open (slotted, perforated, mesh) floor through the product, usually in large batches, usually without moving the product during or after drying.

Bulkhead A partition of wood, concrete, or metal, usually at the end of a container used for drying or storage, to retain the products.

Bulygin Number A dimensionless number, Bu, for representing high intensity heat and mass transfer during evaporation. The number pertains to the amount of heat expended to produce vaporization as related to the heat to bring the liquid to boiling

$$Bu = \frac{L\, c_p\, p_o}{c_q\, (t_c - t_o)}$$

where

L = latent heat
c_p = specific heat
p_o = pressure in the material
c_q = specific heat of the moist material
$t_c - t_o$ = temperature difference between the medium and the material.

Burley Tobacco Drying (Curing) Burley tobacco is used principally for chewing and is grown mainly in the Ohio River area. Burley tobacco is air-cured by ventilation through the storage building. Heat is used when the average relative humidity in the storage cannot be maintained naturally between 65 and 75% during a 24 hr period. Uniform heat must be supplied from a fuel containing a low sulfur content. The heater should have a capacity of 29,300 W (or J/s) (100,000 Btu/hr) for each acre of tobacco stored.

Burner Efficiency See *Dryer Efficiency; Fuel; Safety.*

Bushel A volume measure (often used for grain) which equals 2150.42 in^3 or 1.244 ft^3 in the U. S. One dry bushel is that weight of grain defined by government grain standards as a standard unit for commercial trading, usually at a specific moisture. A bushel of No. 2 shelled corn, for example, is 56 lb at 15.5% wet basis moisture content (Table B-4). A wet bushel is the weight of wet grain which will give one bushel of dry product, as defined above. One bushel is equivalent to 32 quarts (dry), 2150.4 in^3, 0.035239 m^3, 35.238 liters, or 4 pecks. The British bushel is 1.032 U.S. bushel.

Buttermilk Drying Buttermilk may be dried and used in ice cream mix or baking ingredients. The procedure for drying is similar to skim milk drying, using a spray dryer. Also see *Milk Drying*.

Table B-4
Bushel Weights and Densities

Grain or Seed	Approximate Net Weight, lb	Bulk Density, lb/cu. ft.	Grain or Seed	Approximate Net Weight, lb	Bulk Density, lb/cu. ft.
Alfalfa	60	48.0	Lentils	60	48.0
Barley	48	38.4	Millet	48-50	38.4-40.0
Beans:			Oats	32	25.6
Lima, dry	56	44.8	Orchard grass	14	11.2
Lima, unshelled	32	25.6	Peanuts, unshelled:		
Snap	30	24.0	Virginia type	22	17.6
Other, dry	60	48.0	Runners, southeastern	28	22.4
Blue grass	14-30	11.2-24.0	Spanish	30	24.0
Broomcorn	44-50	35.2-40.0	Popcorn:		
Buckwheat	48-52	38.4-41.6	On ear	70	28.0
Castor beans	46	36.8	Shelled	56	44.8
Clover	60	48.0	Rapeseed	50 & 60	40.0 & 48.0
Corn:			Redtop	50 & 60	40.0 & 48.0
Ear, husked	70	28.0	Rice, rough	45	36.0
Shelled	56	44.8	Rye	56	44.8
Green sweet	35	28.0	Soybeans	60	48.0
Cottonseed	32	25.6	Sudan grass	40	32.0
Cowpeas	60	48.0	Sunflower	24 & 32	19.2 & 25.6
Flaxseed	56	44.8	Timothy	45	36.0
Grain sorghums	56 & 50	44.8 & 40.0	Vetch	60	48.0
Hempseed	44	35.2	Walnuts	50	40.0
Hickory nuts	50	40.0	Wheat	60	48.0
Kafir	56 & 50	44.8 & 40.0			

Conversion factors: $1\ lb = 0.45359\ kg$; $1\ lb/ft^3 = 16.018\ kg/m^3$

C

Cabbage Drying Cabbage is shredded and placed on trays and dried in a two-stage tunnel dryer, with the first stage at 82.2°C (180°F) with parallel air and product flow, and the second stage at 62.8°C (145°F) with countercurrent air and product flow with the product dried from 95 to 8% in six hours. Bin drying follows using heated air at 48.9°C (120°F) to dry to 4% in 7-10 hr.

Cabinet Dryer A batch dryer in which products are placed in closed box-like compartments, with heated air circulated or thermal radiation for drying. The products are placed on trays or racks. Natural air circulation can be used but usually forced ventilation is used. Also see *Fish Drying*.

Cacao Drying; Cocoa Cacao is the bean from which chocolate, sometimes called cocoa, is made, during which moisture removal is involved. The processing which precedes drying consists of fermenting and washing the bean upon removal of the coat. Cacao is dried by (1) solar heating of the cacao beans in a tray or on a drying floor, during which time the beans are stirred periodically, (2) limited heating, or (3) heated air drying. Controlled heated air drying using air heated to 60 to 70°C (156 to 211°F) in vertical column or rotary drum dryers.

Caking The uniting of powdered, ground, or crushed material into a larger solid mass. The formation of cakes by pressure may be a desired process, such as for blocks, pellets, and tablets. When it is desired to maintain the powdered dry state, caking is undesirable. Hot bagging or storing of moist or dried materials, such as sugar and dry milk powder, will increase caking and may lead to deterioration of the product.

Calcium Chloride A white salt ($CaCl_2$) which is highly hygroscopic and deliquescent. The salt is used to absorb moisture from air in rooms to reduce the humidity and from air surrounding products to dry products. The salt becomes wet then forms a solution from the absorbed moisture. Over an extended period of several weeks, anhydrous calcium chloride will absorb up to 1-1/4 times its dry weight in a saturated atmosphere. At 20°C (68°F) about 1-1/3 times the dry weight is required to form a solution. Also see *Absorption; Adsorption; Air Drying; Efficiency of Drying Agents*.

Calender A machine for smoothing or dressing cloth or paper by passing over one or more cylinders generally during drying.

Calibration Chart A graphical representation of the value read from an instrument, usually placed on the abscissa, and the true value of the variable, usually located on the ordinate. Calibration consists of comparing an instrument to a known standard for measuring values. The instrument should be calibrated at a sufficient number of points on the scale to assure the proper relationship between measured value and indicated value (Fig. C-1). A calibration chart may be furnished with the instrument by the manufacturer, or may be developed by the user after calibration.

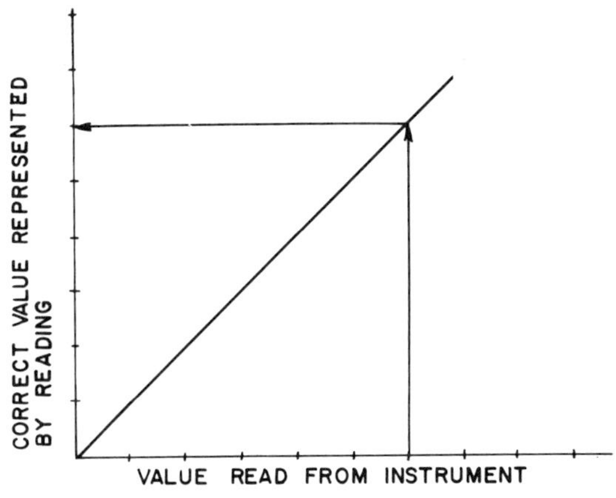

Figure C-1. Calibration chart.

Calorie The quantity of heat required to raise the temperature of 1 gram of water from 3.5 to 4.5°C (38.3 to 40.1°F) known as the small calorie; the normal calorie is defined as the quantity of heat required to raise the temperature of 1 gram of water from 14.5 to 15.5°C (58.1 to 59.9°F). The large calorie, often written with a capital C to distinguish a Calorie, which is 1000 small calories, is also called Kilocalories or kilogram-calories as contrasted to calories or gram-calories. The mean calorie is defined as 1/100 of the amount of heat required to raise 1 gram of water from 0 to 100°C (32 to 212°F); 252 calories equal 1 Btu. One calorie equals 4.187 Joules.

Calorific Value The heat available in fuel by burning or in food by digestion, in calories (gram-calories) per gram, kJ/kg, or Btu/lb (Table C-1). The calorific value of fuels by combustion are commonly referred to as heating values. General calorific values for foods in calories/gram are: cheese, 4-5; eggs, 1.5; milk, 0.7; meat, 1.1; bread, 2.5; cereals, 3.5-4.0; vegetables, 0.5-1.0; fruits, 0.2-1.0; nuts, 6.0-7.5. A 70 kg (154 lb) mature person has a basic minimum requirement of 1700 cal/day and when involved in light work, 2300 cal/day.

Table C-1
Calorific Values, Approximate

	kcal/100 g	kJ/kg
Foods		
Cheese	4.00-4.75	165-200
Egg	1.50	63
Milk	0.71	30
Chicken	1.10	46
Bread	2.6	110
Cereals	3.50-3.75	146-162
Chocolate	5	209
Vegetables	0.50-1.00	21-42
Nuts	6.00-7.50	250-315

Conversion factors: 1 kcal/100 g = 41,840 J/kg = 41.84 kJ/kg.

Calorimeter A device for measuring the quantity of heat (1) produced by combustion, as a fuel, for heating value or (2) transferred by a change in temperature of a liquid or solid, for determining specific heat. A chamber in which the specimen is placed is surrounded by a liquid in a larger chamber. By measuring the temperature change in the outer chamber, the heat produced or transferred is calculated.

Can Dryer A dryer for removing moisture from paper in which the paper passes over steam-heated cylinders and the product is dried by heat conduction and air forced over the cylinders. Also see *Contact Dryer*.

Canning A method of food preservation in which food contained in a permanently sealed container is subjected to an elevated temperature for a definite period of time and then cooled. The spoilage microorganisms are destroyed or their growth inhibited when appropriately heated. Time-temperature exposures have been established for destroying many different organisms. Also see *Dehydrocanning*.

Capacitance The characteristic of flow of electricity in an electric circuit which contains a condenser. Condensers are two parallel surfaces separated by a dielectric material. The parallel surfaces can be in the form of cylinders, spheres, or flat surfaces (plates). One type of indirect moisture measurement uses the principle of capacitance in which the product is placed between the plates, the product serving as a dielectric material. In general the following relationships exist for biological materials: (1) the capacitance does not change greatly with frequency in the range of 170 to 700 kilocycles, and (2) the capacitance varies directly with moisture content. Moisture meters of this type are also called dielectric meters. Also see *Dielectric*.

Capillary Small tubes through which moisture moves in a porous solid body. A tube is considered a capillary if $\ell/h' \ll 1$ where ℓ = mean free path of molecule and h' = maximum height of rise of liquid. As in the rise of liquids in small diameter tubes, flow is the result of surface forces. The capillary theory of drying is used to explain the movement of moisture in the product bed during drying. One theory of drying at high moisture contents is based on the capillary principle. The pore space between granular materials of various sizes provides passages which change during drying. Evaporated moisture permits continuous flow to the liquid surface, which depends on the pore size, surface tension, and density of liquid. Also see *Macrocapillary and Microcapillary*.

Capillary Bound Water A thin layer of moisture adsorbed on the walls of the capillary as distinct from the water contained in the capillaries and bounded by free menisci, which is free moisture. The capillary bound water takes more energy to remove than the free water. Also see *Chemically Combined Water; Free Moisture*.

Capillary Flow The flow of moisture through the interstices of a solid by liquid-solid molecular attraction.

Capillary Number A dimensionless group, Ca, relating the viscous force to the surface tension during two-fluid atomization:

$$Ca = \frac{\mu V}{\sigma}$$

where

μ = dynamic viscosity
V = fluid velocity
σ = surface tension

Capillary-Porous A body in which the pores are considered to be capillaries, and the capillary potential is significantly greater than the gravity potential, so that the action of gravity on the liquid in the capillary can be neglected.

Carbohydrate Includes starches, sugars, and cellulose identified by the fact they all contain carbon, hydrogen, and oxygen. (CH_2O) The possible deleterious effect of the heat and high temperature on carbohydrates when drying food products must be carefully considered. Possible deleterious effects are discoloration, carmelization, physical changes, and nutritional deterioration. Also see *Cellulose; Starch; Sugar*.

Carbon See *Activated Carbon; Combustion*.

Carbon Dioxide A colorless and odorless gas (CO_2) at atmospheric conditions which is a product of combustion of carbon-containing fuels. The amount of carbon dioxide in the product of combustion is one measure of the efficiency of combustion. The carbon dioxide level of flue gas is about 21% for complete combustion of pure carbon in air. Carbon dioxide gas is used to fill containers to replace air (oxygen) to prevent oxidation of contents. Also see *Combustion; Gas Storage*.

Carbon Monoxide A colorless gas (CO) that is a product of combustion and which is highly poisonous. The amount of carbon monoxide present following combustion is a measure of the incompleteness of burning, as measured in the exhaust from a heater in a dryer. Exhaust systems must be properly sealed and ventilated for safety of workers in the area. Incomplete combustion gives carbon monoxide and carbon dioxide plus water. Also see *Carbon Dioxide; Combustion*.

Carmelization Excessive heat treatment may cause a darkening of the sugar in a product, such as of lactose during drying of milk, and other sugars in products. Carmelization usually produces an undesirable effect, but for some foods, such as candies, carmelization may be desirable.

Carotene A chemical compound with orange pigment in leafy portions of green plants, such as alfalfa, carrots, sweet potatoes, and yellow corn. Carotene is changed to vitamin A in human and animal bodies. Pure carotene is oxidized when exposed to air, causing deterioration. The oxidation is accelerated in sunlight and at high temperatures. The carotene loss is greatly reduced if alfalfa, for example, is dried by heated forced air as contrasted to field drying for several days in the sunlight. Carotene loss in vegetables during drying may be reduced by blanching preceding drying, particularly for spinach, sweet potatoes, and carrots. Also see *Blanching; Vitamins*.

Carrier Equation Using the wet-bulb and dry-bulb temperature, the vapor pressure of water vapor can be determined by the empirical relationship:

$$p = p_s - \frac{(B - p_s)(t - t_s)}{2800 - 1.3\ t_s}$$

where

p_s = saturation pressure at the wet-bulb temperature (t_s), in Hg
t = dry bulb temperature, °F
t_s = wet bulb temperature, °F
B = barometric pressure, in Hg
p = vapor pressure, in Hg

Carrot Drying   Carrots are blanched before drying to minimize oxidation of carotene. Carrots are diced and placed on trays for single stage countercurrent tunnel drying or on a belt for continuous drying. Air at 71.1°C (160°F) reduces the moisture content of carrots from 90 to 8% in 7 to 8 hr in a tunnel dryer followed by 7 hr bin drying at 60°C (140°F) to reduce the moisture content to 4%. Using the belt dryer with higher temperatures and controlled air flow, carrots can be dried in about 1 hr.

Carter-Simon   The trade name of a moisture meter in which a sample of finely ground grain is heated at a constant temperature of 155°C (311°F) for 15 min. The sample is weighed on a chemical balance before and after heating.

Cascade   A set-up or arrangement such as in a dryer in which the product being dried flows such that the output from one element enters the next element, etc. In refrigeration, condensers may be arranged in series, to provide a cascade system. Also see *Cascade Dryer*.

Cascade Dryer   A unit for removing moisture in which the product falls by gravity over openings usually triangular or inverted V ducts which conduct heated air through the product for drying and exhaust moist air. These are usually continuous flow units (Fig. C-2).

Case Hardening   In dehydration, or during rapid heating of some products, the formation of a layer in the product that resists passage of moisture movement from the interior to the surface. Case hardening of lumber is avoided or reduced by periodic steaming of approximately 2 hr throughout the drying process. Case hardening of fruits, cereal grains, and beans may be reduced by steaming, using saturated exhaust air from the drying process to heat and wet incoming products. Case hardening of some products is avoided by using an initial high temperature of drying and in other products is avoided by using a low temperature. In the drying of paints and varnishes the surface hardens but much of the remainder remains soft. Case hardening of the outer layers of leather pelts may occur when they are delimed too rapidly.

CDMC   Crop Dryer Manufacturers Council of Farm and Industrial

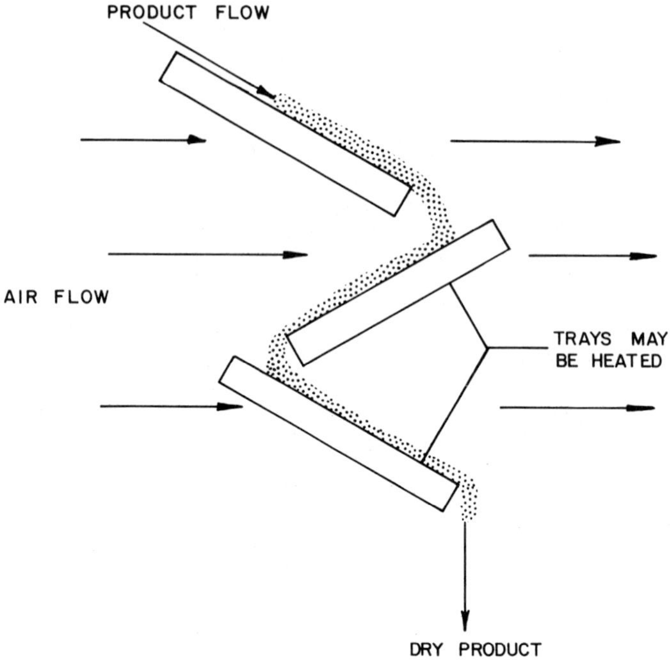

Figure C-2. Cascade dryer.

Equipment Institute (FIEI), 410 North Michigan Avenue, Chicago, IL 60611.

Celery Drying Celery is dried to produce leaf, powder, flakes, or diced products. The celery stalk is diced, blanched, and sulfited, then dried on trays in a two-stage tunnel dryer. Temperatures of the air entering the first and second stages are 82.2 and 54.4°C (180 and 130°F), respectively.

Cellulose A carbohydrate $(C_6H_{10}O_5)n$, a large polymer which has the same general formula as starch. On complete hydrolysis cellulose yields sucrose. Cellulose may contain as many as 3000 glucose units to form a molecule. For cellulose, n may be 5000, as for cotton, and, by comparison, n may be 1000 as for starch. The woody parts and cell walls of plants and textile fibers, cotton and linen, are the major sources of cellulose. Granulated cellulose is dried at 121°C (250°F). Cellulose has an equilibrium moisture content of 2.5%, dry basis, at 25°C (77°F) and 50% relative humidity. Also see *Carbohydrate; Starch; Sugar*.

°Celsius (°C) A unit of representing temperature on a scale of

100 units between the freezing point (0°) and the boiling point (100°) of water at standard sea level atmospheric pressure which compares to 32°F and 212°F or 180 units on the Fahrenheit (Table C-2). Formerly called centigrade.

Table C-2
Celsius (°C) Conversion to Kelvin (°K) and Fahrenheit (°F)

-273°C	0°K	-459.4°F	100°C	373°K	212.0°F
-200	73	-328.0	150	423	302.0
-150	123	-238.0	200	473	392.0
-100	163	-148.0	250	523	482.0
-50	223	-58.0	300	573	572.0
-40	233	-40.0			
0	273	32.0	350	623	662.0
10	283	50.0	400	673	752.0
20	293	68.0	450	723	842.0
30	303	86.0	500	773	932.0
40	313	104.0	550	823	1022.0
50	323	122.0	600	873	1112.0
60	333	140.0	650	923	1202.0
70	343	158.0	700	973	1292.0
80	353	176.0	750	1023	1382.0
90	363	194.0	800	1073	1472.0
100	373	212.0	850	1123	1562.0
			900	1173	1652.0
			950	1223	1742.0
			1000	1273	1832.0

Center Exhaust In a two-stage tunnel dehydrator, the arrangement in which the primary stage is concurrent flow and the secondary stage is counterflow, with the exhaust air from both stages combining at the center.

Centipoise See *Viscosity*.

Centigrade Scale A scale for representing temperature, now officially called the Celsius scale. Also see *Celsius*.

Centrifugal Disc (Disk) Atomizer-Centrifugal Spinning Disc-Centrifugal Spray Head In a spray dryer, a rapidly spinning unit or a pressure unit from which the liquid to be dried is forced by centrifugal action to the edge of the rotating head or wheel. The liquid is dispersed at the periphery over a lip or through slits or orifices to atomize the liquid for drying in a heated air stream in a drying chamber. The spinning disc is particularly useful for making droplets from viscous materials and materials in suspension. Centrifugal spinning discs may be of the following types: radial-vaned with vanes placed between two discs, multiple disc, and bowl-shaped liquid chamber. Small diameter discs of 2 in (5.1 cm) may be driven at speeds up to 50,000 rpm; large diameter discs at lower speed, with a 76.2 cm (30 in) diameter at 3500 rpm, with usual peripheral speeds from 76 to 183 m/s (250 to 600 ft/sec). Also see *Atomizer*.

Centrifugal Fan A centrifugal fan or radial flow fan consists of a wheel with appropriate blades which is rotated in a spiral housing. The air pressure at the center of the rotating wheel is less than the surrounding air pressure. The air pressure is increased in the spiral housing as the wheel rotates. The air flows axially into the wheel and is moved from the center to the periphery of the fan by the centrifugal force of the rotating wheel. It is possible to get higher pressure on a centrifugal fan by increasing the length of the blades because greater centrifugal force exerted. These fans may be further classified into forward-curved, backward-curved, and straight-bladed. The highest mechanical efficiency obtained with these fans is 80% (Fig. C-3). Also see different classes of fans. Also see *Backward-Curved Fan; Forward-Curved Fan; Straight-Bladed Fan*.

Centrifugal Dryer See *Extractor; Sugar Drying*.

Centrifugal Separator See *Cyclone Collector*.

Cereal Grain Drying Cereal grains, often harvested at moisture contents which are too high for storage, are dried before storage. For safe storage, moisture contents should be in the range of 11 to 14%, wet basis, depending on the type of grain, storage conditions, microorganisms associated with the grain, and temperature. Field losses from shattering and weather are reduced by early harvest when the grain is wet. Maximum production is secured when the seed is in the range of 30 to 40% moisture. Mechanical drying can be done by forced air or heated forced air in a bin with appropriate ducts with grain 14 ft or less over the ducts, using 0.013 to 0.134 m^3/m^3s (1 to 10 cfm/bu) of air. Continuous drying can be done in a vertical column or on a horizontal belt in which the grain thickness if 10 to 45.7 cm (4 to 8 in), air temperature 71 to 121°C (160 to 250°F), and airflow of 0.65 to 3.9 m^3/m^3s (50 to 300 cfm/bu). To avoid decreasing germination the grain temperature should be kept below 52°C (125°F); to avoid damaging baking quality, below 60°C (140°F); and to avoid causing reduction in feeding quality for animals, below 82.2°C (180 F). With appropriate control the air temperature for drying may exceed the grain temperature. Also see *Aeration*.

CFM (Cubic Feet per Minute) The most common measure of quantity of air flow in English units. To determine the weight of air it is necessary to know the density of air which is based on temperature and pressure, 1000 cfm = 0.4719 m^3/s.

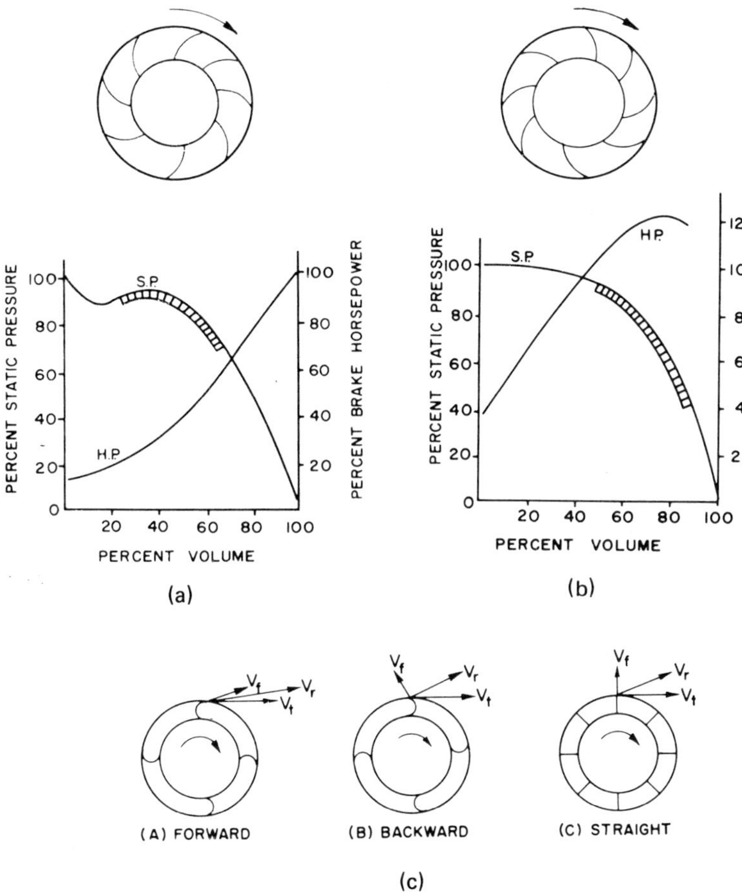

Figure C-3. Centrifugal fan. (a) forward. (b) backward. (c) analysis.

CFM per Bu Used for specifying air flow in cubic feet per minute for each bushel of product in the drying chamber. The value is obtained by dividing the cu ft per min of airflow by the bin capacity in bushels where the capacity is determined on the basis of 1.25 cu ft for 1 bu of small grains and 2.5 cu ft per bu for clean husked ear corn. For drying cereal grains using unheated air, the air flow will range from 0.013 to 0.078 m^3/m^3s (1 to 6 cfm/bu), with the maximum recommended depth of storage from 3 to 4.3 m (10 to 14 ft) (Table C-3).

CFM per Cu Ft Used for specifying air flow in cubic feet per

Table C-3
CFM Requirements for Drying Wheat, Oats, and Shelled Corn with Unheated Air

Grain Moisture Content, %	Recommended Minimum Air Flow Rate, cfm/bu	Practical Grain Depths, ft	Static Pressure,[a] in. Water	Maximum Quantity That Can Be Dried per Fan Horsepower,[b] bu.
		Wheat		
20	3	4	1.2	830
		6	2.3	440
18	2	4	.8	1,880
		8	2.5	600
16	1	8	1.3	2,300
		10	2.0	1,500
		Oats		
25	3	4	0.8	1,250
		6	1.7	590
20	2	6	1.1	1,360
		8	1.8	830
18	1-1/2	8	1.4	1,430
		10	1.0	1,000
16	1	8	.9	3,330
		12	1.9	1,580
		Shelled Corn		
25	5	4	0.7	860
		6	1.6	380
20	3	6	.9	1,120
		8	1.5	670
18	2	6	.6	2,500
		8	.9	1,670
		12	2.2	680
16	1	8	.5	6,000
		12	1.0	3,000
		16	1.6	1,880

[a] Includes 0.25 in. allowance for loss from duct friction.
[b] Air flow (cfm) per horsepower based on 3,000 cfm of air at 1 in. static pressure.

Conversion factors: *1 cfm = 0.0283 m^3/min = 0.00047 m^3/s*
1 ft^3 = 0.0283 m^3
1 ft = 0.3048 m
1 in. H_2O = 249 Pa = 0.249 kPa

minute for each cubic foot of product in the bin or dryer. It is obtained by dividing the cfm of air flow by the total container volume in cu ft. One cfm/ft^3 = 0.01667 m^3/m^{3}s = 1 m^3/m^3 min.

CFM per Sq Ft Used for specifying air flow in cubic feet per minute on the basis of the floor area or area perpendicular to the direction of air flow. It is obtained by dividing the cubic feet per minute of air flow by the cross sectional area in square feet.

For mow drying of hay, 0.076 to 0.10 m^3/m^{2}s (15 to 20 cfm/sq ft) is utilized. For heated air drying, with a temperature rise of 27.8 to 33.3°C (50 to 60°F), the air flow is increased to 0.153 m^3/m^{2}s (30 cfm/ft^2). For heated air grain drying in columns or thin layers the air flow ranges from 0.122 to 0.51 m^3/m^{2}s (24 to 100 cfm/ft^2). In concurrent flow dryers, the air flow may be between 0.5 and 0.9 m^3/m^{2}s (100 and 175 cfm/ft^2). One cfm/ft^2 = 0.00508 m^3/m^{2}s.

CFM per Ton Used for specifying the air flow through forages, bulk root crops, etc., where the total air flow in cubic feet per minute is divided by the tons (2000 lb = 1 ton) of product in the storage or bin to be aerated, cooled, or dried. The capacity in tons is determined on the basis of the volume occupied by crops at the moisture content stored. Thus, for hay, the general recommendation for air flow is 0.155 x 10^{-3} to 0.259 x 10^{-3} m^3/kg (300 to 500 cfm per ton), with a ton of long hay considered as occupying 11.3 m^3 (400 ft^2) of mow space, chopped hay occupying 8.5 to 9.9 m^3 (300 to 350 ft^3), and baled hay occupying 4.95 to 7.08 m^3 (175 to 250 ft^3). One cfm/ton = 0.5188 x 10^{-6} m^3/kg. Also see *Ton*.

CGS Centigrade-gram-second. Also see *Metric System*.

Chamber Dryer A unit used for removing moisture from brick or clay products heated by steam or hot air, in which the air from a partially dried chamber of products can be directed to a chamber with wet products. Four or five chambers may be used with appropriate controls to direct and mix the air for drying. Also see *Clay Drying*.

Charcoal A substance consisting almost entirely of carbon formed by the thermal decomposition of wood. Charcoal may be used for fuel and as an absorbent to remove impurities from gas. Charcoal which is heated with steam to increase its absorbing power is called activated charcoal.

Charge The quantity of a load put into a dryer or storage or of fuel put into a burner.

Charge Kiln See *Compartment Kiln*.

Charge Pan Dryer A unit for removing moisture consisting of shallow cylinder-shaped containers in which the product may move from one pan to the next. The containers are top-loaded and bottom-emptied with an agitator in each container. Heat is supplied through the walls of the containers with the heat source from steam, electricity, or hot air. The unit may be operated under vacuum or atmosphere.

Charles' Law The relationship in which the volume, V, of a perfect gas is proportional to the absolute temperature, T, at constant pressure, $V \propto T$ at constant pressure, or

$$\frac{V_1}{V_2} = \frac{T_1}{T_2}$$

Also see *Boyle-Charles Law.*

Checking As applied to lumber drying, the small cracks which develop during drying as a result of closing of the pores by the hardening of the albumen and resins in the sap before the water constituents have been abstracted. The hardened constituents expand with increasing temperature and cause cracks in the timber. Also applied to grains, particularly corn and beans, in which the seedcoats or seed may check as the result of rapid drying, excessive heating, or rough handling. Also see *Cracking.*

Chemically Combined Water Moisture which is chemically combined as (1) water in the form of hydroxyl ions, OH^-, or (2) water contained in molecular compounds as in hydrates. The water in hydroxyl ions is bound more tightly and requires more energy to remove than molecular-held water molecules. Chemically combined water often is not removed from biological products during ordinary drying operations.

Chemisorption The adsorption of gases and vapors by chemical forces as contrasted to physical forces.

Cherry Drying Whole cherries, split by lye soaking or pit removal, may be dried in a countercurrent tunnel in 8 to 12 hr using 76.7°C (170°F) heated air.

Chicken, Freeze Drying Following slaughter, poultry meat may be dried following slicing and freezing. To dry, the frozen meat is placed in a vacuum chamber where heat is added to assist in the vaporization of water, the frozen water changes to a vapor, which is removed from the meat. The heat is usually added by conduction from hot surfaces to the poultry meat. Dielectric heating or radiation heating may also be used. As the product approaches a low moisture content the temperature is adjusted to a temperature above which the product would be damaged by heat. For long time storage, the product is dried to 1% moisture content; for preventing growth of microorganisms the product is normally dried to below 10% moisture content.

Chili Pepper See *Pepper Drying.*

Chill Factor See *Wind Chill Factor.*

China See *Porcelain.*

CHU See *PCU*.

Circulating Batch Dryer A batch dryer equipped to circulate or mix the product during or between the drying and cooling periods. Also see *Continuous Circulation*.

Citrus Juice Drying Orange, lemon, grapefruit, and other citrus juices are dried to produce a product of 2 to 4% moisture content. If the dried products are to be stored for six months a moisture content of 1% or less is preferred. If dried to 1%, dehumidifying air in the final stage or in-package desiccation is required. Drying is carried out using procedures to assure that the product can be readily reconstituted in cold water.

Dehydrated juice products are usually made from concentrated juice. Atomization is by pressure or centrifugal device. The centrifugal device is used if particles are in the concentrate. Belt or tray dryers using puff drying or foam-mats provide another approach to drying. Modified drum dryers are also used.

The essential oils may be removed from juices in the evaporation process, which is of benefit in minimizing the development of off-flavors. Various procedures may be used to return flavor to the juice. Also see *Doctor Blade; Fruit Juice Drying*.

Clapeyron Equation A relationship representing the change of saturated vapor pressure with the latent heat of vaporization:

$$\frac{dP}{dT} = \frac{L}{(V - \nu)T}$$

where

L = latent heat of vaporization at T
p = saturated vapor pressure
T = absolute temperature
V = specific volume of water vapor at T
ν = specific volume of water at T

Also see *Clausius-Clapeyron Equation; Clausius Equation*.

Classification of Dryers The categorization of dryers according to the product, the method of heating the product or material, the flow of product through the dryer, and the source and flow of heat. Dryers are generally classified as direct or indirect where heating is accomplished by direct contact of the hot gases with the product or heat is transferred through a wall to the product. Dryers are further classified as continuous or batch according to the flow of product. Dryers may be classified according to the heat source such as solar, infrared, dielectric, gas, fuel oil, coal, etc. (Figs. C-4, C-5). Also see following dryers: *Agitated Batch; Atmospheric; Band Through-Circulation; Batch; Belt; Bulk Ventilation; Cascade; Compartment; Concurrent; Conduction; Continuous Band; Conveyor; Counterflow;*

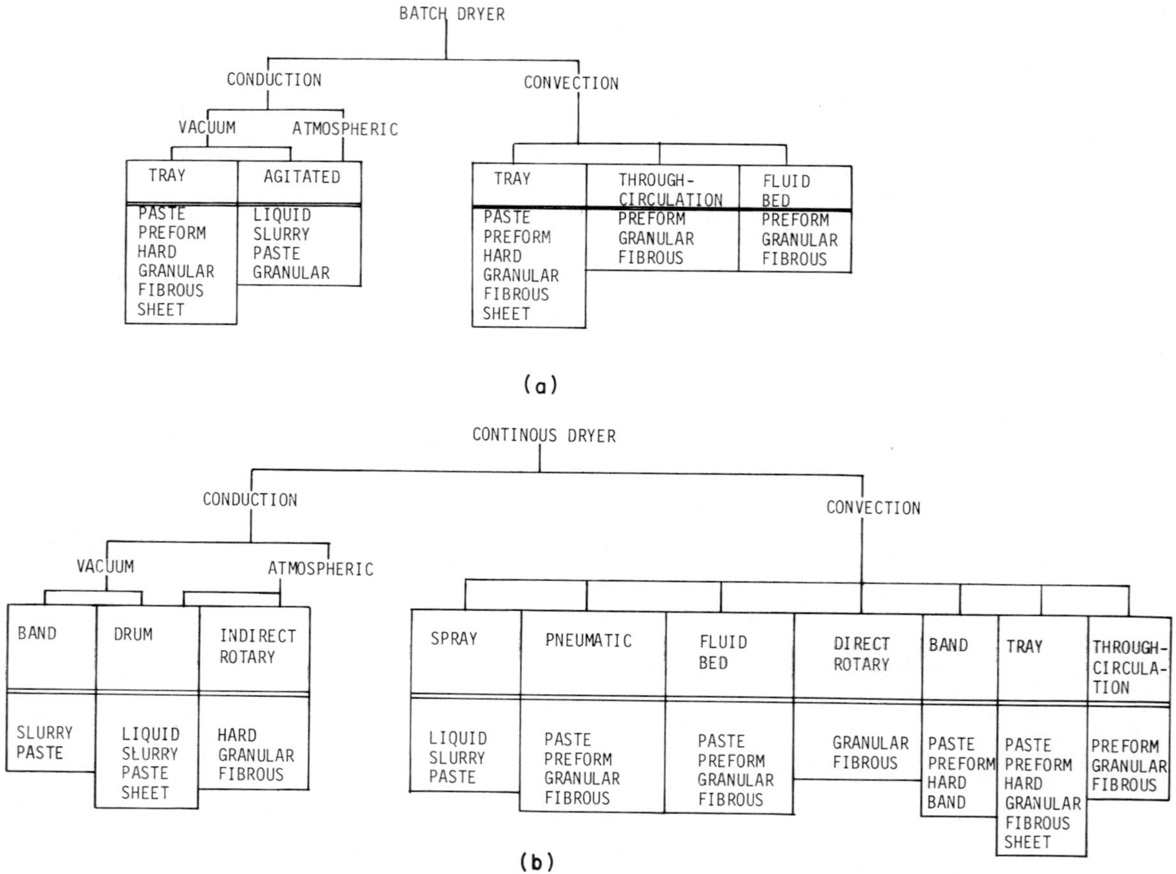

Figure C-4. Classification of dryers according to method of operation. *From Nonhebel, G., and Moss, A. A. H., Drying of Solids in the Chemical Industry (1971).*

LIQUID	SLURRY	PASTE	PREFORM	HARD	GRANULAR	FIBROUS	SHEET
AGITATED BATCH	AGITATED BATCH	VACUUM TRAY	VACUUM TRAY	VACUUM TRAY	VACUUM TRAY	VACUUM TRAY	VACUUM TRAY
DRUM	VACUUM BAND	AGITATED BATCH	CONVECTION TRAY	CONVECTION TRAY	AGITATED BATCH	CONVECTION TRAY	CONVECTION TRAY
SPRAY	DRUM	CONVECTION TRAY	BATCH THROUGH-CIRCULATION	INDIRECT ROTARY	CONVECTION-TRAY	BATCH THROUGH-CIRCULATION	DRUM
	SPRAY	FLUID BED	FLUID BED	CONT. TRAY	FLUID BFD	FLUID BED	CONT. TRAY
		VACUUM BAND	PNEUMATIC		INDIRECT ROTARY	INDIRECT ROTARY	
		SPRAY	CONVECTION BAND		PNEUMATIC	PNEUMATIC	
		PNEUMATIC	CONT. TRAY		DIRECT ROTARY	DIRECT ROTARY	
		CONVECTION BAND	CONT. THROUGH-CIRCULATION		CONT. TRAY	CONVECTION BAND	
		CONT. TRAY			CONT. THROUGH-CIRCULATION	CONT. TRAY	
						CONT. THROUGH-CIRCULATION	

Figure C-5. Classification of dryers based on physical form of feed. *From Nonhebel, G., and Moss, A. A. H., Drying of Solids in the Chemical Industry (1971). Used with permission of Butterworth Publishers, London.*

Cross-Flow; Dielectric; Direct-Fired; Drum; Fluidized Bed; Freeze; Indirect-Fired; Induction; Infrared; Multi-Stage; Pneumatic; Rotary; Rotary Drum; Spray; Through-Circulating; Tray; Tunnel; Vacuum; Vertical Column.

Clausius-Clapeyron Equation A relationship involving the latent heat, pressure, and temperature of a vapor, based on the assumptions that the volume of liquid is negligible as compared to the volume of vapor and that the vapor responds as an ideal gas:

$$\frac{d\ln P}{dT} = \frac{L}{RT^2}$$

where

P = vapor pressure
L = latent heat at T
T = absolute temperature
R = gas constant

Also see *Clapeyron Equation; Clausius Equation.*

Clausius Equation An equation of state of vapor in which the pressure, volume, and absolute temperature are related:

$$p(\nu - b) = RT$$

where

p = vapor pressure
ν = specific volume
T = absolute temperature
R = gas constant
b = constant

Also see *Clapeyron Equation; Clausius-Clapeyron Equation.*

Clay Drying Wet clay containing 15 to 35% moisture is used in a plastic condition for manufacture of tiles, brick, and pottery. Moisture may be removed by (1) air drying or (2) heated air drying (artificial drying) in a continuous kiln or batch kiln. Heating is by direct-fired heaters, steam, or radiation. For a drying temperature of 20 to 40°C (68 to 104°F) the drying time for bricks is from 5 to 10 days. Higher temperatures are used for shorter time of drying. Many clays will not stand a temperature above 121°C (250°F). Also see *Chamber Dryer; Kiln; Leather Hard Moisture Content.*

Cleaning See *Sieving; Screening.*

Clostridium Botulism See *Botulism.*

Coal A solid fuel used for direct but more often for indirect heating for dryers and dehydrators. Coal consists of carbon (68 to 82%), hydrogen, oxygen, nitrogen, and moisture and other components, often undesirable from the aspect of combustion products, such as sulfur. The moisture content of coal varies from 2 to 20%. The higher the moisture content the more heat that is used to vaporize the moisture, so that less heat is available for heating, unless heat is removed from the condensed vapor. After burning, 4 to 20% of the product remains as ash.

Anthracite, also known as hard coal, has a high content of carbon. Anthracite is relatively smoke-free, ignites slowly, burns with a blue flame, and is non-coking. It has a net heating value of 5160 to 5590 kJ/kg (12,000 to 13,000 Btu/lb).

Bituminous, also known as soft coal, is used extensively for commercial purposes. It is generally more volatile than anthracite coal which makes it easier to ignite, will burn more rapidly, will coke, and the ashes will cake more easily. Bituminous coal has a net heating value of 5160 to 6020 kJ/kg (12,000 to 14,000 Btu/lb).

Coal Drying The process of removing moisture from fine or pulverized coal, usually with one of three types of dryers: semi-indirect-fired rotary dryer, cell-type dryer which uses hot flue gas from the stock, or steam heated dryers, in which the temperature inside the dryer is limited to about 121°C (250°F) to avoid overheating of coal. To provide powdered coal for a boiler, the coal is first crushed between rolls to produce 2 to 2.54 cm (3/4 to 1 in) lumps, then dried, followed by pulverizing. Coal is dried to about 8% before pulverizing or during milling. By drying, the capacity of the pulverizer is increased, dry coal packs less in the bin, lower stack losses result because a higher flame temperature is obtained, and the possibility of spontaneous combustion is reduced. Also see *Mill Drying*.

Cocoa Drying See *Cacao Drying*.

Coconut Drying See *Copra Drying*.

Cocurrent See *Concurrent*.

Coefficient of Compressibility A value, C, representing the change in volume, ΔV, of a volume, V, resulting from a change in pressure, ΔP, at constant temperature,

$$c = -\frac{1}{V}\frac{\Delta P}{\Delta V}$$

Coefficient of Performance During the heat pump cycle, the total heat output, Q_h, divided by the heat equivalent of work, W, required to produce the heating effect, and during the refrigeration cycle is Q_c divided by work. Thus,

Cycle	Actual	Maximum (Carnot cycle)
Heating	$COP = \frac{Q_h}{(Q_h - Q)} = \frac{Q_h}{W}$	$\frac{T_h}{T_h - T_c}$
Refrigeration	$COP = \frac{Q_c}{(Q_h - Q_c)} = \frac{Q_c}{W}$	$\frac{T_c}{T_h - T_c}$

To maintain a high coefficient of performance, the system should be operated at a minimum head pressure and a maximum suction pressure. The COP generally ranges from 3.0 to 9.0 depending on refrigerant, head pressure, suction pressure, and superheated conditions. The coefficient of performance of a refrigeration unit:

$$\text{COP} = \frac{778\ Q}{W}$$

where

Q = heat absorbed by the refrigerant, Btu/lb
W = work supplied to compressor per lb of refrigerant, ft-lb.

Coefficient of Rehydration See *Rehydration Ratio.*

Coefficient of Thermal Conductivity See *Thermal Conductivity; Conduction.*

Coffee Drying Coffee can be dried in a single layer depth of 40 to 60 cm (15 to 24 in) in 20 to 50 hr, using drying air temperatures of 37 to 50°C (100 to 120°F), air flow rates of 0.0028 to 0.0068 m^3/kgs (2.7 to 6.6 ft^3/min lb) of dry matter, initially at 50%, wet basis, to 12%, wet basis. To avoid deterioration the coffee beans must be dried to 20%. Most coffee is dried on platforms or floors using solar energy. Artificial drying of coffee is usually done in a drum-type rotary dryer using either a direct or indirect heater. A temperature of drying air temperature of 82°C (180°F), needed to dry in 24 hr, is often considered as excessive for producing a quality product.

Coffee Extract Drying Instant dry coffee is made from coffee extract or liquid coffee. The liquid coffee is made from blended, roasted, and ground coffee beans from which soluble solids are removed by percolators. A vacuum drum dryer may be used to dehydrate the coffee extract. A vertical tower spray dryer using a pressure nozzle at 2068 to 3447 kN/m^2 or kPa (300 to 500 psi) with concurrent flow of extract 10 to 20% solids and heated air at 276°C (530°F) to produce a powder with particles of 60 to 100 microns diameter at 3.0 to 4.0% moisture content.

Coke A fuel consisting of the residue left after the volatile matter is removed from bituminous coal. Coke consists mainly of carbon (85 to 95%) which burns with a clean flame. Coke has a net heating value of 5160 to 6020 kJ/kg (12,000 to 14,000 Btu/lb). A coke burner as compared to a coal burner must have a larger combustion chamber because coke is more bulky than other fuels, otherwise more frequent firing is required.

Column Dryer A unit for removing moisture in which the product being dried moves by gravity through a vertical column, with the heated drying air moving through the column at right angles to the flow of grain. The vertical column dryer is usually a continuous flow crossflow dryer. The speed of discharge of products and temperature and flow of the drying air are regulated or controlled to provide proper drying of the product (Fig. C-6).

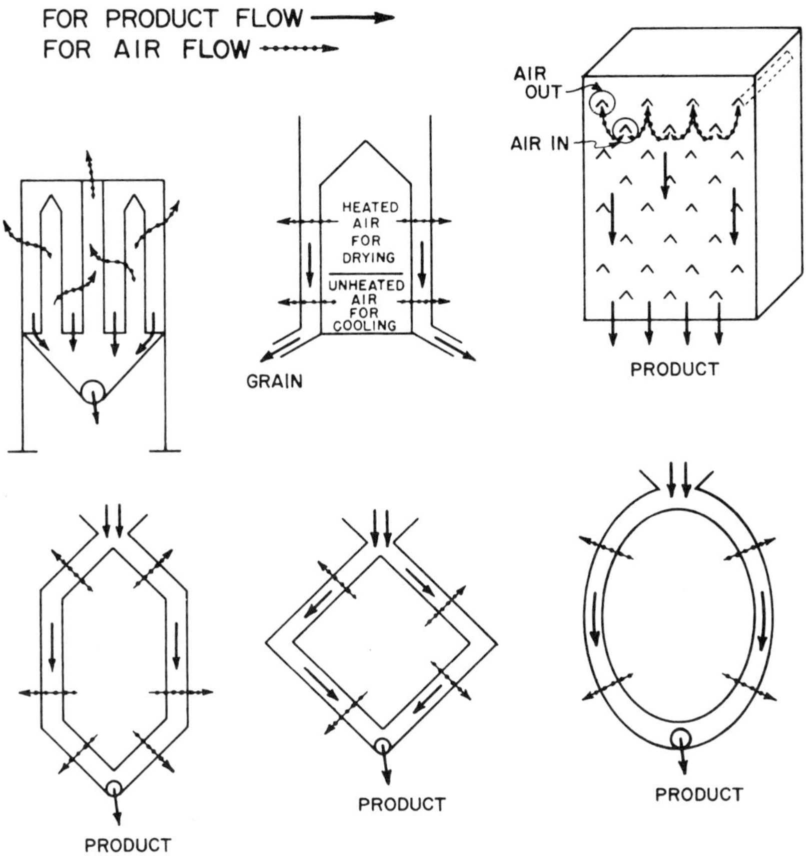

Figure C-6. Column (vertical) dryer types.

Combustion The rapid oxidation of a fuel with the production of heat and light as in the burning of a fuel. Carbon, hydrogen, and sulfur are components of typical fuels which can be burned. With the complete oxidation of fuel in the presence of oxygen, carbon dioxide (CO_2), water (H_2O), and sulfur dioxide (SO_2) are produced. Complete combustion is also known as stoichiometric combustion (Table C-4). The presence of carbon monoxide in the products of combustion indicates incomplete combustion. The quantity of energy released is the higher heating value (HHV) or the lower heating value

Table C-4
Combustion Air Required for Complete Combustion at 1 Atm

	Lb. air/lb. of fuel	Major content of fuel, %	
		Carbon	Hydrogen
Coke	162.1	98	
Coal, hard	144.6	95	2.2
Coal, soft	143.4	87	5.0
Wood, dry	73.4	50	6.0
Natural gas	227.9	68	22.5

Conversion factor: 1 lb = 0.45356 kg.

(LHV). The combustion of one pound of carbon produces 14,600 Btu (6277 kJ/kg); 1 lb of sulfur, 4000 Btu (1720 kJ/kg); and 1 lb of hydrogen, 62,000 Btu (2666 kJ/kg). The amount of air in excess of that theoretically needed to supply the oxygen for complete combustion for stoker-fed coal is 30 to 50%; fuel oil, 20 to 30%, and gas, 15 to 20%. Perfect combustion occurs when all of the carbon is burned without an excess of air. Also see *Calorific Value; Carbon Dioxide; Carbon Monoxide; Heat of Combustion; Net Heating Value; Peat Drying; Stoichiometry; Wood Combustion.*

Commercial Dry Basis A representation of moisture content of products leaving a dryer or in commercial trade on the basis of pounds of water per pound of solid (wet) times 100%. Also known as moisture content, wet basis. Also see *Moisture Content.*

Compaction The process of packing fine particles or aggregates as is purposely done for pelleting, cubing, or wafering; or the packing of products in a drying bin, which may change the air flow-static pressure relationships; or increasing the density of dried food products, such as dry milk or hay, which may be compacted into tablets, blocks, cakes, or pellets following drying to decrease space, improve handling, and reduce oxidation. Also see *Cuber; Packing; Pellets.*

Compartment Dryer A batch dryer or quasi-continuous convection heated air dryer equivalent to two or more cabinet dryers operating in series. Also see *Cabinet Dryer; Tray Dryer.*

Compartment Kiln A building or holding unit which is filled with a product, usually lumber or clay products, the doors are closed, and the drying proceeds, all done in one operation. The product remains stationary. The temperature and relative humidity in the kiln are controlled according to a drying schedule. The unit is also called a box kiln or charge kiln. The design of the compartment kiln is a contrast to the tunnel kiln or progressive kiln, in which the product is moved. Also see *Tunnel Kiln.*

Compressibility As applied to a gas, the negative of the fractional change in volume per unit change in pressure. If the initial and final temperatures are the same, the process is identified as isothermal compressibility. With an increase in pressure, there is a decrease in volume; thus, with a negative sign the compressibility is positive. Where pv^{γ} = constant, with $\gamma = c_p/c_v$, the process is identified as isentropic compressibility; that is, a reversible adiabatic process, with entropy (usually represented by s) constant.

Compressible Fluid A gas or liquid which exhibits a decrease in volume with an increase in external pressure. In many calculations gases and vapors are considered as compressible fluids and liquids as incompressible fluids. In many drying calculations air is considered as an incompressible fluid.

Concurrent Dryer The arrangement of a unit for moisture removal in which the hot air for drying and the wet material to be dried enter at the same end of the dryer and travel through the dryer in the same direction (Fig. C-7). Higher air temperatures, up to 500°F, may be used for cereal grains. Also called cocurrent and parallel flow dryer. Used for grain. The inlet air temperature is high as compared to the crossflow and counterflow dryers. Also see *Energy Efficiency*.

Condensation The change in state from vapor to a liquid accompanied by the release of energy (heat of condensation) of the same magnitude as the heat of vaporization at the same temperature.

Condensation Temperature See *Dewpoint*.

Condensed Milk Water is removed from whole or nonfat milk under vacuum to produce evaporated or condensed milk. Whole milk or milk to which sugar or other solid is added is homogenized ahead of the evaporator. Single or multiple effect evaporators are used.

Evaporated milk is produced by reducing the water in the milk in the range from 3:1 to 2:1. The product is sterilized and packaged aseptically or placed in containers and sterilized at 111.5 to 135°C (240 to 275°F) for 15 min. The term evaporated milk is applied to sterilized condensed whole milk, without additives, except as may be required to standardize the relationship between fat to solids-not-fat.

Condensed whole milk is the term applied to products which have increased solids achieved by removing water under a vacuum. These products are kept in refrigerated storage. Sweetened condensed whole milk and sweetened condensed skimmed milk include the addition of sugar as a preservative and flavor agent. These products are not heat-sterilized. Also see *Dry Milk; Evaporator*.

Condenser An apparatus for changing vapor to a liquid. It carries heat away from the vapor-liquid mixture. Heat is removed from the condenser by air or water. An evaporative condenser

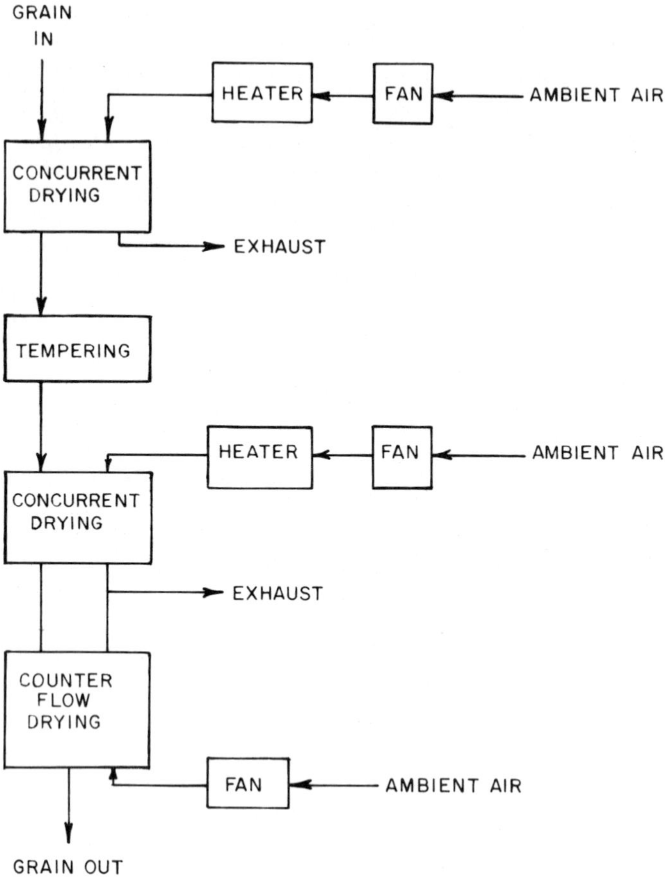

Figure C-7. Concurrent flow (two-stage) dryer with counterflow cooler.
From Brooker, D. B. and Bakker, F. W. (1977). ASAE Paper 77-3017.

utilizes a liquid, usually water, evaporating on the surface of the condenser to remove the heat from the condensing vapor.

Conditioning Using processes to change the physical properties of a batch or of some components of a batch to improve the state of the product for further processing, storage or packaging. Conditioning may be done by wetting with water or steam, by adding a chemical, by drying with forced unheated or heated air, or by equalizing and maintaining a uniform temperature of the product.

Conduction (Heat) The mode of heat transmission characterized by transfer of heat from one part of a solid to another part by molecular activity or in a fluid as distinguished from convection. Conduction is a mode of heat transfer distinguished from convection and radiation. The Fourier equation for thermal heat conduction under steady-state conditions is:

$$q = \frac{kA}{x}(t_2 - t_1) = (k/x)A\ (\Delta t)$$

where

q = rate of heat transfer
k = thermal conductivity of material (see Table C-5)
x = thickness
A = area, perpendicular to direction of flow
Δt = temperature difference along the path of heat transfer

Also see *Fourier; Heat Equation; Thermal Conduction.*

Table C-5
Conductivity Coefficient, k, Conversion Factors

1 Btu-ft/hr ft^2°F	= 12 Btu-in/hr ft^2°F
	= 14.8816 gcal-cm/hr cm^2°C
	= 4.1338 x 10^{-3} gcal-cm/s cm^2°C
	= 1.731 J/msK
	= 1.48816 kcal-m/hrm^2°C
	= 1.731 x 10^{-3} kW/mK
	= 12 pcu-in/hr ft^2°C
	= 1.731 W/mK
	= 17.31 x 10^{-3} watt-cm/cm^2°C
	= 43.931 watt-in/in^2°C

Conduction Solid-Band Dryer An apparatus for moisture removal with a conveyor or solid band holding the product to be dried is moved over a hot surface, such as a steam heated chest. The conveyor and the product are heated by conduction. Drying is often done under vacuum.

Conductivity Coefficient See *Thermal Conductivity.*

Conservation of Energy See *Bernoulli Theorem.*

Constant A quantity which does not vary. A quantity which is the same for the conditions being considered in an equation, formula, or particular situation. Constant coefficients are used in equations. Constants, such as π, g, and e are used in calculations (Table C-6).

Table C-6
Constants

e	2.71828. . .
g	32.17 ft per sec^2; 9.807 m/s^2
π	3.14159. . .
R	1545 ft lb per lb mole R
	8.314 kJ/kg mole K
	1.986 Btu/lb mol R
rad	57.295779. . . degrees
i	$\sqrt{-1}$

Constant Rate Drying An initial phase in the drying of solids during which the rate of evaporation per unit of surface area remains constant. When drying wetted sand, most of the drying occurs at a constant rate. When drying hygroscopic materials, such as grain, very little if any drying occurs during the constant rate period, depending on the initial moisture content of the product. The rate of drying during the constant rate period is dependent on the humidity of the drying air, coefficient of mass transfer, and the velocity of the drying air (Table C-7). Also see *Critical Moisture Content; Equations; Falling Rate Drying.*

Table C-7
Constant Rate Drying Equation

$$\frac{dW}{d\theta} = f_v A \ (p_s - p_a) = \frac{k_f A \ (t_a - t_s)}{h_{f_g}}$$

where $\frac{dW}{d\theta}$ = drying rate, kg/s (lb/hr) of water

f_v = water-vapor transfer coefficient, kg/m^{2}s Pa (lb/hr ft^2 psi)

A = water surface area, m^2 (ft^2)

p_s = water-vapor pressure at t_s, Pa (psi)

p_a = water-vapor pressure in the air, Pa (psi)

k_f = thermal conductance of air film, W/m^{2}K (Btu/hr ft^2 °F)

t_a = temperature of air, °C (°F)

t_s = water temperature, wet bulb, °C (°F)

h_{f_g} = latent heat of vaporization, kJ/kg (Btu/lb)

k_f related to velocity of air

Contact Dryer An apparatus in which a solid or semi-solid product is dried by contact on a hot surface with evaporation occurring at temperatures near the boiling point. The heat is transferred primarily by conduction from a hot surface to the product. Cloth and paper are dried over hot rolls by contact drying. Granular materials, such as sand and grain, can be dried on hot surfaces, but it is not commonly done. Filter cake and slimes can be dried on heated surfaces with the product being stirred during drying. A drum dryer is a type of contact dryer. Also see *Accessory Material Drying; Can Dryer; Drum Dryer.*

Continuous Band or Tray Conveyor Dryer An apparatus for moisture removal in which the material moves continuously, usually in a horizontal position in a band or on a tray, through a heated chamber. The product may be dried by convection heating, or may be moved over a heated surface, in which case the product is heated by conduction, and/or the product may be moved under a radiation heat source, either on a conveyor or in trays on a conveyor (Figs. C-8, C-9). Conveyors may be perforated, permitting air flow, or of a solid-band dryer. Continuous band dryers are used for precipitated inorganic colors, china clay, dyestuffs, molded articles, and some natural fiber textile materials. Dryers of this type consist of *Band Through-Circulation Dryer; Convection Solid-Band Dryer; Conduction Solid-Band Dryer; Continuous Tray Dryer; Radiation Solid Band or Conveyor Dryer.*

Continuous Belt Dryer See *Belt Dryer.*

Continuous Circulation Refers to a dryer or evaporator in which the product is moved continuously during moisture removal. A product is obtained which has a more uniform moisture content throughout the batch being dried, as compared with a unit in which no circulation occurs. Also see *Circulating Dryer; Recirculation.*

Continuous-Flow Dryer An apparatus in which the product being dried is in continuous movement in contrast to a batch operation. Also called continuous operation dryers. The principal continuous flow dryers are spray, fluidized bed, continuous band, pneumatic, continuous rotary (direct and indirect), crossflow, concurrent flow, and continuous tray.

Continuous Tray Dryer An apparatus for moisture removal in which the product to be dried is placed in trays. The product is heated by convection or conduction, usually with air moving through the product on deep trays of uniform materials of granules, pellets, etc., or over the product on shallow trays.

Continuous Web Dryer A machine for moisture removal which moves wet products on a web conveyor in a horizontal direction (similar to a belt or open chain dryer) through heated air, often with countercurrent flow. Used for bulky materials, industrial parts, clay products, leafy and forage crops, and grains (Fig. C-10). Also see *Belt Dryer.*

Control A component of a dryer, or dryer heat source, designed

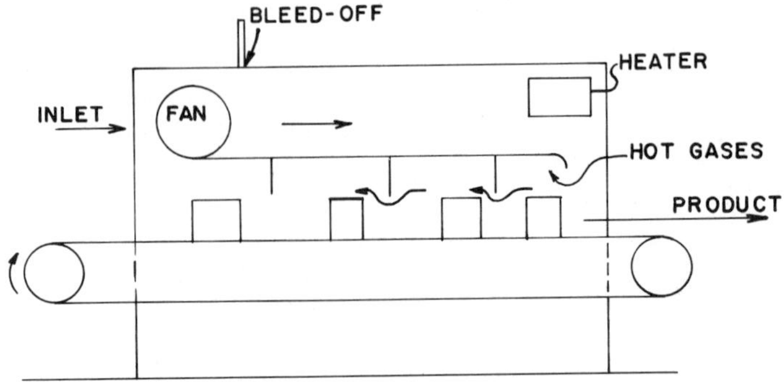

Figure C-8. Continuous band and tray dryer.
From Nonhebel, G., and Moss, A. A. H., Drying of Solids in the Chemical Industry (1971). Used with permission of Butterworth Publishers, London.

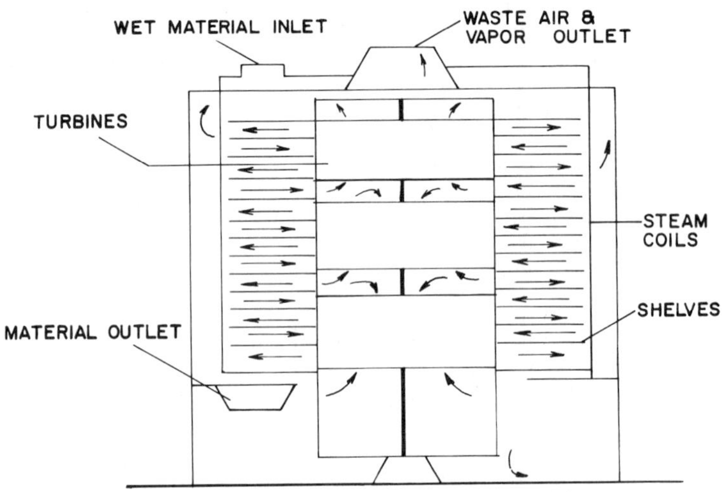

Figure C-9. Continuous agitated tray vertical turbo dryer.
From Nonhebel, G., and Moss, A. A. H., Drying of Solids in the Chemical Industry (1971). Used with permission of Butterworth Publishers, London.

to provide appropriate environmental conditions or limit any normal or abnormal condition of the drying operation. Also see *Safety Control System.*

Controlled Atmosphere (CA) A storage environment which is

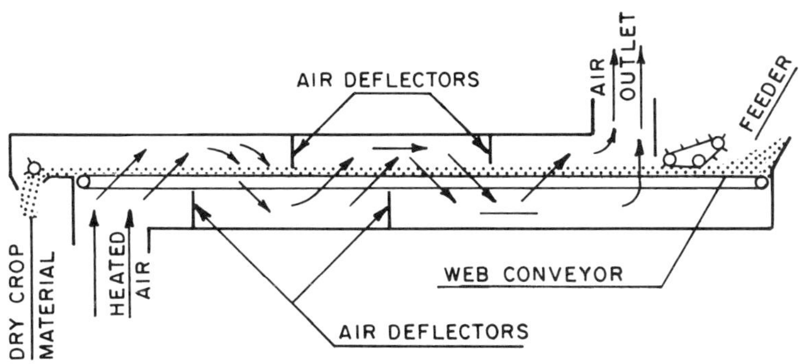

Figure C-10. Continuous web dryer.

controlled or modified as a method of preserving the product or increasing its storage life. The controlled atmosphere storage for apples is used in conjunction with a refrigerated storage. The typical atmosphere surrounding a product in storage is 21% oxygen, 0.3% carbon dioxide, and 79% nitrogen. In the controlled atmosphere storage, the oxygen level is reduced and the carbon dioxide level increased. Several engineering considerations are involved in the design of a controlled atmosphere storage. McIntosh apples, when stored under controlled atmosphere conditions of 3% oxygen and 5% carbon dioxide at 3.33°C (38°F) may have twice the storage life of apples stored in air at 0°C (32°F). Also see *Gas Storage; Modified Atmosphere*.

Convection Heat Transfer Transfer of heat, usually between a solid and a fluid, by moving masses of matter or by currents in gases and liquids caused by differences in density. Convection is the mechanism of heat transfer in which a combination of conduction within the fluid and energy transport is due to the fluid motion, produced by natural flow by density currents, known as natural or free convection; or by forced fluid flow, known as forced convection. Also see *Thermal Convection*.

Convection Heat Transfer Coefficient Represented by h, the transfer of heat by a surface film in convection heat transfer, Btu/hr $ft^2{}^\circ F$ or $W/m^2{}^\circ C$ or J/m^2sK. Also called the film coefficient of heat transfer. Used in the rate of heat transfer equation.

$$q = h A \Delta t$$

where

q = rate of heat transfer
h = convection heat transfer coefficient
A = surface area
Δt = temperature difference

1.0 Btu/hr ft^2°F = 4.886 kcal/hr m^2°C = 5.6783 W/m^{2}K (Table C-8). Also see *Overall Heat Transfer Coefficient*.

Table C-8
Convection Heat Transfer Coefficients, h, Approximate

Condition	Btu/hr ft^2°F	W/m^{2}K
Air, natural convection	1-5	6-30
Air, forced convection	2-100	10-600
Water, natural convection	10-80	50-450
Water, forced convection	20-3000	120-160,000
Water, boiling	500-10,000	2500-60,000
Steam condensing	1000-20,000	5500-110,000
Air inside tubes, 9.75 m/s (32 ft/sec)	8.0	45.4
Air perpendicular to tubes, 2.4 m/s (8 ft/sec)	7.5	42.6
Boiling water on vertical surface	1000	5678
Water inside tubes, 1.5 m/s (5 ft/sec)	1200	6815
Water perpendicular to tube	800	4542
Fluid paste	280	1590
Dry solid	2.5	14.2

Conversion factor: 1 Btu/hr ft^2°F = 5.678 W/m^{2}K.

Convection Solid-Band Dryer An apparatus used to remove moisture in which a solid conveyor belt or band with the heated air blown over the product to be dried.

Covering Power Amount of surface covered by a paint or other covering agent after drying, in terms of area per unit volume or weight of paint.

Conversion Factors The following relationships can be used to convert to different units (exact values are underlined):

Acceleration

1 ft/s^2 = 0.3048 m/s^2
1 g = 9.80665 m/s^2
1 in/s^2 = 0.0254 m/s^2
1 m/s^2 = 3.2808 ft/s^2

Area

1 acre = 0.4047 hectare = 4046.9 m^2
1 cm^2 = 0.155 in^2
1 ft = 92.90 mm^2 = 0.0929 m^2
1 in^2 = 645.2 mm^2 = 6.4516 cm^2 = 0.00064516 m^2
1 km^2 = 0.3861 mi^2

1 m^2 = 10.76 ft^2 = 1550.15 in^2
1 mi^2 = 2.5899 km^2
1 yd^2 = 0.8361 m^2

Density

1 g/cm^3 = 0.0361 g/in^3 = 62.3808 lb/ft^3
1 kg/l = 10.019 lb/gal Imp = 8.3429 lb/gal US
1 kg/m^3 = 0.0624 lb/ft^3 = 0.1603 oz/gal US = 0.133526 oz/gal Imp
1 lb/ft^3 = 16.022 kg/m^3
1 lb/gal Imp = 0.0998 kg/l = 99.81 kg/m^3
1 lb/gal US = 0.11986 kg/l = 119.86 kg/m^3
1 lb/in^3 = 27.6799 g/cm^3 = 2.76799 x 10^4 kg/m^3
1 oz/ft^3 = 1.0035 kg/m^3
1 $slug/ft^3$ = 515.38 kg/m^3

Flow Rate

1 cfm/bu = 0.804 m^3/m^3min = 0.01344 m^3/m^3s
1 cfm/ft^2 = 0.3048 m^3/m^2min = 0.00508 m^3/m^2s
1 cfm/ft^3 = 1.0 m^3/m^3min = 0.01667 m^3/m^3s
1 cfm/lb = 0.00104 m^3/kg s
1 cfm/ton = 0.5188 x 10^{-6} m^3/kg s = 1.12 cfm/tonne
1 ft^3/min = 1.699 m^3/hr = 0.00047195 m^3/s
1 ft^3/s = 0.0283 m^3/s = 28.3198 liter/s
1 kg/s = 2.205 lb/s
1 lb/s = 0.4537 kg/s = 27.223 kg/min
1 liter/s = 0.0353 ft^3/s = 0.220 gal Imp/s = 0.2642 gal US/min
1 m^3/hr = 0.589 ft^3/min = 3.666 gal Imp/min = 4.4027 gal US/min
1 m^3/tonne•min = 0.91 ft^3/bu min
1 m^3/s = 35.3198 ft^3/s
1 tonne/hr = 37 bu/hr

Force

1 kg force = 9.8066 N = 2.205 lb-force
1 lb force = 4.4482 N = 32.17 pdl
1 N = 10^5 dynes = 0.22481 lb-force = 0.10197 kg-force = 7.23065 pdl
1 pdl = 0.1381 N

Length

1 cm = 0.03281 ft = 0.3937 in = 393.7 mil
1 ft = 0.3048 m = 30.48 cm = 304.8 mm
1 in = 0.0254 m = 2.54 cm = 25.4 mm = 1000 mil
1 km = 0.6214 mile
1 m = 39.37 in = 3.281 ft = 1.094 yd
1 mil = 0.00254 cm = 25.4001 microns = 0.001 in
1 mile = 1.609 km = 5280 ft = 1760 yd
1 mm = 0.03937 in
1 yd = 0.9144 m = 91.44 cm

Pressure

1 atm = 760 mm Hg = 33.9 ft H_2O = 14.696 lb/in^2 = 1.0332 kg/cm^2 = 1.013 x 10^5 N/m^2 = 1.013 x 10^5 Pa = 101.325 kPa = 101.325 kN/m^2 = 760 torr = 29.92 in Hg

1 bar = 750 mm Hg = 14.5 lb/in^2 = 10.2 m H_2O = 10^5 N/m^2 = 10^5 Pa = 0.1 MPa
1 in H_2O = 25.40 kg/m^2 = 249.09 N/m^2 = 249.09 Pa = 1.868 mm Hg = 0.0735 in Hg
1 in Hg = 3.3864 kN/m^2 = 25.4 mm Hg = 13.595 in H_2O = 3386.4 Pa
1 kg/cm^2 = 14.223 lb/in^2 = 98066.5 N/m^2 = 98066.5 Pa = 1000 gram/cm^2 = 0.967 atm
1 kg/m^2 = 0.205 lb/ft^2 = 0.03937 in H_2O
1 kN/m^2 = 0.14504 lb/in^2
1 lb/in^2 = 6.8948 x 10^3 N/m^2 = 6.8948 kN/m^2 = 6894.8 Pa = 0.0703 kg/cm^2 = 51.714 torr = 0.068 atm = 2.036 in Hg = 735.5 mm Hg
1 mm Hg = 19.339 x 10^{-3} lb/in^2 = 133.32 N/m^2 = 0.133 kPa = 0.0394 in Hg = 133.32 Pa = 0.535 in H_2O
1 N/m^2 = 1 pascal = 1 Pa
98066 N/m^2 = 14.223 lb/in^2
1 psi = 6897.76 N/m^2
1 torr = 133.322 N/m^2 = 1 mm Hg abs = 0.00136 kg/cm^2 = 0.00131 atm = 0.0193 lb/in^2 = 133.322 Pa

Thermal, Power, Heat, Energy

1 Boiler hp = 33,472 Btu/hr = 9807.3 W = 9.8 kW
1 Btu = 1.0551 kJ = 0.2520 kcal = 778.2 ft lb = 252 gcal = 1055 J
1 Btu/ft hr°R = 1.731 J/m sec K = 1.731 W/mK
1 Btu/ft^2 = 2.712 kcal/m^2 = 1.1356 x 10^4 J/m^2
1 Btu/ft^2°F = 4.886 kcal/m^2°C
1 Btu/ft^2 hr = 3.155 W/m^2 = 0.2712 gcal/cm^2 hr = 2.712 kcal/m^2
1 Btu/ft^2 hr°F = 5.678 W/m^2°C = 5.678 W/m^2 K = 5.678 J/s m^{2}K = 4.882 kcal/m^{2}h°C
1 Btu/ft^2 hr°F/ft = 1.488 kcal/m^{2}h°C/m = 1.731 W/mK
1 Btu/ft^3 = 8.90 kcal/m^3
1 Btu/hr = 0.293 W = 0.293 J/s = 0.298 x 10^{-4} boiler horsepower
1 Btu/ft^2 hr = 11.367 kJ/m^2 day
1 Btu/hr ft^2°F = 5.678 W/m^{2}K = 4.882 kcal/m^{2}h°C
1 Btu/lb = 0.566 kcal/kg = 2.324 kJ/kg = 2324.4 J/kg
1 Btu/lb°F = 4184 J/kgK
1 cal = 4.184 J
1 cal/cm sec°C = 418.4 J/msK = 418.4 W/mK
1 cal/g = 4184 J/kg
1 ft lb = 1.35582 J = 0.1383 kg m
1 ft lb/s = 1.3558 W
1 ft pdl = 0.0421 J
1 gcal/cm^2 = 1 langley
1 hp = 745.7 W = 745.7 J/s = 23.56 centigrade heat units/min = 550 ft lb/s
1 J = 23.73 ft pdl = 0.7375 ft lb
1 J/kg°C = 0.000238 Btu/lb°F
1 J/s = 1 W
1 kcal = 3.9683 Btu = 2.205 metrip hp = 4184 J
1 kcal/100 g = 41.84 kJ/kg
1 kcal/kg = 1.8 Btu/lb

1 $kcal/m^2$ = 0.3687 Btu/ft^2
1 $kcal/m^2{}^\circ C$ = 0.2049 $Btu/ft^2{}^\circ F$
1 $kcal/m^2h^\circ C$ = 0.2049 $Btu/ft^2h^\circ F$
1 $kcal/m^3$ = 0.1124 Btu/ft^3
1 kgm/s = 9.8 W = 0.013 hp
1 kJ = 0.94783 Btu = 0.23884 kcal
1 kJ/kg = 0.4299 Btu/lb = 0.23884 kcal/kg
1 kJ/kg°C = 0.23884 Btu/lb°F
1 kJ/kgK = 0.2388 Btu/lb R
1 kW = 1 kJ/s = 737.6 ft lbs = 1.3410 hp
1 kW/mK = 577.8 Btu/ft hr°R
1 kW/m^2K = 176.1 Btu/ft^2 hr°F
1 kWh = 3.600 MJ = 3415 Btu = 860,445 gcal = 3,600,000 J
1 langley = 1 $gcal/cm^2$
1 langley/day = 4.183 J/cm^2-day = 4.183 x 10^4 J/m^2-day
1 m kg = 7.2333 ft lb
1 m^2/s = 10.7639 ft^2/s = 3600 m^2/hr
1 therm = 10^5 Btu = 25.207 M cal
1 ton refrigeration = 3516.85 W = 3516.85 J/s
1 W (or J/s) = 0.7375 ft lb/s = 3.412 Btu/hr = 0.8598 kcal/hr = 0.00134 hp = 2.843 x 10^{-4} ton refrigeration = 0.102 kg m/s = 0.1019 x 10^3 boiler horsepower
1 W/m^2 = 0.3170 $Btu/ft^2/hr$ = 0.85984 $kcal/m^2/hr$
1 W/m^2C = 0.1761 $Btu/ft^2hr^\circ F$ = 0.85984 $kcal/m^2$ hr°C
1 W/mC = 0.5778 Btu/ft hr°F = 0.85984 kcal/m hr°C

Viscosity

1 ft/s = 0.0929 m^2s
1 kg/ms = 0.00672 lb_m/ft s = 2419 lb/ft hr = 10 poises
1 lb_m/ft s = 1.4882 Ns/m^2
1 m^2/s = 10.7639 ft^2/s = $10^4 cm^2/s$ = 10^4 stokes
1 poise = 0.1 Ns/m^2
1 stoke = 0.0001 m^2/s

Volume

1 ft^3 = 0.0283 m^3
1 gal US = 0.0037854 m^3 = 3.785 mm^3
1 in^3 = 1.6387 x $10^{-5}m^3$ = 16.387 cm^3
1 liter = 1 dm^3 = 61.0246 in^3 = 0.26418 gal US = 0.21997 gal Imp = 0.001 m^3
7.95 lb/gal = 952.5 kg/m^3
1 m^3 = 1000 dm^3 = 35.31 ft^3 = 1.3079 yd^3 = 1 stere
1 oz = 0.00002957 m^3

Also see *Overall Heat Transfer Coefficient; SI Units; Thermal Conductivity.*

Conveying Air Velocity Although exact air velocities for conveying drying products are difficult to prescribe, the minimum vertical velocity in a vertical tube of a dryer should be approximately 2.5 to 3 m/s greater than the free-falling velocity of the largest particles in free air. A velocity of 10 to 30 m/s of the air is used in pneumatic dryers (Table C-9). The electric power used in conveying in a pneumatic dryer is about 4 to 5 kW/1000 m^3/kg.

Table C-9
Conveying Air Velocities for Carrying Products

	ft/min	m/s
Corn	5600	28.56
Cotton	4500	22.95
Dry pulp	4500	22.95
Paper	5000	25.5
Sand	7000	35.7
Sawdust	3000	15.3
Wheat	5800	29.6
Barks, chips	3000-5000[1]	15.3-25.5

For vertical carrying of product, the terminal velocity must be added to the carrying velocity.
Conversion factor: *1 ft/min = 0.0051 m/s*

[1] *Marks, L. S. 1941. Mechanical Engineers Handbook, McGraw-Hill Book Co., N. Y., NY.*

Conveyor Dryer A continuous convection dryer in which the moist material is mechanically moved on a conveyor through a chamber in which heated air is circulated. The air is usually circulated through the material on a mesh, slotted, louvered conveyer, or over the material for products in thin layers. Conveyor dryers are replacing drying trays in tunnels for fruit and vegetable pieces such as apples, carrots, onions, and potatoes. Also see *Band Through-Circulation Dryer; Conduction Solid Band Dryer; Screw Conveyor Dryer.*

Cooling Curve See *Newton's Law of Cooling.*

Cooling Stage The step following heated air drying in which unheated air is moved through the product to reduce the temperature. The exhaust from the cooling stage may be used for heating the product entering the dryer. The term may also refer to the time required for the cooling zone to move through a product mass.

Cooling Zone That portion of the product mass in which the temperature of the product is being reduced during aeration or cooling.

COP See *Coefficiency of Performance.*

Copra Drying Copra is produced by drying the meat portion of the coconut. Copra may be sun-dried, smoke-dried over an open fire on a grate or tray, or artificially dried using a dehydrator (Tapahan). The copra can be dried in the shell and/or removed and dried separately. For sun-drying open halves are placed on trays facing the sun for 4 to 6 days. The meat is removed and further dried either in the sun or using artificial drying. Open halves may be dried using heated air through a platform, drying to 20 to 30% in 4 to 6 hr. The meat is removed and dried to 5%. The copra may be dried to 5% in the shells (half) in 40 hr using 68.3°C (155°F) air for the first 8 hr followed by final heating at 60°C (140°F). Desiccated coconut is prepared in the following steps: shelling, paring, shredding, drying, sieving, and packing. Drying is done on an endless belt dryer where the moisture content is reduced from 50 to between 2.5 and 4.5%. The dry copra or desiccated coconut will have 60 to 65% oil. Coconut oil is made by pressing or squeezing copra. About 7% moisture is added to reconstitute desiccated coconut, which will usually produce a product of 11 to 12% moisture.

Corn Distillers Dried Grain The dried residue resulting from the manufacture of alcohol and distillers liquors from corn or from a grain mix in which corn is the principal constituent.

Corn Drying See *Cereal Crop Drying; Ear Corn Drying; Sweet Corn Drying.*

Correlation Coefficient A number, usually represented by r, representing the relationship of two variables, or two groups of variables, such as x and y, determined by statistical methods. A correlation coefficient of 1.0 represents perfect correlation and 0.0 no relationship. r is defined as:

$$r = \frac{\sum_{i=1}^{n} (x_i - x)(y_i - y)}{\left[\sum_{i=1}^{n} (x_i - \bar{x})^2 \sum_{i=1}^{n} (y_1 - \bar{y})^2\right]^{1/2}}$$

where $\bar{x}$ and $\bar{y}$ are means of the n values of the x and y observations.

Correlation of Drying Data For drying it is sometimes possible to correlate the data by plotting the logarithm of the unaccomplished moisture content change vs. the logarithm of the drying time:

$$\log \frac{M_\theta - M_e}{M_o - M_e} \text{ vs } \log \theta$$

where

θ = time

M_θ = moisture content at time θ

M_o = original moisture content

M_e = final or equilibrium moisture content

These values are plotted on logarithm paper. Also see *Dimensionless Potentials*.

Cotton Seed Drying The seed removed from the cotton during ginning is known as cotton seed, which may be used for animal feed or for producing seed. Cotton seed should be dried to 11% moisture or less for storage. Unheated or heated air drying may be used. Heated air drying is done in a vertical column or rotary dryer, with temperatures less than 43.3°C (110°F) for seed and less than 82.2°C (180°F) for animal feed. Also see *Seed Cotton Drying*.

Countercurrent Flow Dryer See *Counterflow Dryer*.

Counterflow Dryer An arrangement of a convection dryer in which the air flows in a direction opposite to the direction of flow of material being dried. Also called countercurrent flow dryer. Counterflow coolers are used in concurrent flow grain dryers.

Cowl A device utilizing the wind for producing positive or negative pressure to cause air movement. Oscillating cowls may turn with the wind and either (1) face the wind for creating a slight pressure or (2) face in the same direction as the wind in which case a slight vacuum is produced (Fig. C-11). Some cowls utilize rotating turbines to propel the air. The static pressure (S.P.) developed by cowls is as follows:

Theoretical	S.P. = $.000483\ w^2$
Pressure type	= $.00043\ w^2$
Suction type	= $.00017\ w^2$

where w is the wind speed in miles per hour.

The cowl can be used for moving air against small static pressure differences. The pressure cowl needs to be designed and installed to prevent rain and snow entering the structure or the product being ventilated or aerated.

Crackage See *Cracking*.

Cracking The formation of openings in the surface and/or cross-section of a product during drying as a result of rapid heating or drying. For such products as peas, beans, and rice, the cracking results in a reduction of quality and grade. Checking is often a precursor to cracking and breakage. To avoid excessive checking and cracking of cereal grains rapid removal in excess of 3 to 5% moisture per hour is avoided. Also see *Checking; Damage*.

Cranberry Drying While cranberries are sliced before drying for flakes, whole cranberries are perforated before drying. Cranberries are dried on trays in a counterflow tunnel dryer, with an incoming air temperature of 76.6°C (170°F). Cranberries are dried from 88 to 10% in 6 to 10 hr. Flakes and powder at 5% moisture content can be produced from drum-dried cranberry purée.

Critical Humidity The humidity above which a solid hygroscopic salt is wet and below which the salt remains dry, specified at a

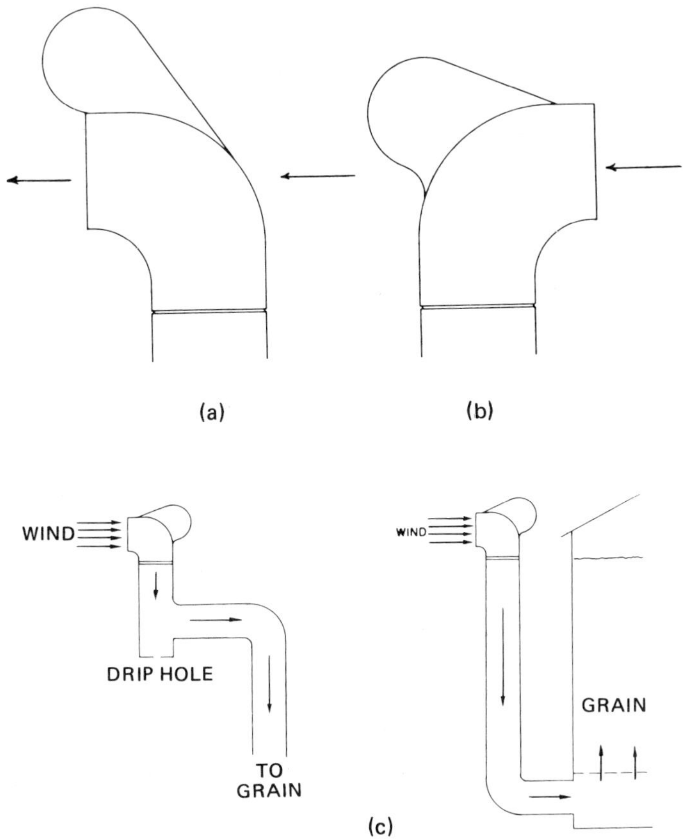

Figure C-11. Cowls. (a) suction. (b) pressure. (c) arrangement to exclude snow.

particular temperature, usually in the range 0 to 100°C (32 to 212°F). The critical humidity of sodium chloride is 75% in the temperature range 0 to 50°C (32 to 122°F); calcium chloride, 44 to 35% relative humidity at 6.7 to 21.1°C (20 to 70°F); lithium chloride is 11.2% at 20.0 to 37.8°C (68 to 100°F); and potassium sulfate is 97% at 10 to 37.8°C (50 to 100°F).

Critical Moisture Content The moisture content of a body at the time during drying when the constant rate drying period changes to the falling rate drying period (or phase) (Fig. C-12). Also used to represent the point between bound and unbound moisture content. Also see *Constant Rate Drying; Falling Rate Drying; Fiber Saturation; Moisture Content Representation.*

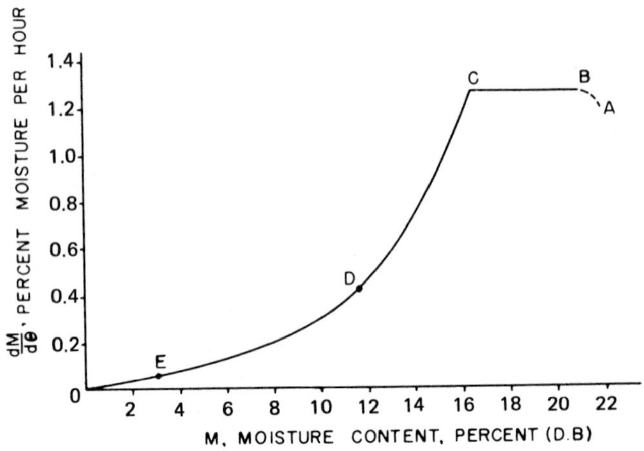

Figure C-12. Critical moisture content: C is the critical moisture content, B-C is constant rate of drying, C-E is falling rate of drying.

Critical Velocity The average linear velocity at which the flow of a liquid at a given temperature stops being laminar (e.g., Re > 2000) for pipe and enters the transition region and moves toward turbulent (> 4000 = Re). The transition velocity varies considerably depending primarily on fluid.

Crossflow Dryer The arrangement of a convection heated air dryer in which the air flows perpendicular to the direction of movement of material through the dryer. Cross flow of the air is also used in the cooling section of a continuous flow convection dryer. These units operate at a temperature of 82.2 to 104.4°C (180 to 220°F), airflow 0.254 to 0.406 m^3/m^2 (50 to 80 cfm/ft^2), and grain thickness 0.3 to 0.46 m (12 to 18 in). Also see *Energy Efficiency*.

Cuber A machine for pressing forage, often ground, or forage and grains, into compact units in the shape of a cube, 1-1/2 in. x 1-1/2 in. x 1-1/2 in. or larger. The product is often dried before compacting. Also see *Wafer*.

Cubical Expansion See *Coefficient of Volume Expansion*.

Cubic Decimeter See *Liter*.

Cubic Feet per Minute See *CFM*.

Curing A form of conditioning as contrasted to simple drying in which a chemical change occurs, such as in tobacco, sweet potatoes, meat, hides, etc., to prepare the product for storage or use. For adhesives, curing is usually applied to processes involving some form of polymerization. Drying of lumber is referred to by some as curing. Also see *Adhesive Hardening; Air Curing; Aging; Seasoning; Ultraviolet Curing; and various products.*

CWT An abbreviation commonly used for one hundred weight, such as 20 cwt lb is 2000 lb or 1 ton. Also see *Ton*.

Cycle A series of changes by a system or a component or the environment in which the change moves away from and back to the normal position, usually repeating the action and motion on a uniform repetitive basis, or frequency (such as, for electricity, cycles per second; four seasons per year; etc.). The complete sequence of values of a periodic quantity which occurs during a period. A process for which the initial and final states are identical.

Cycling A periodic change of the controlled variable, such as the temperature of a dryer.

Cycling Burner Type of operation wherein application of maximum heat is periodic such as cycling between high fire and low fire; cycling between high fire and constant pilot; and cycling from high fire to "off," then restarted to high fire by constant or intermittent ignition.

Cyclone Collector A conical device used to separate solids, normally in the range of 5 to 200 microns, from a swirling air stream. The solids are removed from the bottom of the cone and the air is exhausted from the center at the top (Fig. C-13). Proper design of the collector and air flow is required to remove the materials. The device is also used as a dust collector. Also called cyclone separator. Air with a high velocity enters the cone which has a much larger cross section than the entering duct. The air velocity decreases in the cone permitting the dropping of the solid. Cyclones are commonly used following the spray drying chamber to collect the dried product. The following characteristics pertain to the cyclone separator: (1) The separation is a function of the difference in diameter and density of the product and air with the larger differences providing more efficient separation. (2) The velocity of the inlet is a major factor in controlling efficiency as related to the cyclone diameter. Velocities may approach 30.48 m/s (100 ft/sec). (3) A smaller diameter cyclone at a fixed pressure drop will have a higher efficiency than a larger diameter unit. (4) The diameter of the inlet is usually about one-fourth the diameter of the cyclone. (5) The collection efficiency can be increased by reducing the air outlet diameter, which will increase the pressure drop. A major problem in operation for a unit with these design characteristics is the leaking of air in at the outlet. (6) The pressure drop through the cyclone decreases as product, powder, or dust is incorporated in the air. (7) The pressure drop of the system decreases as the concentration increases for the same airflow in cubic feet per minute.

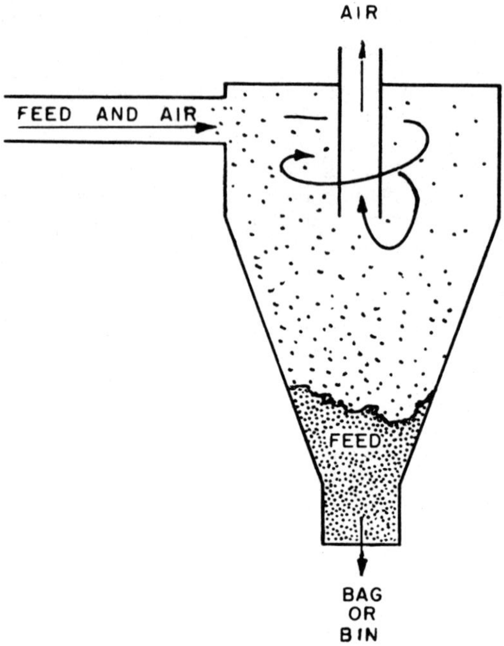

Figure C-13. Cyclone, feed, or dust collector.

D

Dairy Products   See *Condensed Milk; Milk Products*.

Damage   See *Checking; Cracking; Germination; Heat Damage*.

Darcy's Law   An expression which states that the volume of a fluid that will flow through a pipe or a uniform bed of granular material in unit time is proportional to the pressure gradient through the pipe or the bed (Fig. D-1). Also called Fanning equation.

$$F = f \frac{L}{D} \frac{V^2}{2g}$$

where

F = friction loss of flowing fluid
f = friction factor
L = length of path
D = diameter
V = velocity

The friction factor is based on Reynolds number and the roughness of the pipe or channel through which the fluid is flowing. Care must be taken in selecting f, as some sources use $f = 4\ f'$. As a check, in the above equation for laminar or streamline flow, f = 64 ÷ Reynolds number.

Data for Drying and Aeration AMD's (Air Moving Devices)   The following data shall be included on compliance certificates for all fans for drying or aeration:

Model ______________________
AMD Rating ____________________CFM @ 1 in. or 2 in.
WC Pressure at Discharge
Rated Speed _________RPM
AMD Brake HP Max: HP from 0 in. to _________in. WC
Total Electrical Power Requirement _________KW Max
__________AMP Max at ___________Volts Max

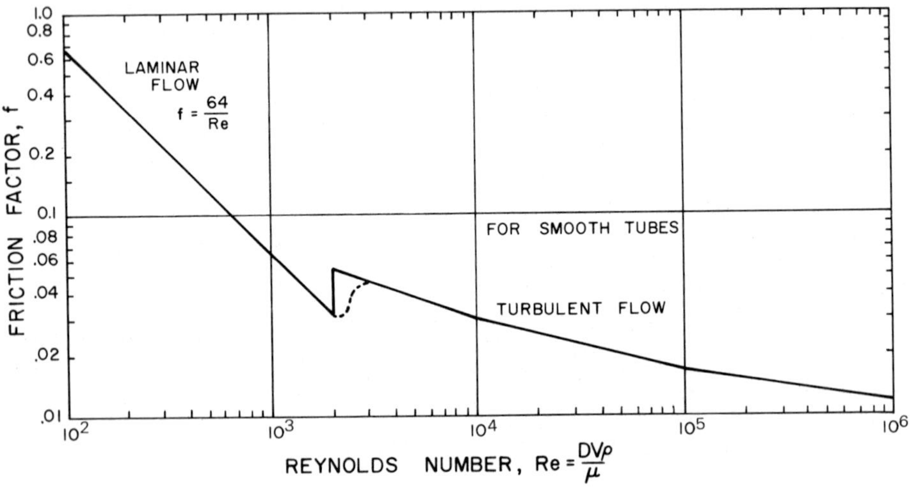

Figure D-1. Darcy's law, f vs. Reynolds number.

D.B. An abbreviation for dry basis moisture content and dry bulb temperature. Usually written as d.b. Also used to represent dry bulb temperature. Also see *Moisture Content; W.B. (w.b.)*.

DBD A process of drying, blanching, drying. See *Apricot Drying; Dry Blanch Dry Process*.

Dead Space The volume of a vessel, tube, container, storage, dryer, etc., not actively involved or used in the containment of the active ingredient or product.

Decay Coefficient The value of c in an exponential function decreasing with time, t, as represented by

$$I = I_o e^{-ct}$$

The decay modulus is 1/c, which is the time for I to diminish to 1/e of I_o.

Decimal A representation of a part of the whole such that 6/10 may be represented by 0.6 or 0,6. The period (.) is used in the USA, Canada, and British Commonwealth countries. The comma (,) is used in Europe, Asia, and South America to represent the decimal point.

Decreasing Rate Drying In deep bed drying, the rate of drying of the bed, bin, or batch decreases when the drying front reaches the top of the bed, when air is moving upward and out of the bed. This terminology refers to deep layer drying in contrast to the term falling rate drying for thin layer drying. Also see *Falling Rate Drying*.

Deep Layer When drying in beds, bins, or other containers in which a depth greater than the dimension of one particle of product is dried. It is usually possible to identify the following locations in deep layer drying: dried zone, drying zone, drying front, and damp undried zone. The initial period of drying in a deep layer, often after a heat-up period, is at the maximum drying rate period, which is followed by a decreasing rate drying period (Fig. D-2).

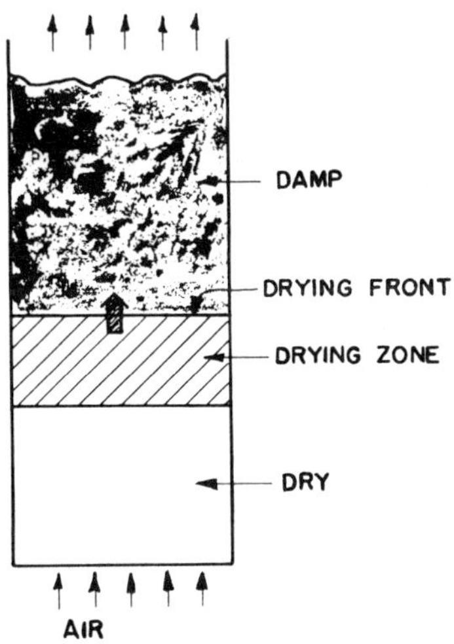

Figure D-2. Deep layer drying.

Degradation of Energy The reduction in energy level to a less readily available state. The order of degradation from most to least available are nuclear, electrical, mechanical, and thermal.

Degrade A drop in lumber grade resulting from kiln drying.

Degree-Day A measure of the departure of the mean outdoor dry bulb temperature from a standard. Standards commonly used are

18.3°C (65°F) for heating or cooling, 4.4°C (40°F) for crop production, and 0°C (32°F) for contractors. The degree-day is calculated by summing the difference between the mean daily temperature and the standard for a period of time. If the maximum daily temperature is 32.2°C (90°F) and the minimum is 15.6°C (60°F), the mean is 23.9°C (75°F) and the degree-day is +10 for that day, based on the 18.3°C (65°F) standard. Should the same mean persist for 30 days, the degree-day would be +300 (based on °F) for that month. Air drying would normally occur more rapidly or less fuel would be required for heated air drying during a time of greater degree-days (Fig. D-3).

Dehisce To open, split, or burst, usually during ripening of seed coats or capsules or vessels, naturally or during drying. Drying may be a natural process or an artificial process by forced air or heated air.

Dehumidify The process of decreasing the water vapor or the humidity of a body of air or gas such as through condensation on a cold surface, diffusion of vapor to an adsorbing agent, chemical reaction, and heating to reduce the humidity.

Dehydrant A dehydrating substance. A substance used to take on water from another substance.

Dehydrate The removal of water from a substance as in dehydration. Removal of hydrogen and oxygen in the same ratio as water (H_2O). Also see *Drying*.

Dehydrating Press A device used to squeeze moisture from pyrocellulose at 25% moisture, used for explosives, at about 17 atm (250 psi). Also see *External Moisture; Powder Drying*.

Dehydration The process of removing water, usually to preserve the properties of substances. Water of crystallization in many salts is eliminated by heat. Alcohols give up water when heated to 400 to 500°C (752 to 932°F) in the presence of alumina. As distinguished from sun drying, dehydration involves controlled artificial heating.

Dehydration Reactions The removal of water, or hydrogen and oxygen in the same ratio of water, as applied to particular chemical reactions, neutralization, esterification, and etherification.

Dehydrator A device used for dehydration of products for preservation. Some early literature uses the spelling dehydrater. Also see *Dehydration; Desiccator*.

Dehydrite Anhydrous, granular magnesium perchlorate which absorbs water to form $Mg(ClO_4)_3 \cdot 6H_2O$, which in turn may be dehydrated to the anhydrous form by heating for 10 hr at 250°C (482°F) in a vacuum.

Dehydrobrining The process of moisture removal and product

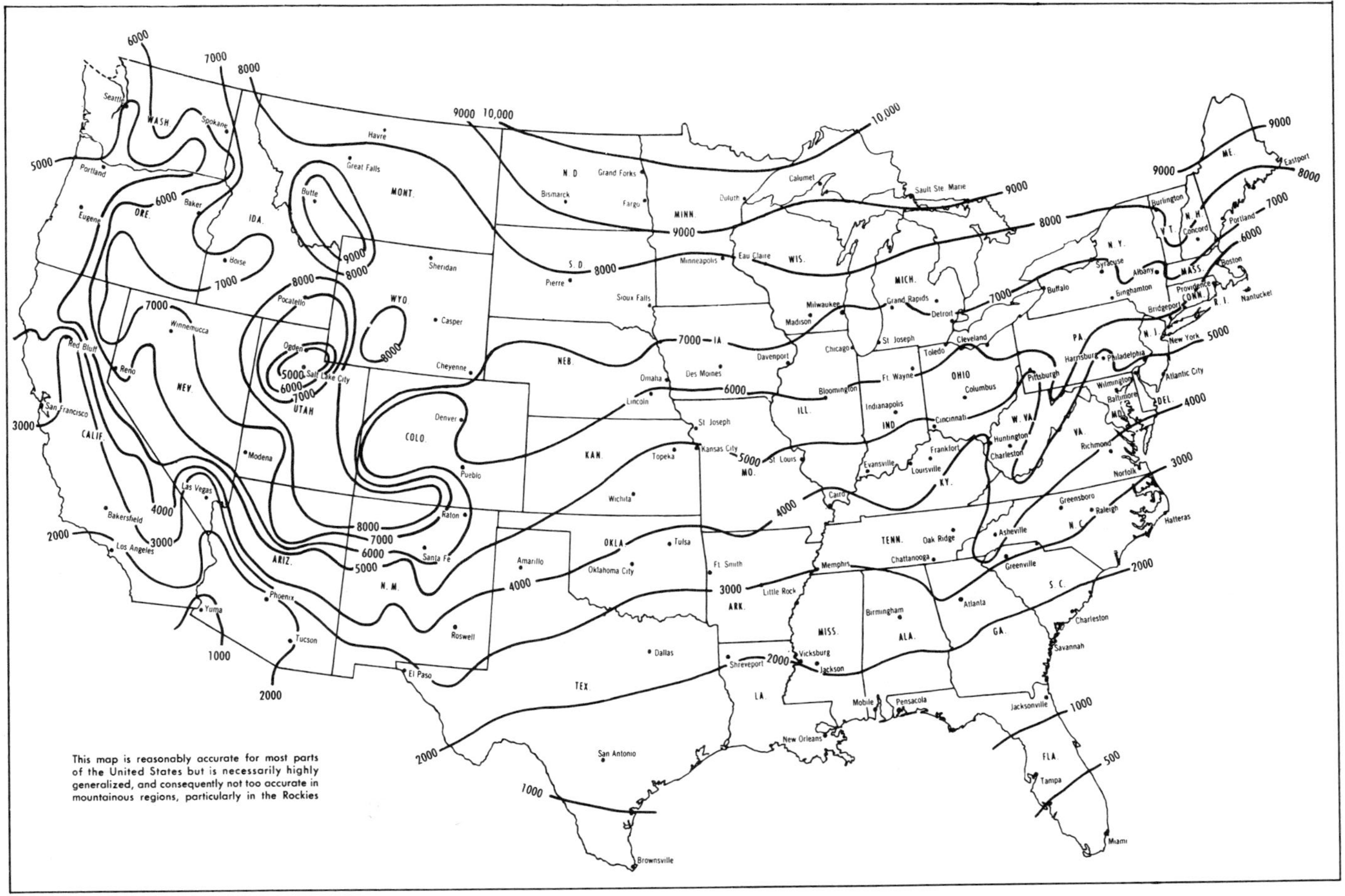

Figure D-3. Degree-day map of the United States.
Reprinted with permission from Handbook of Air Conditioning, Heating and Ventilating, Strock and Koral, Industrial Press, New York, 1965.

preservation by partial drying (90 to 75% moisture content for vegetables) followed by sufficient salt to provide a saturated brine solution with the water remaining in the product after partial drying.

Dehydrocanning   A process of moisture removal and preservation in which about 50% of the water is removed from a food product before canning to save bulk and weight while retaining texture. Also see *Canning*.

Dehydrofreezing   A method of moisture removal in which one-half to two-thirds of the moisture in a product to be frozen is first removed by conventional drying methods using heat, followed by freezing the product using standard or quick-freezing procedures. In some products, such as apples, from 20 to 50% of the moisture can be removed with little expense using normal heat drying, thus greatly reducing the amount of water which would need to be frozen. Considerable saving in shipping and storage of the product results. Product quality is not seriously damaged. Care must be taken to assure that the heating does not decrease the quality.

Dehydrofrozen Apple Slices   Dehydrofrozen apple slices are apple slices dried to half of their original weight (i.e., to about 70% moisture), and then frozen. The slices are prepared by processors located in the eastern United States and are used by some of the large bakeries in the same area. Bakers utilize this product because it is easy to use and is free of "drip" when thawed. The dehydrofrozen apples are dried to one-third or one-quarter of the original weight before freezing.

Deliquescent   The absorption of water vapor by certain solid materials, usually salts, which attract moisture from the air to the extent that a liquid solution is formed, such as with sodium chloride, zinc chloride, quicklime, sulfuric acid, etc. May also be called hygroscopic, but generally deliquescent materials are the more highly hygroscopic materials. Also see *Absorption; Calcium Chloride; Hygroscopic*.

Delmhorst   A commercial moisture detector or meter which measures the electrical resistance obtained by forcing an electric probe into the product. The moisture content of the wettest fibers in contact with the two electrodes near the end of the probe is measured. A device is provided for maintaining uniform pressure of the probe in the product. Also see *Electric Resistance Moisture Meter*.

Denaturation   An irreversible change of property in a protein produced by excessive heating as in cooking, drying, or dehydration, which results in decreased solubility, less emulsifying power, less water holding power, and reduced rehydratability. Also see *Protein*.

Density of Air   The weight per unit volume of air, given in pounds per cubic foot or grams per liter, or kilograms per cubic meter at a specified temperature (Table D-1).

Table D-1
Density of Air, Water and Water Vapor

Air Temperature		Density, lb/ft³ Dry Air at 1 Atm.	Saturated Air at 1 Atm.	Saturated Air at 2 Atm.
32°F	0°C	0.0769	-	-
40	4.4	0.0757	0.0754	0.0821
50	10	0.0742	0.0738	0.0803
60	15.6	0.0727	0.0722	0.0786
70	21.1	0.0713	0.0707	0.0769
80	26.7	0.0700	0.0691	0.0752
90	30.2	0.0688	0.0675	0.0735
100	37.8	0.0675	0.0650	0.0717

Water Temperature		Density gm/cc	Density lb/ft³
39.2°F	4°C	1.00	62.43
50	10	0.99973	62.41
122	50	0.98807	61.68
212	100	0.95838	59.83
392	200	0.8628	53.86

Conversion factor: *1 lb/ft³ = 16.018 kg/m³*

Density of Gas The weight per unit volume of gas, usually expressed in terms of pounds per cubic foot or grams per liter or kilograms per cubic meter, at 0°C (32°F), at atmospheric pressure, unless otherwise specified. Specific gravity values of gases are based on a particular gas, such as air or oxygen, assigning SG = 1 for the reference gas (Table D-2).

Table D-2
Density of Gases at 1 Atm

Gas	g/l	lb/ft³	Specific Gravity Based on Air
Air	1.2929	0.0807	1.00
Carbon dioxide	1.9769	0.1234	1.529
Carbon monoxide	1.2504	0.07806	0.9671
Hydrogen	0.0898	0.00561	0.0695
Nitrogen	1.2505	0.07807	0.9672
Oxygen	1.429	0.08921	1.1052

Density of Water The weight per unit volume of water in pounds per cubic foot or grams per cubic centimeter or kilograms per cubic meter. The temperature of maximum density for pure water is at 4°C

(39.2°F), the density is 1.0 grams/cc. The density decreases as the temperature increases above 4°C (Table D-1).

Density, Relative See *Specific Gravity.*

Depth Factor When drying a product in a bed with forced air, a depth which contains enough product, if all the theoretical heat available for drying could be used, that the depth would dry to equilibrium in a period of time equal to the time required for the fully exposed product to dry half-way to equilibrium, or the time of half response (Fig. D-4). Also see *Equilibrium Moisture; Time of Response; Time Unit.*

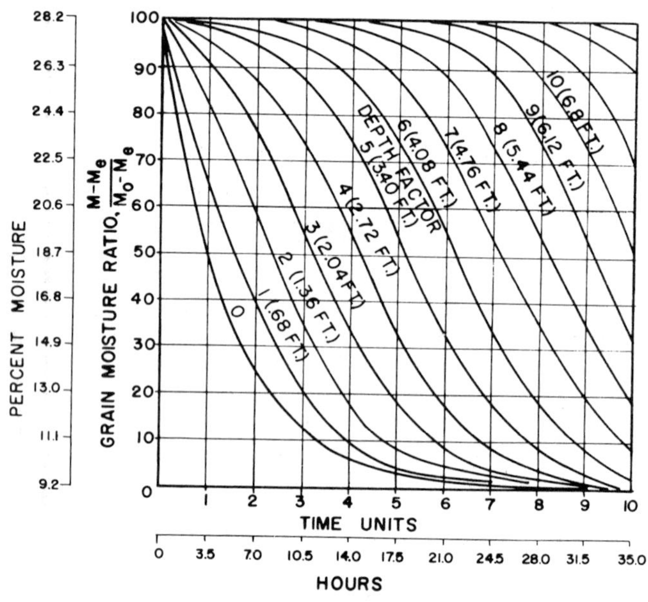

Figure D-4. Depth factor related to drying time and moisture content.

Desiccant A chemical or material which takes on moisture to dry another material, usually a gas, or by drying a gas which has absorbed moisture from a liquid or solid. The desiccant can be used for removing moisture from the product as (1) a means of drying or (2) a method of moisture content determination.

Silica gel and activated alumina are adsorption desiccants which dry by surface attraction. Magnesium perchlorate and calcium sulfate are chemical desiccants, which dry by chemical reaction with water to form hydrates. Magnesium perchlorate is one of the best adsorbers and will adsorb water to about 50% of its dry weight. The vapor pressure of the wet material is higher than that of the desiccant, and the moisture moves from the material to the drying agent as a result of the vapor pressure differences. One standard procedure for moisture determination is to place the sample in a vacuum oven with anhydrous sulphuric acid until constant weight is obtained. This method is particularly useful for materials where dry matter decomposition would be great from heating. With some products, however, the product may decompose, because of the mold and bacteria growth, before the equilibrium vapor pressure is reached. Also see *Adsorption; Calcium Chloride; Efficiency of Drying Agents.*

Desiccation The process of drying and removing moisture from a product.

Desiccator A device used for drying or maintaining dry products with a desiccant. Also see *Dehydrator; Desiccant.*

Desorption The process of removing moisture from a solid. The reverse process of sorption.

Dessichlora A dehydrating agent consisting of barium perchlorate.

Dew Point, Dew-Point, or Dewpoint The temperature at which condensation of water vapor occurs when a mixture of air and water vapor is cooled at constant absolute humidity. The dew point is the condensation temperature. When the relative humidity is 100%, the dew point, wet bulb, and dry bulb temperatures are the same value. When the relative humidity is less than 100%, the dew point is less than the wet bulb temperature which is less than the dry bulb temperature. Also see *Psychrometric Chart.*

Dielectric A substance capable of sustaining an electric field and of undergoing electric polarization. The principle is used in one type of commercial moisture meter. The dielectric constant, K, without dimensions, is a ratio which is a measure of dielectric activity. The dielectric constant $K = Q_1Q_2/r^2F$ where Q_1 and Q_2 are charges; r, the distance between charges; and F, force in dynes of the electric field. Wet materials have a high dielectric constant and dry materials have a low dielectric constant. Water has a dielectric constant of 80 at 20°C (68°F). Grains, hay, wood, etc., have a dielectric constant less than 5 and air in a vacuum a value of 1 (Table D-3).

Dielectric Drying Moisture removal from substances, usually poor conductors of heat, by placing in the field of an alternating current. The alternating electrical field deforms the molecular field of the product, causing production of thermal energy. Heating is in proportion to the dielectric constant of the material, which for water is 80. The theoretical power, kW, using dielectric heating

Table D-3
Dielectric Constant at 40 μc

Product	Moisture Content, %	ε'
Corn	13.7	3.9
Cotton, delinted	7.4	1.9
Peanuts, shelled	4.7	2.2
Pine	6.9	2.1
Rice, rough	15.0	3.8
Tobacco	5.3	1.9
Wheat	12.5	4.0
Soybeans	11.5	3.15
Air at 0°C at 3 million cycles/sec.		1.0005
Steam at 110°C at 3 million cycles/sec.		1.0126
Water at 0°C at 100 million cycles/sec.		88

Wood, at 18°C	% H_2O	1 mcps	10 mcps	100 mcps
Birch	4.65	2.3	2.25	2.15
	12.8	4.4	4.2	3.8
Mahogany	0	1.7	1.7	1.8
	2.6	2.0	2.0	1.95
	12.6	2.5	2.4	2.3
Pine	15	8.1	8.1	
	22	11.3	10.3	
	45	20	15.0	

Sources:

ASAE Data D293

Handbook of Chemistry and Physics. 1967. Chemical Rubber Publishing Co., Cleveland, OH.

Hearmon, R. F. S. 1944. The high frequency properties of wood. Forest Products Lab., Madison, WI.

for drying:

$$P = 2.93 \, M \, c_p \, (t_1 - t_o) \times 10^{-4} + W/3$$

where

M = flow of product, lb/hr
c_p = specific heat
$t_1 - t_o$ = temperature rise of material, °F
W = evaporation rate, lb/hr

Dielectric dryers have an overall drying efficiency of 50 to 60%, providing an evaporation of 0.5 to 1 kg/kWh (1-1/2 to 2 lb/kWh) of electricity. Frequencies in the range of 2 to 100 MHz (megahertz) may be used. The principle may also be used for moisture meters. Also see *Capacitance*.

Dielectric Method of Moisture Determination The dielectric properties of products vary with the moisture content at a constant temperature. The capacity of a condenser changes with the dielectric properties of the material placed between the condenser plates. Wet materials have a high dielectric constant; water has a dielectric constant of 80, and most dry materials have a value less than 5. Air in a vacuum has a value of 1 (Table D-3).

Commonly used dielectric moisture meters are the Steinlite and Motomco. A fixed sample size of grain is placed between the two plates of a condenser. Dielectric meters are less subject to errors from uneven moisture distribution, such as may result from drying or wetting, than resistance meters. Also see *Capacitance*.

Diffusion Coefficient A rate of flow of a material or heat at specified conditions of substance, temperature, vapor pressure, etc., in units of L^2/T, such as cm^2/sec, ft^2/hr, m^2/sec. In drying, the term is used to indicate the flow of heat into a material and flow of moisture or mass out of a material. For a constant rate process, the diffusion is constant; for a falling rate process the diffusion constant normally decreases in value with time. Also known as diffusivity (Table D-4).

Table D-4
Diffusion Coefficients

Wood	[1] Diffusion Coefficient, 10^5 cm^2/sec			
	Temperature			
Specific gravity	50°C	70°C	100°C	120°C
0.4	0.7	1.8	4.1	7.2
0.5	0.4	0.9	2.5	4.4
0.6	0.3	0.5	1.8	2.9

[1] Values are for tangential drying softwoods. The diffusion coefficient for radial drying is approximately 20% greater.

Gases	
Water vapor in air	0.239 cm^2/sec at 8°C
Carbon dioxide in air	0.139 cm^2/sec at 0°C

Values for wood are from: *Stamm, A. J. 1962. Wood and cellulose-liquid relationship. Tech. Bull. 150. North Carolina Agr. Exp. Sta., Raleigh, NC.*

Digital Computer An electrical device which works with numbers, or digits, according to instructions provided, to carry out a computation, either exactly or approximately, to solve problems of all types using the basic manipulations of addition, subtraction, multiplication, division, and logic. The basic components of a digital computer are input, which permits information to be placed in the computer; output, which permits information to be taken from the computer; memory, which permits information to be stored; arithmetic unit, which performs the calculations as instructed and required; and control, which instructs the memory unit to employ the arithmetic unit to take the appropriate action and then call for additional instructions from the memory unit. Digital control of drying systems provides a means of maintaining maximum efficiency. Also see *Microprocessor.*

Dimensional Analysis A procedure for organizing variables into dimensionless groups for evaluation and experimentation. The variables are listed which are independent of each other and which are thought to influence the test. These fundamental variables can be changed independently of the other variables, so are often called independent variables. The variables are then expressed in terms of fundamental units or dimensions, such as L, length, M, mass, and θ, time. The variables are combined either as products or quotients to form dimensionless products of the groups involved (Table D-5). The constants in the equations or relationships of the dimensionless groups are determined by experimentation. Two references which describe the procedure for making dimensional analysis are:

Murphy, Glenn. 1950. Similitude in Engineering. Ronald Press, NY. 302 p.
Bridgman, P. W. 1931. Dimensional Analysis. Yale University Press, New Haven, CT. 113 p.

Also see *Dimensionless Groups; Fundamental Units.*

Dimensionless Numbers or Dimensionless Groups A number without physical dimensions, usually derived by a combination of physical quantities, used to model or represent a physical phenomena. For examples of dimensionless numbers (sometimes called groups) see *Arrhenius; Biot; Bulygin; Capillary; Euler; Fourier; General Equation; Graetz; Kirpichev; Kossovich; Lebedev; Lewis; Lykov; Miniovich; Peclet; Prandtl; Rebinder; Reynolds; Romankov; Schmidt; Sherwood; Spalding-Stantin; von Karman; and Weber numbers used in drying relationships. When checking dimensionless numbers it is important to properly identify and treat separately, lb_f and lb_m.*

Dimensionless Potentials Expressions used to study or relate variables commonly used in drying, in equations and for plotting are for temperature, T; time, θ; moisture, M; pressure, P; and concentration, C. (θ is also often used to represent temperature, with T or t for time.)

$$(T) = \frac{t - t_o}{t_c - t_o} \qquad (P) = \frac{p - p_o}{p_o}$$

$$(\theta) = \frac{\theta_o - \theta}{\theta_o - \theta_p} \qquad (C) = \frac{c - c_e}{c_o - c_e}$$

$$(M) = \frac{M - M_e}{M_o - M_e}$$

Table D-5
Dimensional Analysis

Basic units: M, mass; θ, time; L, length; T, temperature	
A	Area, L^2
c	Specific heat, heat capacity, $L^2/\theta^2 T$
c_a	Concentration, M/L^3
c_b	Specific vapor capacity $L\theta^2/M$
c_m	Mass capacity, L^3/M
D, α	Diffusivity, L^2/θ
E	Entropy, $ML^2/\theta^2 T$
E_a	Activation energy, L^2/θ^2
F	Force, ML/θ^2
f	Frequency, $1/\theta$
g	Acceleration, L/θ^2
h	Heat transfer coefficient, $M/T\theta^3$
k	Thermal conductivity, $ML/T\theta^3$
k_c	Mass transfer coefficient, L/θ
L	Latent heat, L^2/θ^2, or characteristic dimension
M	Moisture constant
P, p	Pressure, $M/L\theta^2$
q	Heat flux, ML^2/θ^3
Q	Heat liberated/unit mass, L^2/θ^2
s	Humid heat, $L^2/\theta^2 T$
V	Velocity, L/θ
v	Volume, L^3
B	Coefficient of bulk expansion, $1/T$
u	Dynamic viscosity, $M/L\theta$
ρ	Density, M/L^3

Dimensions As applied to a physical quantity, the power or exponent of fundamental units of measure. The fundamental units are length, ℓ; time, t; and mass, m. Thus, area is ℓ^2; volume, ℓ^3; velocity, ℓ/t, in which dimensions for ℓ are 2 for area, 3 for volume, and 1 along with K which is -1 for velocity. In SI units the fundamental units are length, meter, m; time, seconds, s; and mass, kilogram, kg. Also see *SI Units*.

Direct-Fired Dryer A type of dryer in which the products of combustion for heating come into direct contact with the product being dried (Fig. D-5). Products which are not unfavorably affected by the products of combustion may be dried with a direct-fired unit. Direct-fired units have a thermal efficiency of 45 to 60%.

Disc Fan A type of air-moving device (AMD) or fan in which there is axial flow of air, parallel to the axis of rotation of the blade. The disc fan is essentially a propeller fan with an enlarged hub. The enlarged hub helps reduce the reverse airflow which may occur opposite the direction of major airflow through the fan. Reverse

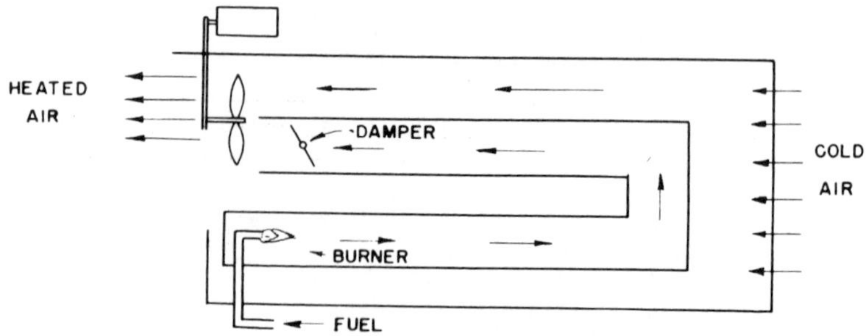

Figure D-5. Direct-fired heater for dryer.

airflow may occur near the center of the fan wheel as the pressure is increased. *Discharge, Coefficient of.* See *Coefficient of Discharge.*

Dispersion Dryer See *Flash Drying.*

Distillation The process of separation of the components of a liquid mixture by partial vaporization of the mixture and separate recovery of vapor and residue. The product is heated to the boiling point; the vapors are condensed and collected.

Doctor Blade or Doctor Knife A knife which removes a sheet of material being dried from the surface of a heated drum. The unit is used with a drum or roller dryer. Usually the doctor blade removes the dried product in less than one revolution of the drum. However, a modified drum dryer may have a control to lift the doctor blade in which more than one revolution of the drum occurs before the product is scraped from the drum.

Double-Drum Dryer An apparatus consisting of two cylindrical, horizontal steam filled drums which turn toward each other, used for continuous drying. Commonly used to dry milk, juices, brewer's yeast, foods, and chemicals. The overall heat transfer coefficients range from 50 to 400 W/m^2K (10 to 70 $Btu/hr\ ft^2{}^\circ F$).

Downdraft See *Updraft.*

Draft Gage An instrument for measuring pressure head, usually of a gas, in which one leg of a U-tube manometer is considerably larger than the other inclined leg. Variations in the level of the inclined leg produce very little level change in the reservoir. The pressure head is the vertical distance between the two levels of the fluid (Fig. D-6). Also see *Manometer, Inclined Tube.*

Drag A term which generally refers to the total force caused by airflow over a body. The drag force F on a particle is

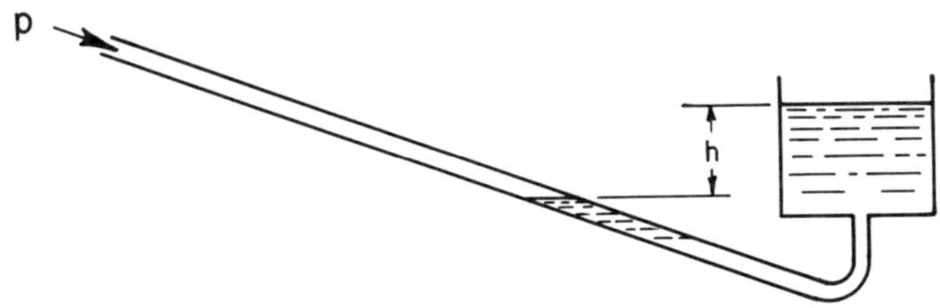

Figure D-6. Draft gage.

$$F = k\rho d^2 v$$

where

k = constant called drag coefficient
ρ = density of the air
d = diameter of the particle
v = velocity of the air

Drag Coefficient See *Drag*.

Dried Cheese Although not of major economic importance, procedures have been developed for making dried cheese from soft natural product. Dried cheese is used in manufactured foods and food products, such as soup, salads, macaroni, and snack foods. A method of drying is by chemical treatment, such as addition of sodium citrate, followed by moisture removal on a drum dryer or in a spray dryer. The cheese may be dried to 2% in a vacuum at a temperature below -7°C (20°F).

Hard cheese, after curing, is grated, surface dried to 15% at a temperature of 24 to 27°C (75 to 80°F), followed by drying the pieces to 5% moisture content using air increased in temperature to 63°C (145°F). Throughout drying, special care must be given to physical properties of whey solids in soft natural cheese and the running of fat in hard cheese.

Drier Comparative form of dry. Used by some to represent an apparatus for drying. Also see *Driers; Dryer*.

Drierite A trade name of a dehydrating agent, consisting primarily of anhydrous calcium sulfate, for drying gases, liquids, and solids. The material may include a color indicator to show the presence of moisture. The chemical may be regenerated by heating at 235 to 250°C (455 to 482°F).

Driers Chemical materials used for increasing the rate of drying paints and varnishes. Driers absorb oxygen from the atmosphere, transfer the oxygen to the paint, accelerating its drying. Driers are oxides of metals and metallic salts of organic acids. Excessive use of driers will destroy film toughness and cause cracking of the paint film. Rapid-acting driers may cause wrinkling of the film. Tung oil, a liquid drier, is classified as a drying oil because of its inherent characteristics, but technically is not a drier.

Driocel A drying adsorbent made from activated bauxite and regenerated by heating at 149 to 260°C (300 to 500°F).

Drip Dry A textile product which dries rapidly using gravity for removal of surface moisture and heat in the normal ambient atmosphere for drying following laundering.

Fibers are treated by covering the surface with resins to repel the water. The resins are primarily aminoplasts. Clothes made from drip dry fabrics are also called wash-and-wear. The usual ironing following washing of drip dry fabrics can be eliminated by hanging the fabrics to dry.

Drop Formation See *Atomization*.

Drosophere An apparatus for maintaining the proper degree of humidity in weaving or spinning mills by a combination of atomizers and airflow.

Drum or Roller Dryer An apparatus in which drying is done continuously on the external surface of an internally steam heated rotating cylinder (Fig. D-7). A thin film is applied at one location and removed at another location, usually after less than one complete revolution of the cylinder. These dryers may be atmospheric or vacuum types and are classified as single drum; double drum, with two drums working together with a nip or sump fed from above; or twin drum, in which the two separate drums function almost as single drums. The single drum may have a dip, splash, or transfer roll feed; the double drum, usually has a nip feed; and the twin drum, the dip or splash feed. Evaporation of moisture in an evaporator concentrating liquids from 2:1 to 3:1 often precedes the drum dryer. Also see *Doctor Blade; Double-Drum Dryer; Textile Drying*.

Dry To remove moisture or liquid from a substance. To desiccate, dehydrate, parch, etc. Free from moisture or at a low moisture level.

Dry, Foreign Word Equivalents The following words represent nearly equivalent words in the various languages:

English	dry	to dry	dried	drying	dryer/drier
Czech	susiti suchy	susit	susehȳ/ā/e	suseni susici	susic̄ susicka
Danish	tør	at tørre	tørret	tørre	tørreri
Dutch	droog	drogen	aedroogd	drogen	droger
Finnish	kuiva	kuivata	kuivattu	kuivaus	kuivuri
French	seche	secher	séché desseché	sechage	secheur sechoir dessecher
German	trocken	trocknen	getrocknet	trocknend trocknung	trockner

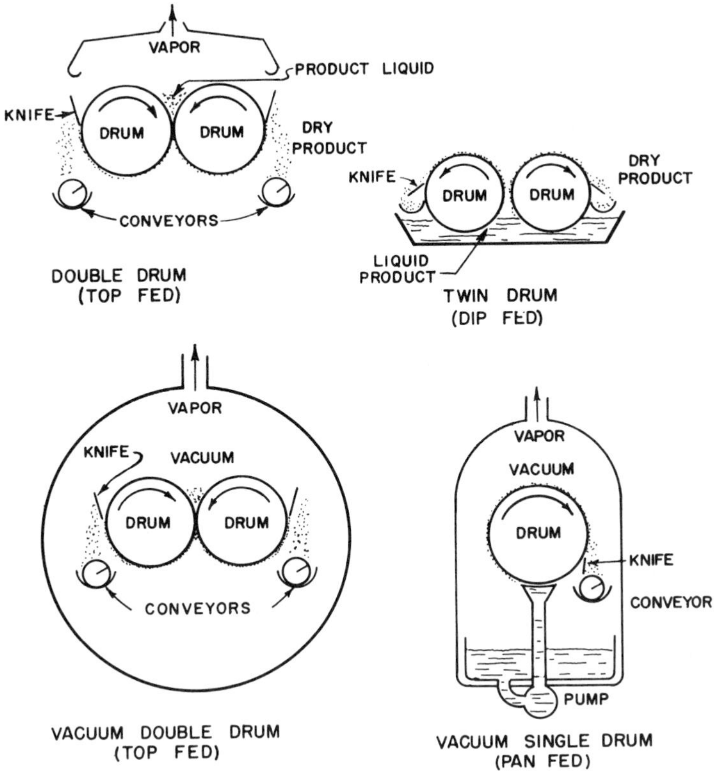

Figure D-7. Drum dryers.
From Hall, C. W., and Hedrick, T. I., Drying of Milk and Milk Products, 2nd edition (1971). Used with permission of Avi Publishing Co., Westport, CT.

Greek	ΖΕΡΟΣ	ΖΕΡΟ̱Σ ζηρōs	ΖΗΡΟ̱Σ ζηρōs	τοταχοστηs	
Hebrew	yavesh	l'yabesh	meubash	yibush	meyabesh
Italian	secco	seccare	seccato disseccato	dissec- camento	seccatoio
Japanese	kansōsuru	ksnsōsu- rukoto	kansosita	ksnsō	kansōki
Latin	siccus	sicco siccare		siccātio	
Norwegian	tørr	å tørke	tørket	tørking	tørke
Polish	suszenie	suszenie		suszenia	suszarki
Portu- guese	seco	secar	seco	secagem	secadora
Russian	suchoi	sushka		sushki	sushilki
Spanish	seco (m) seca (f)	secar (se) desecar	secado desecado	secando desecando	el secador (m) la secadora (f)
Swedish	torr	torka	torkad	torkning	tork

Dry Basis A method of representing the moisture content of a product, either as a decimal or percent, in which the weight of water, W_w, is divided by the weight of dry matter, W_d, in the product, dry basis (Fig. D-8). The dry basis of moisture content is used mainly by researchers, in scientific writings, and in equations dealing with moisture content variations. The moisture content, dry basis, percent = $W_w/W_d \times 100\%$. The dry basis moisture content is usually more acceptable than the wet basis moisture content when making calculations because the dry weight remains constant whereas the wet weight used in the wet basis representation changes during drying. The value of the moisture content on dry basis is larger in magnitude than the wet basis, for the same water content. The relationship between moisture content, dry basis, percent, MC_{db}, and wet basis, percent, MC_{wb}, is given by:

$$MC_{db} = \frac{MC_{wb}}{100 - MC_{wb}}$$

Also see *Moisture Content.*

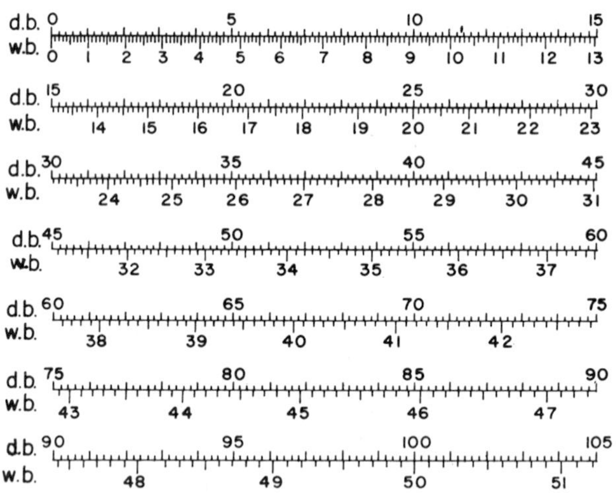

n

Figure D-8. Dry basis to wet basis conversion.

Dry-Blanch-Dry Process A procedure for drying fruit in which the product is dried to 50% moisture at about 80°C (176°F), blanched for a few minutes, then further dried at 70°C (158°F) for 5 to 24 hours until 15 to 20% moisture content. Moisture is removed more rapidly than with sun drying and many products maintain better color and flavor as compared to hot air drying.

Dry Horses   The frames on which goods are placed to dry. The frames are generally movable and used in tunnels or cabinets.

Dry Kiln Automation   See *Automation of Dry Kiln.*

Dry Liquid   A material with 70% liquid which retains liquid characteristics but is dry to the touch. The dry-liquid concept, originally called carrier-dust, such as alumina gel and Fuller's earth used as the carrier-dust, are coated with a liquid while retaining their dust characteristics.

Dry Loft   An area or room in a tannery where leather is hung to dry.

Dry Milk   Whole or skim milk from which most of the water has been removed, down to 3.5% moisture content or less. May be identified as dry whole milk and dry skim milk or dry non-fat milk. Also see *Condensed Milk; Drum Dryer; Evaporators; Milk Drying; Spray Dryers.*

Dry Milk Solids   The solid constituents of milk after the water has been removed.

Dry Shrinkage   The shrinkage which occurs in kiln-dried lumber as water is removed.

Dry Skin   Animal hide or skin preserved after slaughter by air drying.

Dry Vapor or Steam   Steam or vapor containing no free water droplets. Also see *Steam, Saturated.*

Dry Cleaning   Removal of soil from clothes and fabrics using chemical cleaning solvents which are re-used after filtering. Surfactants are added to the solvent. In a dry cleaning still of 0.4 m^3 (100 gal) size, a steam capacity of about 40 kW (4.0 Bhp) at 480 kPa (70 psi) is needed. Clothes are placed in tumble-type washers, excess solvent is removed by centrifuged extractors, and the remaining solvent is removed in heated dryers. Dry cleaning chemicals include petroleum solvents, chlorinated hydrocarbon solvents, and fluorinated hydrocarbon solvents. The excess solvent after centrifugal extraction is removed by evaporation drying in a tumbler at 57 to 71°C (135 to 160°F).

Dry-Cure   The process of curing by drying or moisture removal using a salt or brine.

Dryer A unit which provides the conditions for removing moisture from a product. Also see *Classification of Dryers* and specific kinds of dryers.

Dryer Efficiency The percentage of the total heat supplied by fuel used to evaporate water from a product. Dryer efficiency is the product of the efficiencies of the components and is normally 35 to 70%. Burner efficiency refers to the percentage of the heat in the fuel which through combustion leaves the burner and normally enters the heater. Overall thermal efficiency refers to the heat available in the fuel in relation to the heat used to evaporate water. Dryer efficiency may relate to the efficiency of heat utilization in the space where drying occurs. Also see *Dryer Performance Evaluation Index; Energy Efficiency; Performance Ratio.*

Dryer Performance Evaluation Index (DPEI) The total energy required by a dryer to remove one pound of moisture from the product during drying under a set of specified conditions. The total energy includes the energy to heat the drying air, the energy to move the drying and cooling air, and the energy to move the product. The specified conditions are ambient air conditions (temperature and humidity ratio), the inlet product moisture content, temperature, type and quality of product.

Dryeration A modified drying process involving the portable batch or continuous-flow dryer. In this process, the product is dried with heated air to a moisture level of 16 to 18%, wet basis. The product is transferred immediately without cooling to a temporary storage (dryeration) bin equipped for aeration. The product is allowed to set for a few hours to become tempered before aeration is started. Cooling is accomplished in about 12 hr with airflow rates of about 0.007 m^3/m^3s (1/2 cfm/bu). At this low airflow rate, nearly all the heat in the product is utilized to further dry it. The moisture content may be further reduced 3 to 4% during the 12 hr of cooling. Also see *Equalization; Intermittent Drying; Resting.*

Drying The removal of moisture from a substance involving the simultaneous transfer of heat to the substance and moisture from the substance, known as a unit operation. The heat for evaporation of the moisture in the material is transferred by conduction, convection, radiation, and internal heating, such as by respiration and dielectric heating. The vapor mass is transferred by diffusion or capillary flow to the surface from which the vapor is carried by a gas. As a result of drying, the biological and chemical activity of the product is decreased. Also see *Adhesive Hardening; Dehydration.*

Drying Agent See *Efficiency of Drying Agents; Silica Gel* (Table D-6, Table D-7).

Drying Air The air which is passed through the product being dried in a dryer or drying system.

Table D-6
Drying Capacity of Desiccants, Lb. Water per Lb. of Desiccant, 20°C

Drying Agent	Relative Humidity, percent					
	10	20	30	40	60	80
Lithium chloride (solid)	0.4-1.1					
Lithium chloride (liquid)		1.6	1.9	2.4	3.5	
Calcium chloride (solid)	0.35	0.6	1.0			
Calcium chloride (liquid)				1.5	2.2	3.4
Sulfuric acid (solid)	0.35	0.6				
Sulfuric acid (liquid)			0.9	1.0	1.6	2.8
Glycerine (liquid)	.05	0.1	0.2	0.3	0.6	1.3
Tri-ethylene glycol (liquid)		0.1	0.15	0.2	0.3	0.9
Silica gel (solid)	0.5	0.1	0.15	0.20	0.25	0.30
Activated alumina (solid)			0.15	0.17	0.19	0.20

Based on: *Kirk, R. E., and Othmer, D. F. 1965 (2nd ed.). Encyclopedia of Chemical Technology, vol. 7, Interscience, N. Y., NY.*

Table D-7
Drying Constants, k, 1/hr

Temperature °C	°F	Relative Humidity, %	k, 1/hr	Range of Applicability, % Moisture, d.b.
4.4	40	14	0.075	26.96 to 24.79
			0.0315	24.79 to 20.50
			0.016	20.50 to 16.25
			0.0072	16.25 to ---
4.4	40	60	0.0622	27.01 to 25.2
			0.0325	25.2 to 20.2
			0.0231	20.2 to 18.0
			0.0149	18.0 to ---
15.6	60	76	0.0583	27.2 to 25.20
			0.0396	25.2 to 20.5
			0.0291	20.5 to ---
37.8	100	50	0.1047	23.5 to 18.00
			0.0424	18.00 to 13.00
			0.0143	13.00 to ---
60	140	11.2	0.338	47.0 to 19.0
			0.149	19.0 to 11.5
			0.0503	11.5 to 7.3
			0.0184	7.3 to 5.9
60	140	55	0.271	43.6 to 13.00
			0.139	13.00 to 10.4
			0.074	10.4 to 9.70
			0.052	9.70 to ---

Drying Air Temperature The temperature of the air which enters a dryer.

Drying Capacity As applied to desiccants, the ability of certain chemicals to adsorb moisture from the atmosphere in vapor form (Table D-6). Even though the capacity of certain desiccants, such as silica gel, is low, the efficienty is high as it dries to a lower dewpoint than some other desiccants.

Drying Coefficients Values used to represent the humidity driving force, K_H, or the vapor pressure driving force, K_p, in drying equations to determine the instantaneous drying rate, dW/dt.

$$dW/dt = K_H A\ (H_s - H_g) = K_p A\ 10^{-3}\ (p_s - p_g) =$$

$$h_c A\ (T_g - T_s)/L$$

where

dW/dt = drying rate, g/s
K_p = mass transfer, $g/m^2 s$ bar
K_H = drying coefficient, $g/m^2 s$ unit humidity difference
A = area of material, m^2
H_s = saturation humidity, kg/kg
p_s = saturation partial pressure, bar
p_g = partial pressure inlet gas, bar
h_c = heat transfer coefficient, $W/m^2 °C$
T_g = temperature of drying gas, K
T_s = temperature of drying surface, K
L = latent heat of vaporization, kJ/kg

Coefficients may be used for the constant rate period of drying or the falling rate, in which case the value decreases with time. Also see *Film Coefficients; Mass Transfer Coefficients*.

Drying Constant See *Drying Rate Constant*.

Drying Curve A graphical representation of the moisture content (d.b. or w.b.) of product versus time during moisture removal associated with drying conditions and a particular product. Drying rate curves can be obtained from drying curves (Fig. D-9). Also see *Constant Rate Drying; Drying Rate; Falling Rate Drying*.

Drying Flowers The procedure for drying fresh flowers is to use equal portions of powdered pumice and yellow cornmeal, or equal portions of borax and yellow cornmeal, with 3 tablespoons of uniodized salt to the above mixture. Place flowers in a pan (lined with oiled paper) with stems sticking up. Add small quantities of the mixture to the bunch of flowers with a tablespoon into and around the petals and allow about 1 wk to dry.

Drying Efficiency See *Efficiency of Drying Agents*.

Drying Front The layer which is between the dried and undried

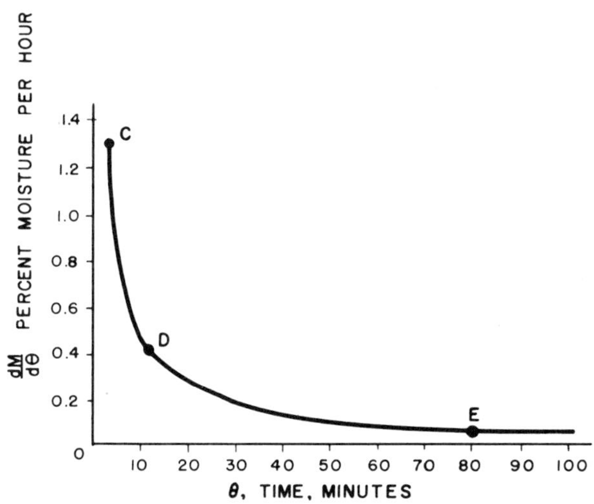

Figure D-9. Drying rate curve for falling rate: C, critical moisture content; C-D, first falling rate; D-E, second falling rate.

products in a drying system. The advancing zone of drying includes the product that has just reached some arbitrary moisture content between the initial and final moisture content. Also see *Deep Layer*.

Drying Loft A room in which paper is dried under carefully regulated atmospheric or environmental conditions.

Drying Oils Vegetable oils, the most common of which is linseed oil, which are easily oxidized in air which assist in producing a paint film. Drying oils have a high proportion of unsaturated acids. The film develops by (1) evaporation of solvent and (2) oxidation of film, enhanced by driers. The drying power of drying oils is represented by their iodine value or iodine number, as the ability to absorb oxygen from the air, which is directly proportional to their ability to absorb iodine. Iodine numbers generally are 125 to 205, with the higher value giving more rapid drying. Also see *Oil, Semi-drying and Oil, Non-Drying*.

Drying Potential The unexpended or unused capacity of a body of air to hold water. The drying potential is represented by the distance from the saturation curve on the line of constant thermodynamic wet-bulb temperature on the psychrometric chart.

Drying Rate The rate or speed at which moist material loses moisture under drying conditions. The drying rate is often divided into constant rate and falling rate periods, expressed in moisture loss per unit time (Fig. D-9). k may be used for the drying constant, or drying rate constant.

$$\frac{M - M_e}{M_o - M_e} = e^{-k\theta}$$

where

k = drying constant, 1/hr
M = moisture content at time θ
M_o = original moisture content, d.b.
M_e = equilibrium moisture content

Values of k for various products are given in Table D-7. Also see *Equations*.

Drying Ratio A dimensionless number representing the ratio of the weight of wet material entering a dryer to the weight of the same material as the material leaves the dryer at a lower moisture content (Table D-8). The number of pounds of fresh product (vegetable, fruit, or meat) prepared for drying required to produce one pound of acceptable dry product. Also see *Rehydration Ratio*.

Table D-8
Approximate Overall Shrinkage, Drying, and Rehydration Ratio of Various Vegetables

	Overall Shrinkage Ratio	Drying Ratio	Rehydration Ratio
Asparagus	30	11	8
Cabbage	21	11.5	10.5
Carrots	12	7.5	6
Celery, stalk	30	17.5	10
Onions, powdered	8	7	5.5
Pimiento	28	10	7
Spinach	13.5	13	9
Tomatoes, flakes	20	14	5

Source: *Van Arsdel, W. B. 1964. Food Dehydration, vol. II, Avi Publishing Co., Westport, CT.*

Drying Temperature The temperature of a product during drying which is related to quality of dried product and rate of drying. Many biological products--pharmaceuticals, foods, seeds, fibers--may be adversely affected by excessive heat treatment. Excessive heat treatment involves a time-temperature exposure of heat, but the limit of exposure is often given by the maximum product temperature reached during the drying period. During drying the evaporation of moisture may keep the product cooler than the drying air. The temperature of the product approaches the temperature of the drying air as the product dries (Table D-9).

Table D-9
Drying Temperatures, Maximum Air Temperature at the Termination of Drying/Dehydration

Fruits and Vegetables	Temperature °C	°F		Temperature °C	°F
Apples	74	165	Grapes	71	160
Apricots	71	160	Onions	88	190
Cabbage	63	145	Peaches	66	150
Carrots	79	175	Pears	63	145
Cherries	71-77	160-170	Potatoes	66	150
Figs	71	160	Prunes	74-77	165-170
Grain					
Feed	82	180			
Seed	52	125			

Drying Time The time required to reduce a product from one moisture content to a lower moisture content based on experimental or practical conditions (Fig. D-10). Many procedures are used to estimate the drying time of various products under different conditions. The drying time, t_h, based on heat transfer to accomplish heated air drying is approximated by

$$t_h = \frac{\Sigma L}{\Delta T_{\ell m} (h A_u)}$$

where

L = latent heat of vaporization
$\Delta T_{\ell m}$ = log mean value of temperature difference
h = film coefficient of heat transfer
A_u = area over which heat is transferred

The drying time, t_m, based on mass transfer to accomplish moisture removal is approximated by

$$t_m = \frac{(M_1 - M_2)\ 10^{-3}}{\Delta p_{\ell m} K A_u}$$

where

$M_1 - M_2$ = difference in moisture content of product

$\Delta p_{\ell m}$ = log mean value of pressure difference
K = mass transfer coefficient
A_u = area over which the moisture is transferred

Also see *Constant Rate; Drying Coefficients; Equations; Falling Rate.*

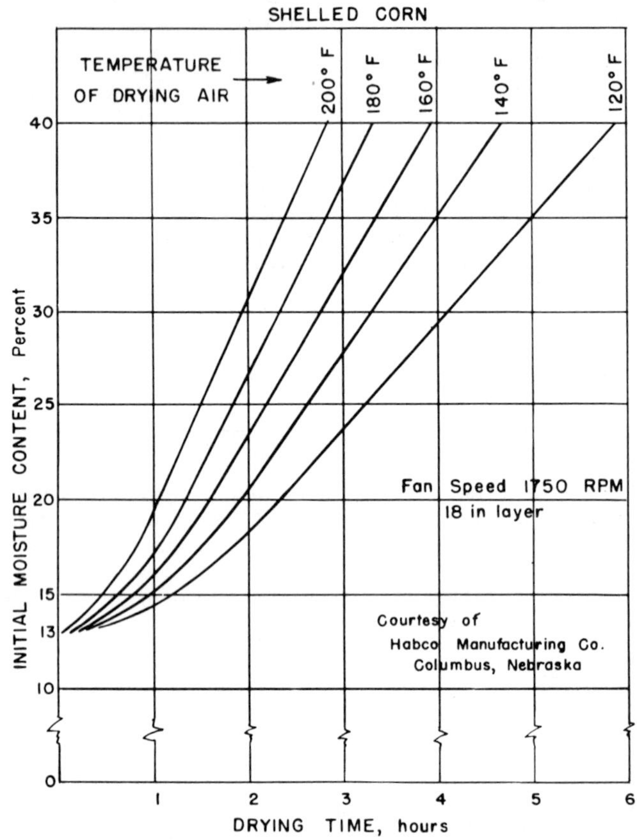

Figure D-10. Drying time chart for drying shelled corn to 13% moisture with heated air.

Drying Time, Maximum Permissible The maximum elapsed time that may be used to complete the drying of any portion of the product without an undesirable change in quality.

Drying Tube A tube or pipe filled with a hygroscopic material (such as calcium chloride) used to dry gases.

Drying Zone The band or layer of product in a bin or bed in which most of the drying occurs. The leading edge of the drying zone is called the drying front (Fig. D-2). Also see *Drying Front*.

Dryly; Drily The condition of a material without moisture.

Dryness The condition of a substance being dry or having a low quantity of water.

Dryo A prefix meaning dry.

Duct A pipe, tube, or channel for directing air to and from a fan, to and through a dryer, and/or to and through the drying bed. To avoid excessive pressure drop, an air velocity of 5.08 m^3/m^2s (1000 ft/min or 1000 cfm/ft^2) to 10.1 m^3/m^2s (2000 ft/min) is used (Table D-10). Duct systems include the following, based on shapes: triangular, rectangular, wire mesh, perforated, and louvred. Also see *Barn Hay Dryer; Flow of a Fluid*.

Dust Dust concentrations greater than 0.4 gram per standard cubic foot are usually considered excessive. Atmospheric dusts are 1.0 to 500 microns in diameter; industrial dusts up to 4000 microns. Also see *Air Pollution; Fines; Maximum Average Contaminants; Safety*.

Dust Chamber See *Cyclone Collector*.

Dust Collector See *Cyclone Collector*.

Dust Filter See *Filter*.

Dynamic Equilibrium A system in which the equilibrium value (such as equilibrium moisture content) is constant, while the evaporation of the liquid from the substance, and the condensation of vapor on the substance, are equal. Equilibrium moisture content determined in a moving air stream may be called dynamic equilibrium moisture content. Also see *Equilibrium Moisture*.

Table D-10
Friction of Air in Pipes, In. of Water per 100 Ft.

cfm	6	9	12	15	18	21	24	27	30	33	36	39	42	45	50	55	60
1000	8.3	1.3	.28	.084	.035												
5000			5.0	1.7	.68	.32	.17	.09	.052								
10,000				6.0	2.5	1.1	.58	.31	.14	.12	.07	.05		.026			
20,000					9.1	4.1	2.2	1.2	.72	.38	.28	.18	.14	.09	.056	.033	.023
40,000							8.0	4.3	2.7	1.5	1.0	.68	.47	.34	.20	.12	.055
80,000														1.3	.72	.46	.30

Conversion factors: *1 cfm = 0.00047* m^3/s
1 in. water = 0.248 kPa

E

e The base of the natural logarithm which has a numerical value of 2.71828. e is the sum of the infinite series

$$1 + 1/1! + 1/2! + 1/3! \ . \ . \ . + 1/n!$$

Ear Corn Drying Ear corn is dried with natural air moving through the ears on piles or narrow open slotted cribs. Ear corn can be stored at a higher moisture content (25%) than shelled corn (13%) without spoilage unless forced air drying is used.

Forced natural air and heated air drying of wet ear corn are used to reduce the moisture to a safe level in cribs exceeding 1/2 to 2 m (4 to 6 ft) width in the major corn growing areas. Depths or distances of air flow in ear corn of 4.5 m (15 ft) provide a static pressure drop of 0.1 to 0.2 kPa (1/2 to 3/4 in. w.c.). An air flow of 0.065 to 0.13 m^3/m^3s (5 to 10 cfm/bu) heated to 43.3°C (110°F) is used for ear corn drying. Moisture must move from the cob as well as the kernel (Fig. E-1). Also see *Corn Drying*.

Ebullition See *Boiling*.

Earthenware Drying The removal of moisture from earthenware in plaster molds. Drying may be accelerated by circulating heated air at a temperature not to exceed 70°C (160°F). Degradation of the plaster from excessive heat must be avoided. Pottery is dried from 30 to 20% moisture to permit removal from the mold. Depending on the product, mold, thickness, shape, and time, considerably more moisture may be removed, perhaps down to 5%. Also see *Porcelain Drying; Pottery Dryer*.

Effective Heat Utilization Factor The effective heat utilization is the ratio of the total heat expended (Btu) to the heat in the water evaporated (Btu). In order to facilitate the comparison of

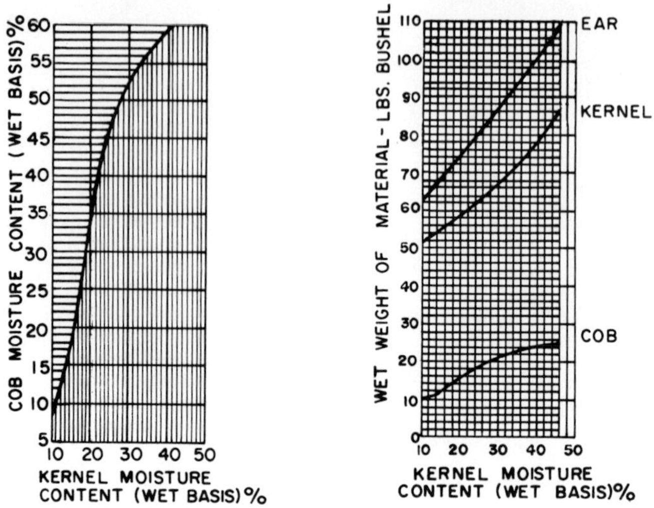

Figure E-1. Ear corn to produce 56 lb. of shelled corn at 15.5% moisture.

drying tests the effective heat utilization should be determined on the basis of the following, or similarly defined base data:

1. Net or low heating value of fuel.
2. Outside air temperature, T_1 = 40°F (or some other base).
3. Temperature of product after drying the same as before drying.

$$EHU = \frac{(LHV)\ f - (h_{40} - h_t)Q - (W - w)\ s\ (T_f - T_s)}{Lw}$$

where

LHV = low (net) heating value, Btu/ft^3 (Btu/gal)
f = quantity of fuel used, ft^3 (if gas, converted for temperature and pressure)(gal)
$h_{40} - h_t$ = enthalpy difference of 40°F and ambient air, Btu/lb air
Q = quantity of air used for drying, lb
W = initial weight of product, lb
w = weight of water removed, lb
s = specific heat of product leaving dryer, Btu/lb°F
T_f = temperature of grain leaving the dryer, °F
T_s = temperature of product entering the dryer, °F
L = latent heat of vaporization of moisture, Btu/lb

The expression $(h_{40} - h_t)Q$ is the correction for outside air temperature which varies from the base. The expression $(W - w)s(T_f - T_s)$ is the correction for product temperatures before and after drying. Also see *Energy Efficiency; Heat Utilization Factor*.

Efficiency of Drying Agents The ability of chemicals used for drying air to remove moisture from the air. Chemicals which have a low vapor pressure at saturation are described as having a higher efficiency as a drying agent. Thus, phosphorus pentoxide, P_2O_5, is relatively more efficient than calcium chloride ($CaCl_2$) as a drying agent. Using P_2O_5 the residual water in 40,000 liters is less than 1 mg whereas with $CaCl_2$, in 1 liter is about 0.2 mg. The efficiency as used here does not necessarily relate to the quantity of chemical required. See Table E-1.

Table E-1
Efficiency of Drying Agents

Vapor in Dried Air, mg. per liter	
P_2O_5	<1 mg. in 40,000 liters
$Mg(ClO_4)_2$ anhydrous not weighable in 210 liters	
BaO	0.00065
KOH	0.002
CaO	0.003
H_2SO_4	0.003
Al_2O_3	0.005
$CaBr_2$	0.18
$Ba(ClO_4)_2$	0.82
$ZnCl_2$	0.85
$ZnBr_2$	1.16
$CaCl_2$ granular	1.5
$CuSO_4$ anhydrous	2.8

Source: *Handbook of Chemistry and Physics. 1967. Chemical Rubber Publishing Co., Cleveland, OH.*

Effusion The passage of gases under pressure through small holes in which the rate of flow is inversely proportional to the square root of the density. Effusion occurs through holes larger than those to which diffusion is applicable.

Egg Drying Eggs may be dried whole or separated into whites and yolks, which are dried. Whole eggs and yolks are dried in the same manner. Rehydrated eggs have 47% protein and 43% fat.

Egg white drying (Fig. E-2) includes the following steps: (1) remove some of the free glucose by yeast, enzyme systems, or bacteria, to prevent the Maillard reaction; (2) adjust the pH with a food acid, as lactic acid and citric acid; and (3) the liquid egg white is spray dried in a spray dryer or tray dried in a cabinet or tunnel system at 54.5°C (130°F).

Egg yolk drying consists of: (1) pasteurization at 61.7°C (143°F) for at least 3 min; (2) storing after cooling to 4.4°C (40°F), or spray drying immediately; and (3) stabilization of yolks with a glucose-oxidase enzyme system.

Whole egg drying is similar to egg yolk drying. Carbohydrate

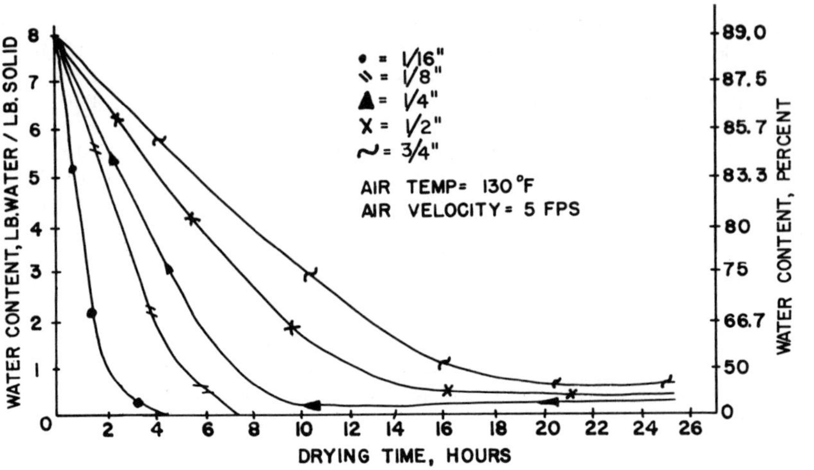

Figure E-2. Egg white drying rates with different depths on pans.

products, such as sucrose and corn syrup solids, are often added prior to spray drying to reduce foaming and to minimize emulsifying abilities of liquid eggs.

Excessive heat, above 57.2°C (135°F), may have an adverse effect on the whipping properties of egg whites. An adverse effect from excessive heat is more likely to occur during tray drying than spray drying. Also see *Browning Reaction; Maillard Reaction.*

Electrical Resistance Moisture Determination The resistance, R, of a material to the flow of electrons, usually expressed in ohms, or multiples thereof. The voltage or electromotive force, E, divided by the amperage, I, is the resistance in ohms. The electrical resistance may vary with temperature, strain, material, and moisture content. The electrical resistance of hay decreases with an increase in pressure. Alfalfa hay at 17.3% w.b., has a d-c resistance of 750 kilohms with slight pressure, 180 kilohms with a 5 psi pressure, and 90 kilohms with 10 psi pressure. The 60 cps d-c resistance is slightly higher. The electrical resistance is greater with current flowing perpendicular to the stems than with current flowing parallel to the stems. The electrical resistance of grain decreases when the pressure is increased. For 15.2% shelled corn, with a pressure of 100 psi, the d-c resistance is 1.0 megohms.

Above 17% moisture content, there is a parabolic relationship between the moisture content and the logarithm of the electrical resistance. Most resistance meters do not give readings below 7% moisture where there is very little change in the electrical conductivity. Immediately after drying with heated air lower readings than the actual moisture content of the product are obtained. This occurs because the tendency of resistance meters is to measure the resistance on the surface. If moisture has been added to the grain, the readings are higher than the average moisture content of the product. Values of electrical resistance of wood are given in Table E-2. Correlations for variations in temperature must be made for tests conducted at temperatures other than those at which the calibration is provided. Also see *Delmhorst; Electric Resistance; Moisture Meters.*

Table E-2
Resistance (Conductivity) of Wood

Wood	Water, %	Electric, Ω^{-1} cm^{-1} x 10^6, at 18°C 1 mcps	10 mcps	100 mcps
Birch	4.65		1.0	6.5
	12.8		1.5	17.5
Mahogany	0		0.15	2.5
	2.6		0.3	4.0
	12.6		0.7	10
Pine	15	0.17	2.8	
	22	1.1	5	
	45	7	58	

Sources: *Hearmon, R. F. S. 1944. The high frequency properties of wood. Forest Products Lab., Madison, WI. Kollman, F. F. P., and Cote, W. A. 1968. Principles of wood science and technology. Springer-Verlag, N. Y., NY.*

Electrodialysis A process using electric current to cause migration of saline cations and anions through a solution, and a set of water (or solvent) permeable membranes and to either the cations or anions to remove them from water (or a solvent) (Fig. E-3).

Electroosmosis The phenomenon whereby when an electric potential is placed across a porous material containing a liquid suspension, and the fluid flows across the solid. The process may be used for dewatering operations and as an assist to mechanical dewatering. The phenomenon is applied to drying materials normally difficult to dry, such as glue-like or jelly-like substances.

Emissivity As applied to radiation heat transfer, the rate of emission of radiant energy from an opaque material due to the temperature, divided by the rate of emission by the same area of a black body at the same temperature (Table A-1). The total radiation from a black body is

$$q = \sigma A T^4$$

where

q = radiation, Btu/hr
σ = Stefan-Boltzmann constant, 0.173×10^{-8}

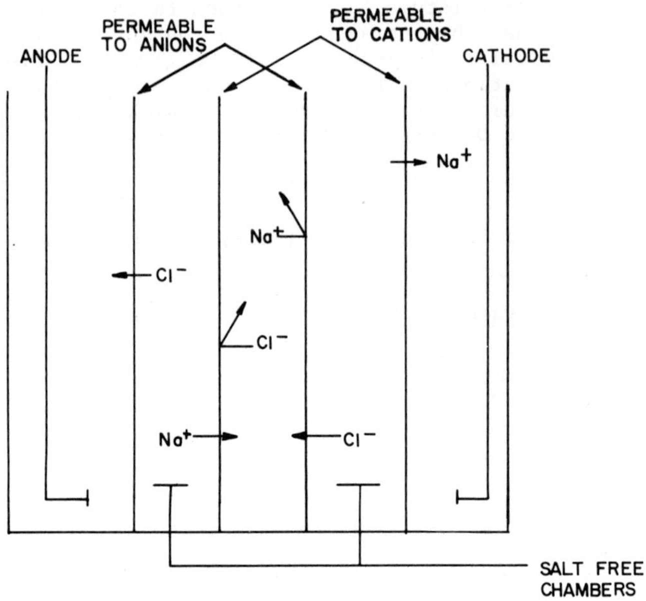

Figure E-3. Electrodialysis.

0.173×10^{-8} Btu/ft^2 hr°R^4 = 5.71×10^{-5} ergs/cm^{2}sec°K^4 = 56.7×10^{-12} kw/m^{3}K^4

A = area

T = absolute temperature

Emissivity is dependent upon the material, its surface characteristics, and its temperature. Also see *Absorptivity; Stefan-Boltzmann Law*.

Empirical Refers to an equation, formula, curve, analysis, or number based on observation and experimental values which is not necessarily supported by established theories.

Endosmosis The flow of water by osmosis through a semi-permeable membrane into a cell containing a solution.

Energy The capacity or ability to do work. Energy may be identified on the basis of type such as kinetic energy, potential energy, internal energy, flow energy, or according to source such as electrical energy, chemical energy, mechanical energy, biological energy, solar energy, etc. Energy cannot be created nor destroyed but can be converted to different forms.

Kinetic energy may be referred to as velocity energy or velocity head and represents the energy a body or substance possesses because of its motion. Kinetic energy is calculated from KE = $1/2\ mv^2$ and for a rotating body, KE = $1/2\ I\ \omega^2$, where m = mass, v = velocity, I = inertia, and ω = angular velocity.

Potential energy is possessed by a body or substance because of its position. PE is the head of substance above a datum, and may also include the flow energy due to pressure of a fluid in a pressurized vessel, where 1 atm of water is 14.7 psi, and the pressure head is lb/ft^2 divided by the density in ft^3/lb. Also see *Calorific Value; Radiation.*

Internal energy is that energy associated with the temperature and state of a body or substance

$$u_2 - u_1 = C_v(T_2 - T_1)$$

Energy Efficiency As is now used in drying, the amount of energy used to evaporate a unit of water, kJ/kg (Btu/lb), with the latent heat of vaporization as a value to be approached. A rule-of-thumb is that approximately two to three times the latent heat value is required in energy to accomplish drying. Some typical values are:

Concurrent commercial dryer	4180-5120 kJ/kg (1600-1900 Btu/lb)
Cross flow commercial dryer with recirculation	5800 kJ/kg (2200-2500 Btu/lb)
Cross flow commercial dryer without recirculation	7400 kJ/kg (3200 Btu/lb)

Energy for Drying Crops Artificial drying of crops gained commercial significance in the 1950's. In 1975 about 85% of the U. S. corn production was artificially dried. Artificial drying is also used to dry rice, sorghum, wheat, soybeans, peanuts, and forages. On-farm drying is used for 50 to 55% of the corn crop, mostly with LP gas. The 20% off-farm drying is done with LP or natural gas in about equal amounts. Corn harvested at 30% moisture will require 25 gal of LP per acre for drying (at approximately 85 bu/acre), which is two or three times the fuel required in field operations to produce corn.

Enthalpy Represented by H or h, the thermal energy of a substance with respect to an arbitrary reference level. The enthalpy is the sum of the internal energy, u, and the flow energy which is the product of the pressure, P, and the specific volume v ($h = u + Pv$). For an ideal gas, $\Delta H = m\ C_p\ \Delta T$. In drying, calculations are usually concerned with the changes in enthalpy. The enthalpy tables for dry air are based on -17.8°C (0°F) for steam, 0°C (32°F), and for refrigerants, -40°C (-40°F).

Entropy Represented by S or s, is a measure of the degree to which energy of a system is unavailable. The entropy of a body increases when its free energy decreases or when the disorder increases. When energy is lost by heat transfer, entropy increases. For thermal systems, entropy is defined by $\Delta S = \Delta Q/T$ for a reversible process where ΔS is a small change in entropy, ΔQ is a small change in heat transfer to or from the system, and T is absolute temperature at which ΔQ takes place. Also see *First Law; Free Energy; Second Law.*

Equalization The procedure of holding products so that moisture or temperature will attain a uniform temperature throughout. Also see *Dryeration; Resting; Sweating; Tempering; Turning.*

Equation of State An equation which connects the relationship of pressure, volume, temperature, and mass of a substance.

Equations, Drying Many equations are available for representing the drying rate. Two general equations are given below:

(a) Constant rate period

$$\frac{dw}{dt} = K_g A (p_s - p_v) = \frac{h A \Delta T}{L}$$

where

$\frac{dw}{dt}$ = rate of moisture loss
K_g = mass transfer coefficient from wet surface
A = surface area
p_s = saturation vapor pressure
p_v = partial pressure of vapor

The above equation is modified so the rate of transfer of moisture depends on the air velocity, v:

$$\frac{dw}{dt} = K'A (p_s - p_v) v^{0.8}$$

where

h = film heat transfer coefficient
ΔT = surface to air temperature difference
L = latent heat

(b) Falling rate period

$$\frac{dM}{dt} = -A k (M - M_e)$$

$$\frac{dM}{dt} = -\frac{\pi^2 D_L}{4\ell^2} (M - M_e)$$

where

M = moisture content, decimal, d.b.
M_e = equilibrium moisture content, d.b.

D_L = diffusion coefficient
ℓ = radius or half-thickness of material
k = rate of drying per unit area to moisture content

Also see *General Equation; Mass Transfer.*

Equilibrium The condition in which a balance exists between opposing forces or effects, such as moisture equilibrium, chemical equilibrium, temperature equilibrium, pressure equilibrium, etc. Also see *Le Chatelier Principle.*

Equilibrium Moisture A condition or state of relationship of moisture in a product and its surroundings. A product is in moisture equilibrium with its environment when the rate of moisture loss from the product to the surrounding atmosphere is equal to the rate of gain of the product from the surrounding atmosphere. The atmospheric conditions are described by the temperature and relative humidity or absolute humidity (Table E-3). The moisture content of the product when the product is in equilibrium with the surrounding atmosphere is known as the equilibrium moisture content or hygroscopic equilibrium. The relative humidity of the surrounding atmosphere at the same condition is the equilibrium relative humidity at the particular temperature. Equilibrium moisture content values may be determined in an air stream and identified as dynamic equilibrium moisture content, or in a closed container and identified as static equilibrium moisture content. The graphical plots of equilibrium moisture and equilibrium moisture content are called isotherms. Also see *Adsorption Isotherm; B-E-T; Harkins-Jura; Henderson Equation; Hysteresis; Hydrophilic Fiber.*

Equilibrium Relative Humidity See *Equilibrium Moisture.*

Erg A unit of work in the cgs system of units in which one erg of work is done when a force of 1 dyne is exerted upon a body through a distance of 1 cm. Because the erg is a small unit, the joule (10^7 ergs) is more commonly used. Also see *Joule.*

Error The difference between the indicated value and the true value of a measurement. In experimental errors are included illegitimate, random and accidental, and systematic or fixed. The illegitimate errors include blunders, mistakes, computational, and copying; random errors are unbiased and result from judgment, variation of conditions, and definition; systematic errors are biased and include technique, human, experiment, and calibration (Fig. E-4). The standard error of estimate (S.E.) is a statistical means of representing the variation of readings. Also see *Correlation Coefficient; Standard Error.*

Euler Number A dimensionless parameter, Eu, used to represent the relationship between pressure drop due to friction and inertia forces in the flow stream:

Table E-3
Equilibrium Moisture (Adsorbed) with Air of Various Humidities at Room Temperature, 20°C (77°F)

	Moisture Content, w.b., %						
Relative Humidity, (%)	15	30	45	60	75	90	100
Barley	6.0	8.4	10.0	12.1	14.4	19.5	26.8
Buckwheat	6.7	9.1	10.8	12.7	15.0	19.1	24.5
Corn, shelled, YD	6.4	8.4	10.5	12.9	14.8	19.1	23.8
Corn, shelled, WD	6.6	8.4	10.4	12.9	14.7	18.9	24.6
Corn, shelled, Pop	6.8	8.5	9.8	12.2	13.6	18.3	23.0
Flaxseed	4.4	5.6	6.3	7.9	10.0	15.2	21.4
Oats	5.7	8.0	9.6	11.8	13.8	18.5	24.1
Rice, rough	5.6	7.9	9.8	11.8	14.0	17.6	--
Rice, undermilled	5.9	8.6	10.7	12.8	14.6	18.4	--
Rice, polished	6.6	9.2	11.3	13.4	15.6	18.8	--
Rye	7.0	8.7	10.5	12.2	14.8	20.6	26.7
Sorghum	6.4	8.6	10.5	12.0	15.2	18.8	21.9
Soybeans	--	6.2	7.4	9.7	13.2	--	--
Wheat, white	6.7	8.6	9.9	11.8	15.0	19.7	26.3
Wheat, Durum	6.6	8.5	10.0	11.5	14.1	19.3	26.6
Wheat, soft red winter	6.3	8.6	10.6	11.9	14.6	19.7	25.6
Wheat, hard red winter	6.4	8.5	10.5	12.5	14.6	20.1	25.3
Wheat, hard red spring	6.8	8.5	10.1	11.8	14.8	19.7	25.0

$$Eu = \frac{\Delta p}{\rho v^2}$$

where

Δp = pressure drop due to friction
ρ = density
v = velocity of flow

Evaporated Milk See *Condensed Milk*.

Evaporation The change of a liquid to a gas which occurs at the surface of a liquid. The evaporation occurs as excited molecules at the surface of a liquid gain energy to escape. The rate of evaporation for a particular liquid depends only on the temperature of a fluid. Pressure vessels are used to change the temperature of evaporation. The latent heat of vaporation represents the quantity of heat required to change the state from liquid to gas at a particular temperature. Boiling may be considered as evaporation throughout the liquid. Evaporation is a synonym for *Vaporization*.

Evaporative Efficiency The amount of evaporation of water (liquid) which occurs in a direct dryer or evaporator compared to the amount of evaporation if the leaving air is saturated adiabatically.

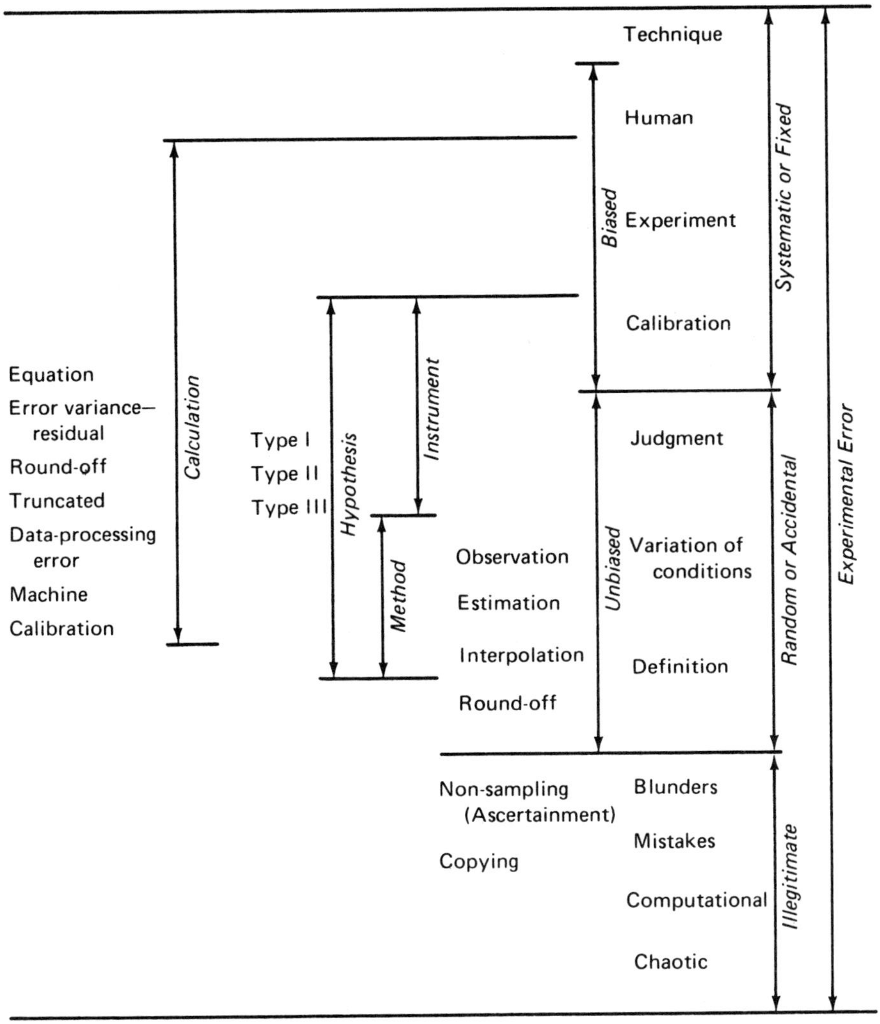

Figure E-4. Errors.
From Hall, Carl W., Errors in Experimentation (1978). Used with permission of Matrix Publishers, Inc., Portland, OR.

Evaporative Limit The maximum amount of moisture that can be evaporated adiabatically by a mass of moist air. The amount of moisture is proportional to the distance from the saturation curve along a line of thermodynamic wet-bulb temperature, which relates to the ability of air to dry and carry away moisture. Also see *Drying Potential*.

Evaporator A device for concentrating a solution or solid-liquid mixture by boiling. The temperature difference for evaporation is usually taken as the difference between the saturated temperature of the stream (neglecting superheating and subcooling) in the steam space and the temperature of the boiling liquid in equilibrium with vapor at the pressure of vapor-liquid. The evaporator usually refers to a vacuum pan or boiler for food or other heat sensitive products (Fig. E-5). May apply to a kiln or other dryer such as is used for fruits, with the term referring to an evaporator for vaporizing moisture from these products. Present literature, however, usually refers to the latter example as a dehydrator or dryer. The evaporator is often used to remove the bulk (50 to 80%) of the moisture in a product preceding a dryer operation. The cost of moisture removal from a liquid is less in an evaporator than with a dryer. Also see *Condensed Milk; Horizontal Tube; Long Vertical Tube; Multiple Effect; Vertical Tube.*

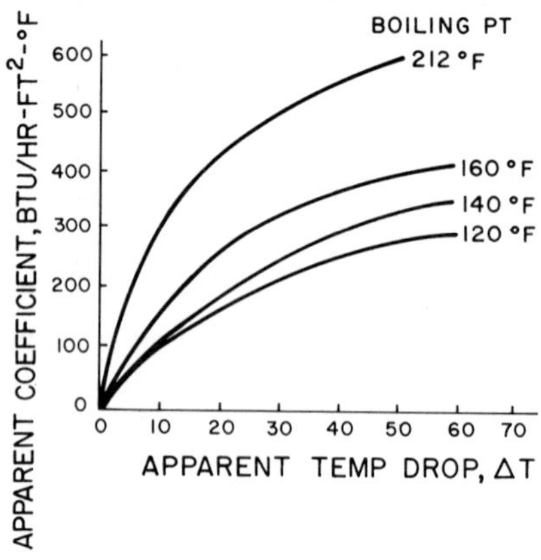

Figure E-5. Evaporator, apparent coefficient of heat transfer versus temperature drop.

Evaporites The material remaining after evaporation of moisture from a liquid product.

Excess Air See *Combustion.*

Excess Flow Valve A check valve which permits flow of fluid in either direction but which limits excessive flow in one direction. If the designated flow rate is exceeded the valve automatically closes.

Exhaust Oven A unique method of drying moisture samples using the exhaust gases from a tractor engine. The unit consists of an open-ended cylinder which contains a sample and is dried with the exhaust gases. A scale is used to determine the weight of the sample before and after drying. The temperature of the sample is controlled at 140°C (282°F) by adjusting the engine speed. The sample is turned about once each minute to prevent burning. About 6 min are required to dry a sample of grass of 15.8% moisture and 10 min to dry a sample of oats at 27%. Also see *Oven Method of Moisture Determination*.

Exosmosis The flow of water by osmosis through a semi-permeable membrane out of a cell.

Exothermic A term used to identify chemical reactions in which energy is liberated during the reaction. The burning of carbon as in coal gives off carbon dioxide and heat. The burning of one gram molecular weight of 12 grams of carbon gives off 22.6 kJ (94,400 cal) and is an exothermic reaction. Also see *Combustion; Endothermic; Net Heating Value*.

Expansion There is an increase of volume of a substance during wetting, which may occur when humid air is circulated through a product. The lateral pressure of a stored product may increase as a result of expansion during wetting of a product. The lateral pressure of shelled corn in storage increases from 0.0246 to 0.1336 kg/cm^2 (0.35 to 1.9 psi) at a 1.56 m (5 ft) depth as the moisture increased from 13 to 15%; when increased to 22% moisture, the lateral pressure is 0.22 kg/cm^2 (3.2 psi).

Explosion The rapid combustion of a substance with a violent reaction which may be desirable, such as for internal combustion, or undesirable as in an atmosphere where the appropriate concentration of particles is ignited with a spark or hot surface. The possibility of explosion must be considered around and in dryers. Three basic requirements for an explosion are: (1) a suitable mixture of dust of fine particles and oxygen; (2) a source of ignition, such as a spark or heat; and (3) a confined area. Explosions are avoided by proper ventilation, dust removal equipment, dustproof electrical equipment and lights, and grounding of equipment. The lowest concentration or lower explosive limit of grain and flour dusts is generally assumed to be 20 g/m^3 (0.02 oz/ft^3). The upper limit for flour dust is 1 kg/m^3 (1 oz/ft^3), and for grain dusts 2 kg/m^3 (2 oz/ft^3). Associated ignition temperatures are 410 to 430°C (770 to 806°F) at 30 to 90% humidity. Also see *Electrostatics; Safety*.

Exsiccant A material capable of drying or being dried.

Exsiccate To dry up.

Exsiccator A means of extracting moisture, such as dehydrate, to dry, mainly applied to paints, varnishes, etc.

External Moisture   In the explosive powder industry, the name used to represent the surface moisture on powder. The external moisture is determined by heating the product for a specified time and temperature, such as 60 to 100°C (140 or 210°F) for 4 to 6 hr. Also see *Dehydrating Press; Free Moisture; Powder Drying; Surface Moisture.*

Extractor   A device for removing excess moisture from clothing and textiles and some food products, such as sugar, by centrifugal action without the addition of heat. Further moisture removal is usually by tumble dryers.

F

Fahrenheit Scale A scale for temperature with 180 equal divisions between the freezing (32°) and the boiling point of water (212°) at atmospheric pressure, designated as degree (°F). Named after G. D. Fahrenheit. Also see *Celsius*.

Fail-Safe Control A control designed so that a malfunction will stop the operation of the device or equipment, such as a burner in a dryer.

Falling Rate Drying The period of drying in which the rate of drying, the change of moisture (M) with time (t) is decreasing with time, represented by dM/dt. Usually applied to thin layer drying, whereas decreasing rate drying is applied to deep layer drying. Also expressed in terms of moisture loss per unit time per unit of exposed surface. The falling rate period follows the constant rate period of drying. The constant rate period of drying is not always present. Field harvested grains dry entirely in the falling rate period. Also see *Decreasing Rate; Drying Rate; Equations*.

Fan A mechanical device for moving air with a revolving wheel or blade driven by a power unit. Fans may be classified as axial flow or radial flow. A standardized designation of a fan is AMD (air moving device). The performance of fans is represented by fan laws. The mechanical efficiency of a fan is the ratio of the fluid or air horsepower at the output divided by the brake horsepower at the input to the fan. The mechanical efficiency is in the range of 40 to 70%. The fluid horsepower is the work output of the fan, in ft/lb/min divided by 33,000. If the volume of flow is in pounds per minute, the head must be in feet of flowing fluid, which for a fan is air. The static pressure is usually given in inches of water which must be converted to feet of air by multiplying by density of water/density of air. Also see *Centrifugal Fan; Fan Laws; Fan Rating;*

Fluid Horsepower; Forward-Curved Fan; Mechanical Efficiency; Radial Fan; Static Efficiency.

Fan Laws Expressions which relate the fan operating procedure and performance. The usual relationships are: the capacity varies directly as the speed ratio; the pressure (static, velocity, total) varies as the square of the speed ratio; the horsepower varies as the cube of the speed ratio.

Fan Rating Refers to the specification of the operating characteristics of a fan. The rating includes speed, horsepower, and volume of flow at various heads. Inasmuch as the output of a fan and its characteristics vary with atmospheric conditions, the fan is rated at 1 atm (29.42 in Hg), 20°C (68°F), and 50% relative humidity with readings taken as specified by the Air Moving and Conditioning Association, Inc., 30 West University, Arlington Heights (Chicago), IL 60004. Fan manufacturers provide the rating of their fans (Fig. F-1). The fan output must match the system needs for air.

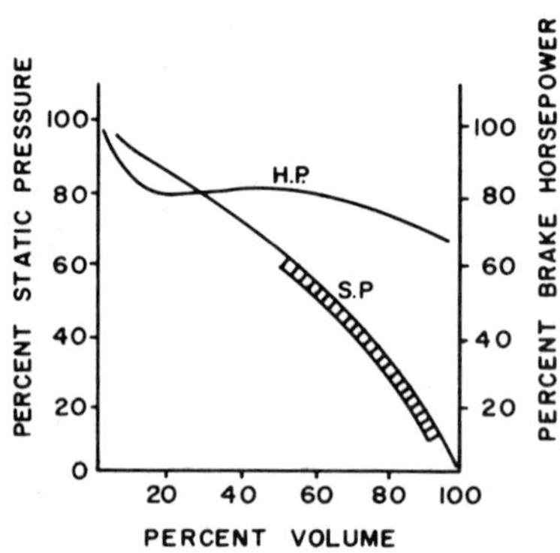

Figure F-1. Fan specifications of high speed propeller.

Fatty Acid See *Acid Number.*

Feed Collector See *Cyclone Collector.*

Fermentation The process of anaerobic respiration which is equivalent to oxidation in the absence of oxygen with the production of carbon dioxide and other products such as alcohol. These chemical changes occur by the action of yeast or bacteria in materials containing starch or sugar. Changes may occur before or after drying depending on the substance, temperature, moisture content, presence of yeast or bacteria, and presence of oxygen. Also see *Tobacco Curing.*

Festoon Dryer A device for removing moisture from paper materials in which the material is kept tension-free and is continuously fed in loops on supporting poles carried by a conveyor chain, passed through forced warm air convection (air flow downward). Also called a loop dryer or a tensionless dryer.

FFA (Free Fatty Acid) See *Acid Number.*

f-Factor, f-Friction Factor The constant in Darcy's equation of liquid flow for determining the friction head of flow of a liquid.

$$\text{Friction Head} = f\left(\frac{L}{D}\right)\left(\frac{V^2}{2g}\right)$$

where f in laminar flow is 64/Re; and for turbulent flow is taken from a figure such as that included under Darcy's law. Also see *Darcy's Equation; Reynolds Number.*

Fiber Saturation Point The moisture content of wood and other cellulose above which free moisture exists and below which the water is bound. For other materials this point is often called the critical moisture content. The fiber saturation point, depending on the species, is between 20 and 35% dry basis. Other parameters related to the fiber saturation point are electric conductivity, dielectric constant, elastic and strength properties. For wood, the fiber saturation point is the moisture content reached when all the fibers are completely swollen and saturated with colloidal water and no liquid or free water exists in the coarser capillary structure. Typical values for wood are:

Willow, poplar, birch, beech	32-35%
Fir and spruce	30-34%
Pine, larch, Douglas fir	26-28%
Locust, chestnut, oak, ash, cherry	22-24%

Also see *Critical Moisture Content.*

Fick's Law A mathematical statement which states that the rate of diffusion of moisture is proportional to the cross sectional area times the concentration gradient:

$$\frac{dM}{dt} = -DA\frac{dC}{dx}$$

where

M = moisture transferred
t = time
D = diffusion coefficient
A = area of cross section
C = concentration
x = distance in direction of diffusion

The law is also applied to mass transfer and dialysis.

Fig Drying Figs are halved and sulfured before drying. Drying is done on trays in a tunnel countercurrent dryer. An air inlet temperature of 73.9°C (165°F) is used to dry to 5% moisture in 8 to 12 hr.

Film A layer, usually thin, of a gas or liquid on a surface which may predominate in heat transfer calculations. The film, considered a conducting or an insulation layer, is a component in convection heat transfer. The rate of heat transfer is increased by agitation in which the thickness of the film is decreased. Heat transfer by boiling or condensation is related to the film characteristics. The film may also relate to the layer next to the surface or wall along which there is a moving fluid. Also see *Film Coefficient*.

Film Coefficient See *Convection Heat Transfer Coefficient; Drying Coefficient*.

Filter Device used to remove components, usually undesirable, from gases and liquids. The size of the pores in a filter determines the size of particles retained (Table F-1). The air supply to a controller is filtered so that particles will not cause malfunction. Fuel to the burner is filtered. Air entering the spray dryer for food drying may be filtered. Air for ventilation may be filtered. Absolute filters will remove 99.9% of 0.3μ particles and will remove microorganisms as well as larger particles. Filters are classified as roughing, medium efficiency, high efficiency, and ultra high or absolute (Table F-2). In general, the greater filtration that is required the larger the pressure drop across the device and the more expensive the cost of operation. Coarser filters may be called strainers.

Table F-1
Filter Pore Sizes and Products Retained

Pore Size, μm	Retention
0.2	Retain all bacteria
0.45	Retain all coliform groups
0.8	Retain airborne particles
1.2	Retain nonliving particles considered dangerous in fluids intravenously administered
3	Yeast cells
5	Retain all significant cells from body fluids
75	Diameter of human hair

Table F-2
Characteristics of Air Filters to Handle 1200 cfm

Filter Type	Initial Pressure Drop, In. H_2O	Approximate Operating Pressure Drop, In. H_2O	Approximate Horsepower of Fan Motor
Roughing	0.09	0.4	0.15
Medium efficiency	0.25	0.6	0.23
High efficiency	0.43	0.75	0.3
Ultrahigh efficiency	0.94	1.5	0.56

Conversion factors: *1 cfm = 0.00047* m^3/s
1 hp = 0.746 kW
1 in. water = 0.248 kPa

Source: Hall, C. W., et al., 1971. Encyclopedia of Food Engineering. Avi Publishing Co., Westport, CT.

Final Moisture Content   A representation of the moisture control of a product leaving a dryer or other process. The lowest moisture content of a product being dried is not necessarily the best moisture content. The optimum final moisture content depends on mechanical properties to avoid damage for handling (grain), excessive loss of leaves (forages), likelihood of rapid regain of moisture by some highly hydroscopic materials, crumbling of sheets or flakes (paper), splitting (lumber), and loss of stability (food products).

Dehydrated food products appear to be most stable with minimum oxidation where there is a monomolecular layer of water internally (rather than a lower moisture content) which for milk and protein is 3.5%; potatoes, 6.5%; high sugar foods, near 0%; and nuts (walnuts), 3.0%.

Fines   Small diameter particles such as those which form from larger particles during handling and drying. The fines may be carried by the drying air. These fines may be removed by a filter, centrifugal separator, electronic precipitator, washing, etc., to avoid air pollution by the exhaust. Fines may accumulate in drying bins under spouts or conveyors. These fines, often of high moisture content, may obstruct uniform flow and distribution of drying air. Fines may include dust, with sizes primarily in the range of 1 to 400 micron diameter, as well as particles much larger.

Finish Drying   The product leaving a particular dryer often is not at the desirable low moisture content desired. If the dryer were used to remove the last 2 to 5% moisture, excessive heat effects (too long, or too high a temperature or both), there could be a deterioration in organoleptic properties. Finish drying is done to remove the final 2 to 5% moisture in a product, without destroying the physical structure, without producing undesirable chemical changes, and without permitting biological contamination. Several procedures are used for finish drying: (1) dehumidified air, at a low temperature moved slowly through the product, (2) air forced through a deep bed (called finishing bins), (3) intermittent heating, permitting moisture diffusion; and (4) tumbling product through controlled air environment. Although applicable to many products, finish drying is more commonly used for wood particle drying.

Fire Curing Consists of curing the tobacco in the presence of the smoke from the burning or smoldering wood. The tobacco leaves remain in the barn for about three months until they are as brown as dark mahogany. The U. S. fire-cured tobacco is produced in Kentucky and Tennessee. Also see *Air Curing; Bright Leaf; Burley; Flue Curing; Shade Grown; Tobacco Curing.*

Fire Point The temperature of an oil at which enough vapor is released to burn continuously. Also see *Flammable; Flash Point; Ignition Point.*

First Law of Thermodynamics The law which states that during a process energy is conserved. That is, although heat may be transferred from one form to another, there is no net gain or loss in total energy. For a system which undergoes a cycle, the net heat transfer across the system boundary is equivalent to the net work done by the system. Also see *Second Law.*

Fischer (Karl) Method A procedure for determining the moisture content of a product using chemicals often used as a standard. The material to be checked for moisture is finely ground. Chemical solvents are used which penetrate the cells and provide rapid removal of moisture. Anhydrous methyl alcohol is used to extract the moisture. As compared to the oven method of moisture determination, a higher moisture content is indicated for hygroscopic materials because the procedure removes some of the water of hydration which might not otherwise be removed. As much as 2.5% higher moisture content may be obtained using this method as compared to thermal methods.

Fish Drying Fish may be dried by several methods, of which salting is one method, usually utilizing the fish available in excess of the fresh market needs. Gutted fish, placed on screened or slotted shelves or trays, can be dried in a cabinet dryer with heated air. Air temperatures should be kept low enough to avoid dripping of the fat. Pickle curing is another method in which 17 lb of salt are used per 100 lb of fish, the salt and fish are layered, with 1-1/2 weeks required. Tunnel dryers, trays or belt roller dryers have been used, with forced air at 23.9°C (75°F), 40% relative humidity, to dry fish to 40%. Sun drying of shrimp is widely used, following boiling in brine. A depth of 5 to 7 cm (2 to 3 in) of shrimp is spread in the sun and turned every half hour, taking 3 to 4 days to dry.

Freeze-drying cubes or strips of fish meat can be used as a method of moisture removal. Fish flour can be prepared by spray drying a wet slurry or by azeotropic drying, using dichloroethane solvent. Moisture contents of dried fish are summarized in Table F-3. Also see *Azeotropic Drying*.

Table F-3
Approximate Moisture Content of Cured Fish, Percent

Fish, fresh	60-75
Smoked haddock	75
Smoked herring	61
Salted cod	57
Salted herring	55
Tunnel dried	35
Dried cod	12
Freeze-dried fish	5

Flakes Slotted frames constructed of laths which are placed outdoors upon which products such as fish are placed to dry. Fish dry to 40% moisture in 4 to 5 days.

Flame The visible effect from radiation during combustion. The flame results from burning of the soot particles and the gaseous combustion products. Simple controls are on-off flame sensing devices. The temperature of a flame varies as shown in Fig. F-2. The temperature and color of flames vary with the gas or other element involved (Table F-4). A sodium flame in oxygen is yellow; copper, green; phosphates, blue.

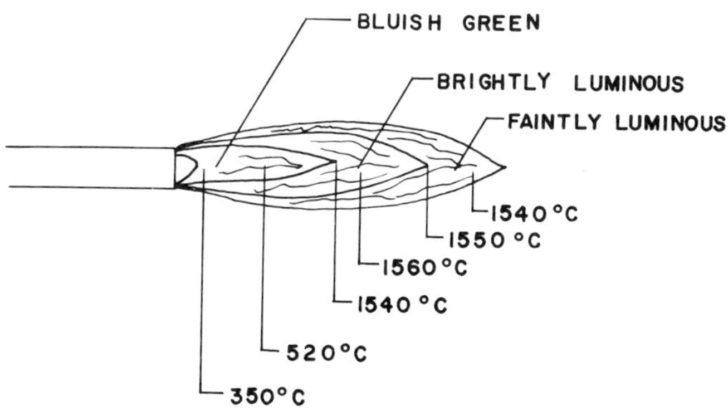

Figure F-2. Flame-color and temperature distribution.

Table F-4
Flame Temperature of Different Gases, Approximate

	°C	°F
Butane, C_4H_{10}	1875	3440
Candle	1500	2732
Carbon monoxide, CO	2400	3800
Coal gas and air	2000	3632
Hydrogen, H_2	2350	3700
Methane, CH_4	1850	3400
Natural gas	1900-2300	3700
Oxyacetylene	3340	6000
Propane, C_3H_8	1950	3500

Flammability See *Limit of Flammability*.

Flammable Combustible or ability to support combustion such as represented by fuel. Because of the possibility of misunderstanding the meaning of inflammable, a term previously used for this purpose, flammable is now used to a greater extent commercially.

Flash Drying In some processing or unit operations, such as in size reduction, energy for size reduction is decreased if the product is dry. Heat may be added during the grinding or crushing operation to reduce energy requirements. Flash drying may accompany size reduction of solid fuel, pneumatic conveying of solids, etc. Flash drying is also called dispersion drying and pneumatic drying.

Flash Point The lowest temperature at which a liquid produces enough vapor which when mixed with air will burn when ignited at a given pressure. Important for storage, transportation, and safety considerations for fuels. Flash point also refers to the temperature at which lubricants for compression refrigeration units, such as used in freeze dryers, vaporize (Table F-5). Also see *Ignition*.

Table F-5
Flash Temperatures of Different Fuels and Oil

Benzene, C_6H_6	-11°C (12°F)
Ethyl alcohol, C_2H_5OH	12°C (54°F)
Fuel oil	
No. 1	38-74°C (100-165°F)
No. 2	43-88°C (110-190°F)
No. 3	43-110°C (110-230°F)
No. 5	55°C (130°F)
No. 6	65°C (150°F)
Lubricating oils	163-218°C (325-425°F)

Flaxseed Drying Flaxseed should be 8% moisture content or below for safe storage because of a high fat content. Flaxseed can be dried with unheated or heated air. Flaxseed with 18% moisture can be dried with 79.4°C (175°F) air at a velocity of 80 ft/min without detrimental effect on germination. If the moisture content is from 9 to 11% it is more economical to use 65.5°C (150°F) air for drying.

Florite Desiccant An activated bauxite selected specifically for drying or desiccating which is used extensively for dehydrating natural gas.

Flow See *Laminar Flow; Rotameter; Turbulent Flow.*

Flowers See *Drying Flowers.*

Flue Curing A high heat is provided through a central heating system or flues on the floor of the barn for the drying of tobacco. Hot air, usually heated by gas, oil, or coal, is evenly distributed throughout the barn with a blower or fan. Flue-cured tobacco is kept in the barn for about a week until a honey yellow color develops. Virginia-type tobacco, whether grown in Virginia, North Carolina, South Carolina, South America, Africa, or Asia is flue cured. Also see *Curing; Fire Curing; Tobacco.*

Flue Gas A mixture of gases emerging from a stack, chimney, or flue, resulting from combustion, containing oxides, carbon monoxide, carbon dioxide, water vapor, sulfur dioxide, particulates, etc. An analysis of the flue gas, as with an Orsat apparatus, provides information for evaluating the efficiency of combustion. Also see *Combustion; Pollution.*

Fluid Bed Dryer See *Fluidized Bed Dryer.*

Fluid Horsepower A representation of power derived from the basic definition of horsepower:

$$\text{Fluid horsepower} = \frac{\text{ft-lb per min}}{33{,}000}$$

The numerator may be written as lb x ft per min, ft x lb/min, or ft-lb/min. For air, as with a fan, the flow is in pounds per minute and the head must be in feet of the flowing fluid. The fluid horsepower of an air fan at 20°C (68°F) is:

$$\text{Fluid horsepower} = \frac{(\text{cfm})(\Delta P_t)}{6350}$$

where cfm is cubic feet per minute and ΔP_t is the total operating pressure (although the static pressure is usually used in calculations). Also see *Air Horsepower; Fan; Horsepower.*

Fluidized Bed Dryer A unit for drying in which the product is in suspension, or the product is moved through the dryer by the drying and/or the cooling air (Fig. F-3). The spouted bed dryer is a special type of fluidized bed dryer. Also see *Spouted Bed Dryer*.

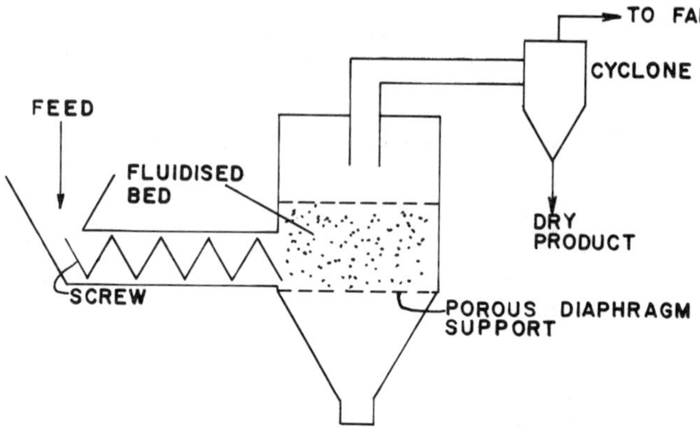

Figure F-3. Fluidized bed dryer.

Flux The rate of flow through an area, such as Btu/hr ft^2; lb/min ft^2; cfm/ft^2; kJ/m^{2}s; kg/m^{2}s; m^3/m^2.

Foam Spray Dryer Products which can entrain air or a gas, such as certain food and biological materials including skim milk, cream, and juices, are foamed then forced through an atomizing nozzle in a conventional spray dryer. Air or nitrogen are gases commonly used. Foam spray-dried products have less heat-induced changes, such as flavor and discoloration, as compared with conventional spray-dried products. A product with density less than conventional is obtained which is about equal to the density of instantized or agglomerated powder (Fig. F-4).

Forage Drying See *Barn Drying; Hay Drying*.

Force The physical agent which brings about a change in the momentum of a body and is directly proportional to the product of the mass, m, and acceleration, a, of a body, F = ma, where m = w/g.

Force units
F = m x a

System	Force	Mass	Acceleration
cgs	1 dyne	1 gram	1 cm/sec^2
mks	1 newton	1 kg	1 m/sec^2
fps	1 poundal	1 slug	1 ft/sec^2
SI	1 newton (N)	1 kg	1 m/s^2

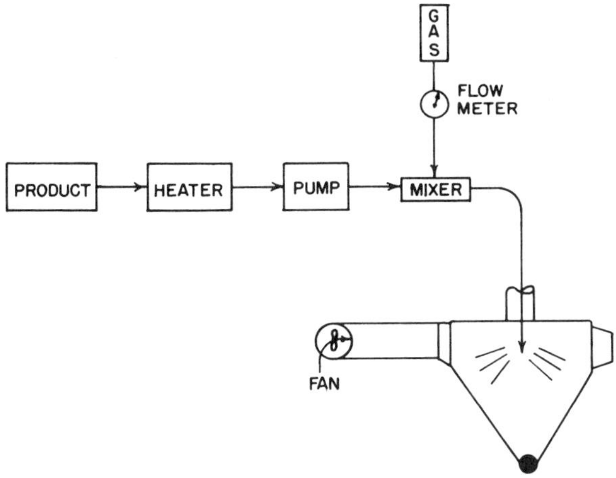

Figure F-4. Foam spray dryer.
From Hall, C. W., and Hedrick, T. I., Drying of Milk and Milk Products, 2nd edition (1971). Used with permission of Avi Publishing Co., Westport, CT.

Forced Air Drying A system for removing moisture from a product in which heated or nonheated air is moved through the product with a fan or air moving device (AMD). Also see *Convection*.

Formula A statement representing a relationship of variables in symbolical form. Thus, F = ma represents the force which equals mass times acceleration. An empirical formula is based on experimental results, not necessarily based on a physical law.

Forward-Curved Fan A centrifugal or radial flow fan constructed with curved blades mounted on the outer periphery of a power driven revolving wheel. There are usually 20 to over 66 curved blades in this type of fan with the curved blades so arranged that the leading edge of the blade is on the outer edge of the periphery. Under the best operating conditions, this fan has a mechanical efficiency of about 80%. The horsepower of the forward-curved blade fan increases as full 100% volume of flow is approached and where the static pressure is low. Proper design must consider the possibility of overloading. Usually used for static pressures in the range of 2 in. water column or less where there is a fixed, nearly constant load at the outlet. Also see *Backward-Curved Fan; Centrifugal Fan*.

Fourier Law A basic relationship for conduction heat transfer stating that the steady-state rate of heat transfer by conduction, dq, is proportional to the cross-sectional area at right angles to the direction of heat flow, dA, and to the temperature gradient on the conduction path, -dt/dL. The proportional constant, k, is the thermal conductivity of the substance through which heat is being transferred:

$$q = -kA \frac{\Delta t}{L}$$

$$dq = -k\ dA\ (dt/dL)$$

Units of k must be consistent with values in equation. The amount of heat, Q, Btu, transferred over a period time is Q = q x time. Also see *Conduction*.

Fourier Number A dimensionless number, F_o, in unsteady heat or mass transfer which characterizes the connection between the rate of change of temperature, the physical properties, and the dimensions of a product or body. The Fourier number for heat transfer:

$$F_o = \frac{k\ t}{\rho\ c_p\ \ell_m^2}$$

where

k = thermal conductivity
t = time
ρ = density
c_p = specific heat
ℓ_m = distance, V/A, volume/surface area, may be midpoint to surface but can vary depending how the data are assembled

The Fourier number for mass transfer:

$$F_o = \frac{D\ t}{\ell^2}$$

where

D = diffusivity
t = time
ℓ = distance

Fourier Series A mathematical relationship representing periodic occurring phenomena which is a series consisting of sine and cosine functions. Also known as the Fourier expansion. The general form of the Fourier series is:

$$\frac{1}{2} a_o + \sum_{n=1}^{\infty} (a_n \cos nx + b_n \sin nx)$$

with

$$a_n = \frac{1}{\pi} \int_{-\pi}^{\pi} f(x) \cos nx \, dx$$

$$b_n = \frac{1}{\pi} \int_{-\pi}^{\pi} f(x) \sin nx \, dx$$

Fractional Distillation The procedure whereby the distillates which boil off a product in different temperature ranges are collected separately.

Frame Dryer See *Tenter Dryer.*

Free Energy The difference between the enthalpy and the product of entropy and absolute temperature of a thermodynamic system. Also called Gibb's free energy and thermodynamic potential.

$$\text{Free energy, } G = H - TS$$
$$= U - TS + pV$$

where

H = enthalpy, H = U + pV
U = internal energy
T = absolute temperature
S = entropy
p = pressure
V = volume

Also see *Enthalpy; Entropy; Gibb Function; Helmholtz.*

Free Fatty Acid See *Acid Value.*

Free Moisture or Free Water The water in a substance not held by hygroscopic forces. The water can be removed by mechanical methods or gravity, such as draining, centrifugal force, and pressing. Some authors refer to free moisture as that moisture in a product above the equilibrium moisture content, $M - M_e$. Some authors define free moisture content as including unbound water. Also see *Moisture Content Representation.*

Freeze Dryer A unit for removing moisture in which vacuum is usually maintained around a product. The vacuum is maintained such that the water is frozen in the substance, and moisture is removed by sublimation of the solid ice to water vapor. The freeze dryer is used for drying products which otherwise would be damaged if higher temperatures were used. Also see *Accelerated Freeze Drying; Freeze Drying; Pressure Freezing-Air Drying; Vacuum Contact Dehydration.*

Freeze Drying A method of drying heat-sensitive materials in which the product is frozen, heated under carefully controlled conditions, and moisture removed by sublimation, that is, as a vapor from the solid state, usually under vacuum. (Fig. F-5). The porous structure of the product is maintained during drying by carefully controlling formation of solid ice crystals and carefully applying heat (Table F-6). Freeze-dried products usually can be readily reconstituted because of the maintenance of the cell structure. Also called freeze concentration. Also see *Accelerated Freeze Drying; Freeze Dryer; Triple Point; Vacuum Contact Dehydration.*

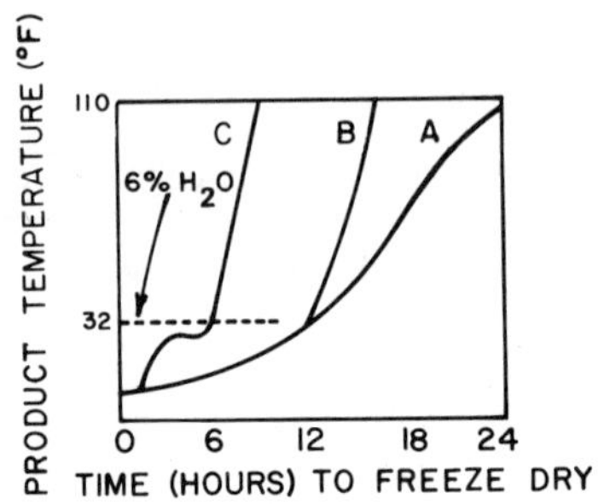

Figure F-5. Freeze drying temperatures during drying. *From Desrosier, N. W., Technology of Food Preservation, 3rd edition (1970). Used with permission of Avi Publishing Co., Westport, CT.*

Freezing Point The temperature at which a liquid changes to the solid state, which for pure water is 0°C (32°F). The freezing point of a solution of water is decreased as material (solute) is dissolved into solution. At the freezing temperature both liquid and solid exist in equilibrium. The depression of the freezing point of water is 18.5°C (33.3°F) with the dissolving of one gram molecular weight in 100 grams of water. Also see *Raoult's Law.*

Freundlich Isotherm A relationship for adsorption of gas on the surface of a material such that

$$\frac{x}{m} = k\,p^{1/n}$$

where

x = amount of gas adsorbed
m = mass of the absorbent
k and n = constants
p = pressure of the gas

Table F-6
Freeze-Drying/Dehydration Compared to Conventional Drying/Dehydration

Conventional Dehydration	Freeze-Dehydration
Successful for easily dried foods such as fruits, seeds, and vegetables	Successful for most foods but usually limited to those not successfully dried by other methods
Meat generally not satisfactory	Successful on cooked and raw animal products
Continuous processing	Batch processing
Temperatures between 100° and 200°F. generally used	Temperatures sufficiently low to prevent thawing used
Usually at atmospheric pressure	Pressures below 4 mm Hg used
Drying time may be short, usually less than 12 hr	Drying time generally between 12 and 24 hr
Evaporation of water from food surface	Moisture loss by sublimation from boundary of ever receding ice crystal zone
Solid dried particle	Porous dried particle
Higher density than original food	Lower density than original food
Odor frequently abnormal	Odor usually natural
Color usually darker	Color usually natural
Slow rehydration, usually incomplete	Rapid, complete rehydration possible
Flavor may be abnormal	Flavor generally natural
Storage stability good, tendency to darken and become rancid	Storage stability excellent
Costs generally low, in the order of 2 to 7 cents a pound water removed	Costs generally high, in the order of four times more than conventional dehydration

From Desrosier, N. W., Technology of Food Preservation, 3rd edition (1970). Used with permission of Avi Publishing Co., Westport, CT.

Also see *Adsorption Isotherm; Equilibrium Moisture Content.*

Friction The resistance to movement of one material to another material or to itself. For liquids see friction factor. For solid bodies, the coefficient of friction between two surfaces is the ratio of the force required to move one over another to the normal force of one body against the other. The coefficient of friction relates to calculations involving handling and storing of liquids and granular materials. Small grains on smooth steel have a coefficient of friction of about 0.41 (Table F-7). The coefficient of friction is the tangent of angle of friction, ϕ. One body will just slide over another at the angle of friction, ϕ (Fig. F-6). Values of the coefficient of friction will change depending on temperature, moisture content, and surface condition of product and the opposite surface. The coefficient of friction for initiating movement, known as the static coefficient, is larger than the sliding or kinetic coefficient of friction. Also see *Angle of Repose.*

Table F-7
Friction, Coefficients of (Estimates Only)

	Steel	Concrete	Wood (Smooth)
Ashes	0.60		
Beans	0.37	0.44	0.32
Coal, hard	0.28	0.50	0.47
Coal, soft	0.32	0.70	0.70
Corn	0.41	0.44	0.36
Oats	0.41	0.47	0.37
Rice	0.41	0.52	0.44
Sand, dry	0.32	0.57	0.57
Sawdust, wet	0.60		
Steel, dry	0.15-30	0.30-0.7	0.20-0.62
Wheat	0.37	0.42	0.31-0.425
Wood	0.50-0.62	0.4	0.20-0.54

Friction Factor See *Darcy's Equation; Duct; Friction.*

Friedholm See *Adsorption Isotherm.*

Froude Number A dimensionless number, Fr, used in fluid flow which represents the ratio of inertial force and gravity force in a homogeneous fluid:

$$Fr = \frac{v^2}{g\ell}$$

where

v = velocity

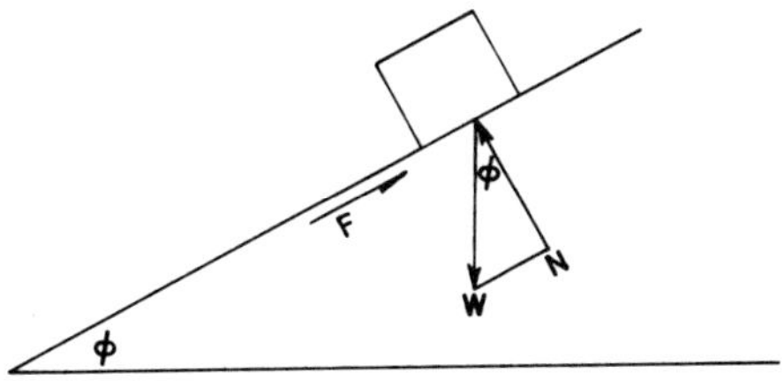

Figure F-6. Friction diagram.

g = gravity
ℓ = distance

Fruit Juice Drying The principal non-citrus fruit juices dried are apple, grape, and strawberry. Drying is preceded by concentration in an evaporator. Excessive heat treatment in the evaporator must be avoided to prevent off-flavors. Generally, for fruit juices (including berry), a falling-film evaporator operating below 55.6°C (130°F), usually less than 2 hr, does not produce excessive off-flavor. Of these products, strawberry is the most heat sensitive. Heat damage in the evaporator is carried to the dryer. Spray drying is most commonly used. Puff drying, foam-mat drying, and foam spray drying offer other methods. Products leaving the spray dryer are normally at 2 to 4% moisture. To dry further, a dehumidified air for final drying, or in-package drying, is used. Sweetened (50% of juice solids in sugar) fruit drinks (apple, grape, and tart cherry) of an instantizing quality can be made in a turbulent film evaporator in a vacuum, which produces small flakes upon cooling, the sugar being a drying agent. Also see *Citrus Juice Drying*.

Fuel A substance used for producing thermal energy by combustion of fossil fuels or organic compounds. An excess of air over that theoretically required for complete combustion is provided (Table F-8). Also see *Combustion; Heating Value; Installation; Peat Drying; Vapor Pressure; Vaporization Burner*.

Fuel System Dryer A liquid chemical, usually retailed in 0.355 liter or 340.9 cm^3 (12 fl oz) containers, which may be added to the automotive or burner fuel system (for gasoline or diesel) to absorb moisture to prevent ice formation in winter and corrosion of metals from moisture condensation. The contents of one is added to 10 to 20 gal of fuel depending on the concentration of the drying agent and extent of moisture to be absorbed. The dryer may consist of (1) isopropyl alcohol, (2) methanol (at least 45%), and (3) petroleum distillate.

Table F-8
Fuel, Excess Air Supplied for Combustion in Burners

	Percent
Bagasse	25-35
Fuel oil	5-10
Natural draft	50-60
Natural gas	5-10
Pulverized coal	15-20
Wood	20-25

Fugacity The fugacity is a pseudo-pressure used to describe the behavior of real substances in solutions compared to ideal solutions. For gases the fugacity coefficient is the ratio of the fugacity to the partial pressure. For liquids the fugacity is the mole fraction of the constituent multiplied by the fugacity of the pure substance at the pressure and temperature of the solution.

Fundamental Units The primary dimensions of mass, M, length, L, and time, T. Some authors use θ for time instead of T. The fundamental units of measurement are listed in Table D-5. These fundamental units are used in dimensional analysis of physical systems. Also see *Dimensional Analysis; Units*.

Funicular A condition in drying a porous body where moisture in the pores forms a continuous connection. The funicular condition precedes the pendular condition. Also see *Pendular*.

Fur See *Leather Drying*.

Furnish The name given to the wood materials, such as wood chips, fed into a dryer, or other process. Also see *Particleboard Furnish Drying*.

Fusion The process of changing a solid to a liquid, as in melting, by the application of heat. Heat is absorbed by the solid as it changes from a solid to a liquid. The quantity of heat required to change a fixed amount of solid to a liquid, without a change in temperature, is the latent heat of fusion. For changing ice to water at 0°C (32°F), the latent heat of fusion is 333 kJ/kg, 79.7 cal/gram, or 144 Btu/lb at one atmosphere pressure. Also see *Latent Heat*.

G

Gage, Gauge A device which gives a reading or value of a primary measuring element. Applied to a point of measurement, such as a pressure gauge. Also see *Absolute Pressure Gage*.

Gage Pressure The basis of representing air or gas pressure in which the atmospheric pressure is equivalent to zero, identified as zero gage pressure, or 0 psig, pounds per square inch gage. At 1 atm pressure the absolute pressure is 14.7 psia, with the letter a representing absolute.

Galileo Number A dimensionless parameter, Ga, which is a measure of the relationship between the force of molecular friction and the force of gravity in fluid flow, particularly for a viscous liquid:

$$Ga = \frac{\ell^3 g \rho^2}{\mu^2}$$

where

ℓ = length dimension
g = gravitational constant
ρ = density of fluid
μ = dynamic viscosity

The Galileo number is related to the Grashof (Gr) number by:

$$Gr = \beta \Delta t (Ga)$$

where

β = coefficient of bulk expansion, 1/T
Δt = temperature difference

See also *Grashof Number*.

Gantry Drying An open framework used to support materials, or vessels containing materials, in a convenient manner for drying.

Garbage Drying Removing the moisture from organic garbage may be done with (1) direct heat dryer, followed by grease extraction in percolators, and (2) an indirect heat dryer where the heat may be provided by steam in a jacket surrounding the garbage, or in tubes inside a rotating steel drum through which the garbage passes. In the direct method, some of the materials may be scorched. Vapors from the drying process must be treated, commonly with water spray and chlorination.

Garlic Drying The garlic bulb is sliced and dried on trays in a cabinet or tunnel, similar to onions. Garlic is dried to about 6.5% moisture and is made into powder or granules. Also see *Gladiola Drying; Onion Drying*.

Garner Bin A holding container placed ahead of a dryer which receives wet grain. The garner bin discharges grain to the wet grain chamber of the dryer, either by gravity or augers.

Gas A fluid in a physical state in which the molecules are free to move widely and which has neither definite shape nor volume. In contrast, a vapor is a substance in a gaseous state but is involved as a part of the vapor-liquid relationship such that if the pressure on the vapor is increased some of the vapor returns to liquid.

Gas is also a fuel used for drying. The name is also a shortened form of the word gasoline, another fuel, used primarily for engines. Also see *Fluid; Installation; LP Gas; Natural Gas; Vapor*.

Gas Constant The constant of proportionality in the law of ideal gases represented by R (Table G-1), in the equation

$$pV = MRT$$

which is the general gas law, where

p = absolute pressure
V = volume
M = mass
T = absolute temperature

Also see *Boyle's Law; Charles Law; Ideal Gas; van der Waals Equation*.

Gas Drying The process of removing moisture from gas. Three methods used for drying are (1) refrigeration, in which air is cooled below the dew point, in which moisture is condensed and removed; (2) compression, in which the saturation humidity is decreased with an increase in pressure (Fig. G-1); and (3) drying agents, which absorb or adsorb moisture from the air. Also see *Absorption; Adsorption; Drying Agents*.

Gas Law See *Gas Constant*.

Table G-1
Gas Constants

	R		Molecular Weight,	Specific Heat	
	J/kgK	ft lb/lb°R	M	c_p Btu/lb°R	c_v Btu/lb°R
Air	287	53.35	28.9	0.24	0.17
Carbon dioxide, CO_2	189	35.1	44	0.21	0.16
Carbon monoxide, CO	296	55.1	28	0.18	0.25
Hydrogen, H	4124	767	2.02	3.42	2.44
Nitrogen, N	296	55.1	28.02	0.247	0.176
Oxygen, O	260	48.3	32	0.217	0.155
Steam	462	85.8	18.02	0.41	0.36

Conversion factors: *For R (specific) gas constant, 1 ft lb/lb°R = 5.38 J/kgK.*
For c_p and c_v, 1 Btu/lb°R = 4.188 kJ/kgK.
Molecular weight in SI is called molecular mass

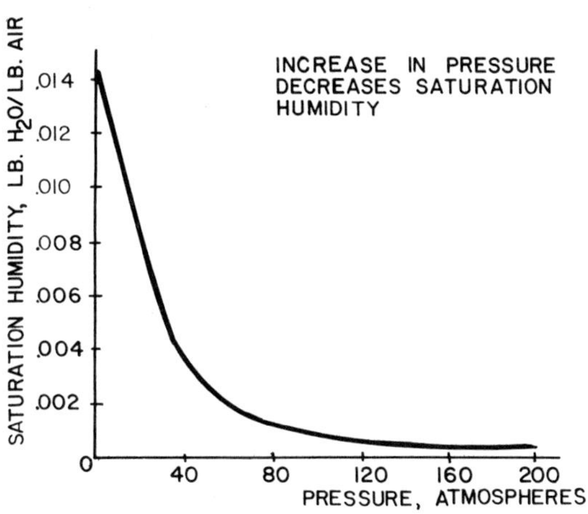

Figure G-1. Gas drying.

Gauge See *Gage; Gage Pressure.*

Gay-Lussac Law See *Charles Law.*

Gel A viscous emulsion that will solidify or set to a jelly upon cooling. Colloidal suspensions may change from a gel, that is, a quasi-solid pliable condition, to a sol state identified by greater fluidity. A gel may convert to a sol by addition of water, increased temperature, change in pressure, change in pH, and other physical or chemical phenomena. Many of the conversions are reversible, such as change from a sol to a gel by drying. Also see *Silica Gel*.

General Equation (Drying) The general equation of drying can be expressed in dimensionless form:

$$Ki_q(\tau) = Ki_m(\tau) \cdot Lu \cdot Ko\ (1 + Rb)$$

where

$Ki_q(\tau)$ and $Ki_m(\tau)$ = Kirpichev numbers for heat and mass transfer, respectively
Lu = Lykov number
Ko = Kossovich number
Rb = Rebinder number

Also see *Equations, Drying; Kossovich Number; Lykov Number; Rebinder Number*.

General Gas Law See *Gas Constant*.

Germination Refers to the ability of a seed to grow. The germination properties of a seed depend upon the moisture content, previous heat treatment, age, and physical treatment. Germination can be impaired by mechanical abuse. Germination is lost by storage in a sealed container. In general, most seeds will not germinate after heating to a temperature of 51.7°C (125°F). Higher temperatures of drying air can be used, if the product does not reach 51.7°C (125°F). To avoid the difficulty of control of heating based on temperature of the product rather than on air temperature, it is common practice to limit the temperature of drying air for seeds to 43.3 to 46.1°C (110 to 115°F). Grains stored at high moisture contents will lose their ability to germinate due to microbial action. Germination will also be reduced if wet grain is frozen before the moisture is removed. Corn at a moisture content of 25% exposed to 0°C (32°F) for 24 hr or -15.6°C (4°F) for 4 hr will lose germination ability.

Gibb Function Represented by G, an expression which represents the maximum useful work which can be obtained, the relationship is:

$$G = U - TS + pV = H - TS$$

where

U = internal energy
T = absolute temperature
S = entropy
p = pressure
V = volume
H = enthalpy

The decrease in the Gibb function of a system sets an upper

limit to the work, exclusive of pdV work, in any process between two equilibrium states at the same temperature and pressure. Inasmuch as the decrease equals maximum useful work that can be obtained in a process, the function G has also been called the free energy. Also see *Free Energy; Helmholtz.*

Gladiola Drying The curing of gladiolas is very similar to curing of onions. By curing at 32.2 to 34.9°C (90 to 95°F) with relative humidity above 80% a disease-resistant surface layer is developed and some rot fungi are killed. Good results have been obtained by curing at 26.7 to 32.2°C (80 to 90°F) for 48 to 72 hr with a relative humidity between 50 and 80%, using air velocities up to 1 m/s (200 ft/min). It is a common practice to store the corms at 4.4 to 10°C (40 to 50°F) after curing and cleaning. Also see *Garlic Drying; Onion Drying.*

Gradient The rate at which a variable quantity changes in value, such as °F/hr, Δp/sec, m/s, and ft/sec. The term also refers to the gradient of a function in vector notation, written as grad f or ∇f where

$$\nabla f = if_x + jf_y + hf_z$$

Graetz Number A dimensionless parameter, Gz, which characterizes forced convection of a fluid for streamline flow relating thermal capacity of the fluid and convection heat transfer:

$$Gz = \frac{\omega c_p}{k\ell}$$

where

ω = mass flow rate
c_p = specific heat
k = thermal conductivity
ℓ = length dimension

which has been modified to:

$$Gz = \frac{4\ell/d}{Re \cdot Pr} = \frac{4\ell/d}{Pe}$$

where

Re = Reynolds number
Pr = Prandtl number
Pe = Peclet number

Also see *Peclet Number; Prandtl Number; Reynolds Number.*

Graham Law The physical relationship which expresses the relative rates of diffusion of gases as being inversely proportional to the square root of the densities of the gases when considered under the same conditions.

Grain A unit of weight of 7000 grains is equal to one pound. Often used to represent the quantity of moisture in vapor form in air. The unit was originally based on the weight of a grain of wheat and equal to 0.0648 grams.

Grain Dryer A device which reduces the moisture content of grain, with or without heated air, but including an air moving device, appropriate ducts for air distribution, a chamber for drying, often with conveying or materials handling devices to move grain to and from the dryer. Also see particular type of dryer.

Grain Drying Recommendations A general compilation of recommendations for drying various crops is given in Table G-2. These recommendations are representative and can be modified with appropriate design to carry out the drying process.

Table G-2
Grain Drying Recommendations

	Ear Corn	Shelled Corn	Small Cereal Grains	Rice
Maximum moisture content				
with natural air, %	30	25	20	25
with heated air, %	35	35	25	25
Maximum relative humidity of natural air entering grain to dry to safe storage, %	60	60	60	60
Preferred depth of grain for batch drying with heated air, m	1.5-6	0.4-0.6	0.4-0.6	0.23-0.45
ft	5-20	1-1/3-2	1-1/3-2	3/4-1.5
Range of airflow to dry crop, Natural air, cfm/bu	5-3	5-2	4-1	4-1
Maximum temp., rise of 8°C (10°F), cfm/bu	5-3	6-2	5-1	4-1

Based on early CDMA recommendations.

Gram A unit of mass in the c-g-s system, with 453.59 grams per pound, 1000 grams per 1 kilogram, and 15.432 grains per 1 gram. In the SI units, the basic unit of mass is the kilogram (kg).

Gram-Calorie See *Calorie*.

Granulator Device used in cane and beet sugar manufacture to remove the unbound moisture from the sugar by dryers and coolers. Moisture is removed in the dryer by blowing hot air through a stream of cascading sugar or through a bed of wet sugar. Upon heating and drying the sugar crystallizes and must be cooled after the dryer.

Grape Drying See *Raisin*.

Grashof Number A dimensionless number which represents the ratio of buoyance forces to viscous forces in natural convection:

$$Gr = \frac{\rho^2 \, g \, \beta \ell^3 \, \Delta t}{\mu^2} = \beta \, \Delta t \, (Ga)$$

$$= \frac{\text{Galileo number}}{\text{Gay Lussac number}}$$

where

ρ = density
g = local acceleration due to gravity
β = volume coefficient of expansion
ℓ = length dimension
Ga = Galileo number
Δt = temperature difference
μ = dynamic viscosity

Also see *Galileo Number*.

Grass Dryer See *Green Crop Dryer*.

Gravel See *Rock and Sand Drying*.

Green Bean Drying Snap beans or green beans are used for edible pods or seeds in the pods which can be dried. Split green beans are dried in a tray or on a belt at a temperature of 62.7 to 65.5°C (145 to 150°F) for 6 hr, followed by bin drying at 46.1°C (115°F) for 2 days to produce a product at 2 to 5% moisture content. Higher temperatures can be used if the time is reduced. An improved quality of product can be obtained by freezing the beans prior to drying.

Green Crop Dryer A device used to remove moisture from forage crops, which is identified as a low temperature green crop dryer if the hottest air in contact with the crop does not exceed 176.6°C (350°F). The device is known as a high temperature green crop dryer for hotter temperatures of air in contact with the crop, which usually exceed 315.5°C (600°F). Also see *Forage Drying*.

Green Lumber Freshly sawed lumber, or lumber that has received no intentional drying; also known as unseasoned lumber. The term does not apply to lumber that becomes wet through waterlogging.

Green Pea Drying Green peas are edible peas which are difficult to dry and rehydrate by conventional methods. Peas are slit prior to blanching, sulfiting, and dehydrating. Tunnel, cabinet, and continuous belt dryers are used for green peas. Temperatures of 82.2°C (180°F) and 71.1°C (160°F) are used in the first and second stages of the tunnel dryer, with peas on trays. An initial temperature of

87.7 to 93.3°C (190 to 200°F) is used for a continuous belt dryer. Peas are removed from the dryer at 8% moisture and finished to 4% in a bin dryer using air at 48.9°C (120°F).

Green Stage Floor   The part of the oasthouse where green hops in pokes are received and stored before drying. Also see *Hop Drying; Oast*.

Grinder Drying   See *Mill Drying*.

Grog   A liquor of an unsweetened mixture of spirit and water. Special dryers are available for drying grog ware.

Gross Drying Rate   A term representing the loss of weight from moisture per unit time ignoring the change in drying area.

Gross Heating Value   See *High Heating Value*.

H

h See *Convection Heat Transfer Coefficient.*

Hack Drying The drying of stacked raw clay bricks in open air on a platform. The bricks are stacked in hacks about 1.25 cm (1/2 in) apart. The first layer is partially dried so a second layer can be supported and partially dried until the stack is eight bricks high. Bricks may be restacked after partial drying, with drying continued in the air for about a month before placing in the kiln. Also see *Clay Drying.*

Hair Dryer An apparatus for drying hair, of the fixed in-place or hand-held portable types. Air is directed over the head of wet hair. Units for commercial establishment have at least 1000 w heating elements with five or more control levels. Home units are usually rated at 500 to 1000 watts. Hand-held units have heating elements with 250 to 750 watts, with two or three levels of heating. Except for small portable travel hair dryers, a duct directs the heated air from the dryer to the head.

Hair dryers may also consist of a brush and drying comb with an electric heating element, which may include a nozzle for directed air flow. The unit may be powered with 300 to 700 watts, the 500 to 700 watt units having two settings of intensity, identified as "dry" and "style." Hair is dried much more quickly using these conduction/convection heated dryers than with air flow alone.

Half-Life A term used to represent the time at which a radioactive substance has decayed until one-half of the original material, activity, or strength remains (Fig. H-1). The relationship has been applied to many physical and chemical and biological phenomena. For drying a similar term is time of one-half response. Also see *Time of One-Half Response.*

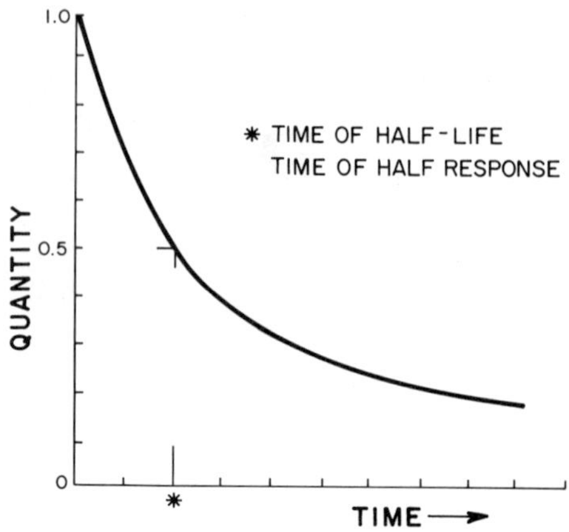

Figure H-1. Half-life.

Hammer Mill A machine used to comminute products which may be used preceding drying to provide a more uniform small size of product, to increase the surface area, and to decrease the length of path of moisture flow for drying. Energy added during grinding and with air flow may appreciably reduce moisture content of the material. Also see *Mill Drying*.

Hand Dryer An electric heated forced air device, usually fixed to a wall, for drying of hands and face, used in place of towels and napkins after washing. Usually operated on a time switch for a definite time of less than one minute. A 2000-watt heating element is common in commercial units.

Hand-Held Dryer See *Hair Dryer*.

Hardness The resistance of a material to crushing, indenting, or abrading when impacted, measured on a Moh scale of 1 to 10, soft to hard, or as given by a time-force-physical change plot. In general, products become harder upon drying. Hardness may be used as a means of rapidly determining indirectly the moisture content. For metals, the Brinnell scale represents the hardness.

Hardness also refers to water containing the salts of calcium, magnesium, and iron which affects the sudsing property. Water hardness is generally expressed in terms of equivalent calcium carbonate in parts per million or grains per gallon. According to the U. S. Geological Survey standard for water hardness: soft is 0-60 ppm; moderately hard, 60-120 ppm; hard, 120-180 ppm; and very hard, over 180 ppm.

Harkins-Jura A method of plotting equilibrium moisture content data or isotherms which gives plots for products (proteins and grains) in the range of 50 to 95% relative humidity. The H-J equation is:

$$\text{Ln}\,\frac{p_g}{p_s} = B - A/M^2$$

where

p_g = vapor pressure of product
p_s = saturated vapor pressure
M = moisture content
A and B = constants

Also see *Adsorption Isotherm; Equilibrium Moisture Content; Henderson Equation.*

Hay Drying The removal of moisture from long, chopped, ground, or baled hay, using forced unheated air or forced heated air. The product is dried from green cut at 60 to 80% moisture, wet basis, using heated air, up to 150 to 815°C (300 to 1500°F), often in a rotary dryer with chopped or ground forage, reducing the moisture content to 5 to 10%. This process is often referred to as dehydration. In a belt or conveyor dryer the temperature of the air is usually limited to 176.6°C (350°F).

Partial drying of cut long forage in the field, often crushed or lacerated, reducing the moisture to 30 to 50% in 1 to 3 days, followed by forced unheated air drying of chopped, long, or baled hay in storage is often practiced (Table H-1). Drying must occur rapidly enough to avoid mold formation, depending on the moisture content (Fig. H-2). An air flow of 0.15 to 0.26 m^3/tonne·s (300 to 500 cfm/ton) (one ton of long hay is about 400 ft^3) at about 0.25 kPa (1.0 in) static pressure is needed (Figs. B-2; H-3; H-4). Supplemental heat can be used for these systems. A major problem in drying is maintaining uniform air flow through the stored hay. Also see *Barn Dryer; Green Crop Dryer.*

Head See *Friction; Impact; Potential; Pressure; Static; Velocity; Work.*

Heat A form of thermal energy which substances have as a result of movement of the molecules and atoms. A change in thermal energy is associated with calories, temperature, change of state, expansion or contraction, and electrical effects.

Heated-Air Drying The use of forced ventilation of air using an air moving device with the addition of heat for removing moisture from materials and products.

Heated-Air Unit A basic heated-air-producing unit including an air moving device, burner system, and control system (Fig. H-5). The unit is usually coupled to a drying structure by means of a flexible duct. The unit may have a transport chassis and wheels for portability. Also see various types of dryers.

Table H-1
Comparison of Duct Systems for Forage

Type of Duct System	Form of Storage	Widths of Mow for Which Recommended ft	Maximum Width of Mow ft	Depth Above Duct ft	Depth Above Floor ft	Capacity tons	Maximum Length ft
Triangular	long	20 to 36	36	13	21	60	40
center main	chopped		36	13	21	60	40
	baled		24	9	17	60	50
Rectangular	long	30 to 36	36	15	23	120	70
center main	chopped		36	13	21	120	75
	baled		28	11	19	120	90
Side main	long	20 to 40	40	16	17	120	60
with lateral	chopped		40	13	14	120	60
ducts or slatted floor	baled		40	12	13	120	60
Center main	long	32 to 44	44	16	17	120	55
with lateral	chopped		44	13	14	120	55
ducts or slatted floor	baled		44	12	13	120	55

Approximately 7 to 10 tons can be dried per horsepower.
Static pressures encountered are usually less than 1 in. water.

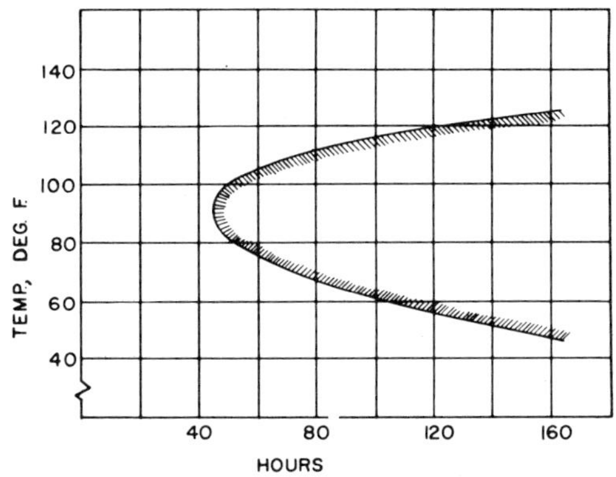

Figure H-2. Hay relationship of temperature and mold for high moisture conditions.

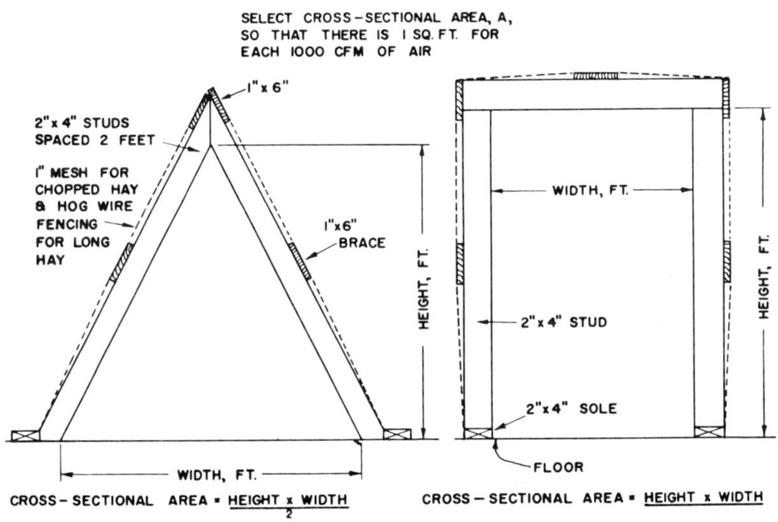

Figure H-3. Hay drying center main duct without lateral ducts.

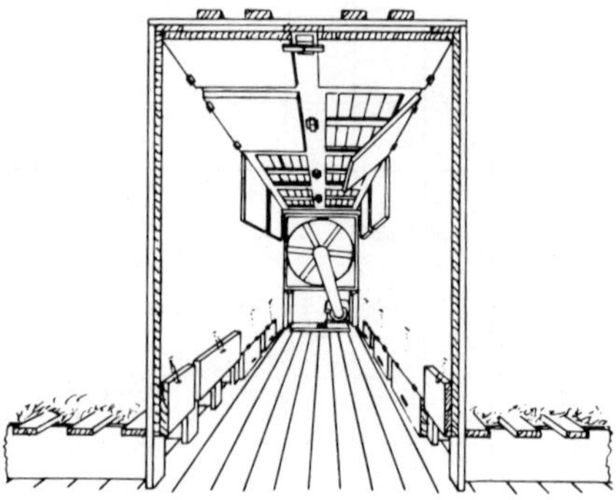

Figure H-4. Hay drying main duct with doors and slatted floor duct.

Heated Forced Air See *Forced Air*.

Heating Value See *Gross Heating Value; Net Heating Value*.

Heat, Latent See *Latent Heat*.

Heat, Mechanical Equivalent See *First Law of Thermodynamics*.

Heat, Radiant See *Radiation; Electromagnetic Radiation*.

Heat, Specific See *Specific Heat*.

Heat Capacity The quantity of heat necessary to raise the temperature of a substance by one degree, either in calories/°C or Btu/°F.

Heat capacity, C, is defined as $C = dU/dT$ where U is the change of internal energy of the system. The heat capacity of water is 1.00803 cal/g°C at 0°C at 1 atm. For a gas the specific heat at constant pressure is larger than the specific heat at constant volume. The heat capacity and specific heat are the same. Also see *Specific Heat*.

Heat Conductivity See *Conduction, Heat; Thermal Conduction*.

Heat Content See *Heat Capacity*.

Heat Damage A general term which might apply to any of several undesirable changes in a food or biological material during drying. The most common example is browning of food products by overheating. Other examples are checking, splitting, cracking, charring, and burning. Browning occurs as a result of a reaction between the free amino groups and lactose in milk, which can be delayed or reduced by storing at low moisture content. For dehydrated potatoes the rate

Figure H-5. Heated air portable heater for drying.

of browning is greatest at 15 to 20% moisture. Browning increases with an increase in temperature. The deterioration of quality of a product is a result of excessive heat exposure. The heating may occur as a result of excess moisture, in the field or in storage, or during drying. Heat damage is manifested by discoloration, darkening, checks, etc., and may result in lowering the grade for cereal grains (Table H-2). Also see *Checking; Cracking; Maillard Reaction.*

Table H-2
Heat Damage, Kernel Damage and Grade, U.S.

Product	Grade Number	Heat Damage, %	Maximum Allowable Kernel Damage, %
Corn	1	0.1	2
	2	0.2	3
	3	0.5	4
Sorghum	1	0.2	4
	2	0.5	8
	3	1.0	12
Wheat	1	0.1	2
	2	0.2	4
	3	0.5	7

Heat Equation The parabolic second order partial differential equation applicable to heat conduction:

$$\frac{\partial u}{\partial t} = \frac{k}{c\rho}\left(\frac{\partial^2 u}{\partial x^2} + \frac{\partial^2 u}{\partial y^2} + \frac{\partial^2 u}{\partial z^2}\right)$$

where

$u = u(x,y,z;t)$ denotes temperature at location (x,y,z)
t = time
k = thermal conductivity
c = specific heat
ρ = density

Heat Exchanger A device used to transfer heat from one fluid to another. A heat exchanger is used in an indirect fired dryer. Also see *Indirect Heater.*

Heat Flow The quantity of heat flow per unit time, or equivalent, per unit cross-sectional area. Units and conversions are: 1 Calorie/sec cm^2 = 4.186 watts/cm^2 = 3600 Calories/hr cm^2 = 13,272 Btu/hr ft^2 = 318,531 Btu/day ft^2 = 41,873 W/m^2.

Heat of Combustion   The quantity of heat produced by burning a certain quantity of material, expressed in calories/gram molecular weight, Kilogram-calories/mole, Cal/gram, cal/gram, Btu/lb, or kJ/kg. Also see *Combustion; Fuel; High Heating Value; Low Heating Value; Wood Combustion.*

Heat Pipe Dryer   A dryer using a heat pipe as a means of recovering heat from the exhaust air. A heat exchanger pipe transports heat by evaporation and condensation of a liquid in an enclosed pipe (Fig. H-6). Using a heat pipe in a concurrent (heating)-counterflow (cooling) dryer, a 20% saving in energy is reported; for a fixed bed dryer, a 10% saving in energy is reported.

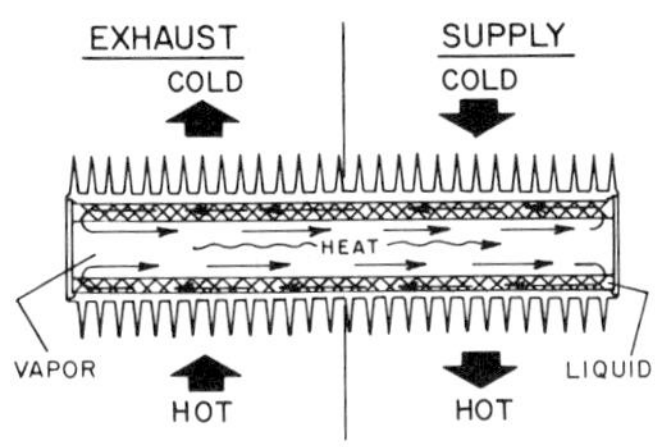

Figure H-6. Heat pipe.

Heat Pump Dryer   A heat pump is a refrigeration system which can be used for heating or cooling designed in a manner which enables a surface to deliver usable heat to a space during heating and drying or to cool the same space (Fig. H-7). When operating as a heating system, the evaporator (cooling coils) absorbs heat from an outside space and delivers it with the heat from compressing the refrigerant, to the condenser which releases heat to the space being heated. When operating as a cooling system, the evaporator absorbs heat from the conditioned space and rejects it to the outside media, basically. The heat pump is similar to a conventional refrigeration system. A system making use of rejected heat cannot properly be called a refrigeration system because cooling is not the primary objective of the cycles. For drying, heat is added at the condenser and moisture is removed at the evaporator. The heat pump dryer is used for temperature sensitive products.

Heat Stable Toxin   A poisonous compound which is produced by a microorganism which is not usually destroyed by prolonged heating or drying.

Heat Transfer   The movement of thermal energy through a body or across a space as a result of temperature difference by one or a combination of the three basic modes of heat transfer: (1) conduction

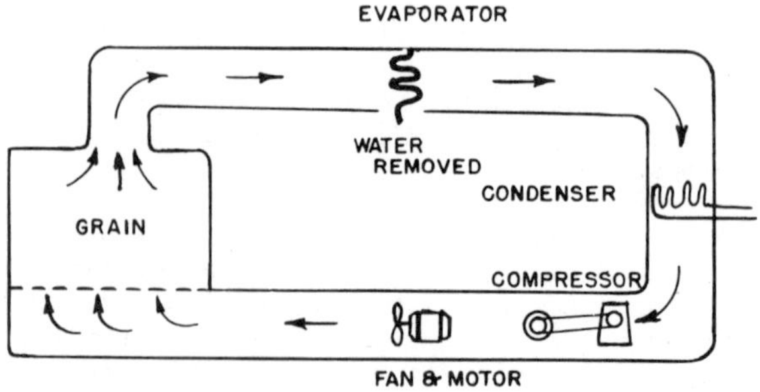

Figure H-7. Heat pump dryer-closed circuit refrigeration.

(molecular vibration depending on the temperature), (2) convection (transfer of a fluid mass from one to another location depending on density), and (3) radiation (electromagnetic, depending on the surface characteristics and the absolute temperature). Also see *Conduction; Convection; Conversion Factors; Heat Equation; Radiation.*

Heat Treatment During dehydration, drying, and evaporation, the thermal effects of the heat treatment may be similar or as effective as thermal processes identified as:

1. Thermization. A treatment less severe on the product and associated microorganisms than pasteurization.
2. Pasteurization. A treatment of sufficient severity to kill pathogenic organisms and inactivate certain enzymes.
3. Commercial sterilization. A treatment of sufficient severity to kill all the pathogenic organisms and most of the remaining organisms.
4. Sterilization. A treatment of sufficient severity to kill all of the organisms in the product.

Also see *Pasteurization; Radiation Treatment; Sterilization.*

Heat Utilization Factor (HUF) The ratio of temperature decrease due to cooling of the air during drying (evaporative cooling) and the temperature increase due to heating the air.

$$\text{Heat Utilization Factor} = \frac{\text{air temperature (dry-bulb) decrease during drying}}{\text{air temperature (dry-bulb) increase during heating}} \times 100\%$$

$$\text{HUF} = \frac{t_2 - t_3}{t_2 - t_1} \times 100\%$$

where

t_1 = original dry-bulb air temperature
t_2 = temperature of air after heated
t_3 = dry bulb temperature of air exhausted from the dryer

The heat utilization factor is usually less than 100%, but can exceed 100% in numerical value. Also see *Effective Heat Utilization; Energy Efficiency; Performance Ratio.*

Heat Wheel A device for recovery of heat, also known as a Ljungstrom regenerator. The device consists of a wheel or rotor which turns slowly (1 to 3 rpm) in a housing made of two compartments. One compartment is for the cold air being heated; the other is for the hot air being cooled. In one revolution the rotating wheel is heated by the hot air or gas and gives up heat to the incoming cold air or gas. The wheel may rotate in a vertical or horizontal plane. The wheel has a large heating surface formed by plates, tubes, or cells.

Heating Tendency The ability of stored products, such as grains and solid fuels, to heat spontaneously, which can be reduced by drying. Also see *Spontaneous Combustion.*

Helmholtz Function The property of a thermodynamic system defined by F, also known as free energy, and represented by the equation:

$$F = U - TS$$

where

U = internal energy
T = absolute temperature
S = enthalpy

Thus, the decrease in value of the Helmholtz function of a system sets an upper limit to the work done in any constant volume process between two equilibrium states at the same temperature during which a system exchanges heat only with a single reservoir at this temperature. Also see *Free Energy; Gibb Function.*

Henderson Equation An empirical relationship derived from Freidholm isotherm equation representing the equilibrium moisture content:

$$1 - RH = e^{-cTM_e^n}$$

where

RH = relative humidity, decimal

T = absolute temperature, °R
M_e = equilibrium moisture content, percent, dry basis
c,m = constants varying with the material (Table H-3)

The equilibrium moisture content is zero at zero relative humidity and approaches 100% as the moisture content approaches infinity. The slope of the curve is much steeper as the moisture content approaches infinity and zero. Also see *Adsorption Isotherm; Equilibrium Moisture Content; Freundlich Isotherm; Harkins-Jura*.

Table H-3
Henderson Equation, Value of Constants

	c	n
Cotton	4.91×10^{-5}	1.70
Corn	1.10×10^{-5}	1.90
Soybeans	3.20×10^{-5}	1.52
Flaxseed	6.89×10^{-6}	2.02
Wheat	5.59×10^{-7}	3.03
Sorghum	3.40×10^{-1}	2.31

Herbage Drying See *Forage Drying; Green Grass Drying; Hay Drying*.

Hertz A term for designating frequency, one cycle per second.

HFC An abbreviation for high frequency current, which for electricity is a frequency between 3 and 30 x 10^6 cycles per sec. Also see *Radio Frequency*.

Hide or Skin Drying The removal of moisture from raw hides or skins which contain 50 to 70% water under conditions where drying does not occur very rapidly. Fur, which is skin or hide with hair, is treated the same way with treatment on the side opposite the fur. The raw skin is washed with cold water to remove the blood and dirt before drying. Hides are dried to prevent deterioration. Bacteria will not usually develop if the moisture content is below 30% and mold will not usually develop if the moisture content is below 12%. Hides should be dried in the shade. Also see *Leather Drying*.

High Dried A term once applied to fish which have been dried throughout so as to store long periods as distinguished from those dried on the surface for keeping a short time.

High-Frequency Dielectric Heating Heating wood or other product with an electric field oscillating at 1 to 30 million cycles/sec, applied to the wood or other product dried between metal plate electrodes. Microwave heating and processing are done with energy at shorter wavelengths than dielectric heating. The rate of heat generation is

$$E = 1.4 \left(\frac{V}{t}\right) f\ e$$

where

E = energy generation rate, watts/cu in
V/t = voltage gradient
f = frequency, megacycles/sec
e = loss factor = power factor x dielectric constant

Also see *Dielectric Drying*.

High Heating Value (HHV) The high heating value, also known as the gross heating value, is the amount of heat produced from burning a solid fuel, in kJ/kg or Btu/lb (Table H-4). The high heating value is obtained in a calorimeter by assuming that energy (latent heat) is removed from the combustion products until the water produced has been liquefied. The complete combustion of fuel including carbon, hydrogen, and sulfur in the presence of adequate oxygen gives carbon dioxide, water, and sulfur dioxide. The presence of carbon monoxide in the exhaust indicates incomplete combustion. Also see *Low Heating Value*.

Table H-4
High Heating Values

	Heating Values	
	Btu/lb	MJ/kg
Butane	21,293	49.4
Cane sugar bagasse	7,900-9,100	18.3-21.1
Carbon	14,086	32.7
Carbon monoxide	4,343	10.1
Cotton ginning waste	7,900	18.5
Fuel oil	18,500	42.9
Gasoline	20,200	46.9
Hardwood, dry	8,100-8,900	18.8-20.6
Hydrogen	60,950	141.5
Kerosene	19,900	46.2
Methane	23,860	55.5
Propane gas	21,650	50.2
Softwood	8,800-9,800	20.4-22.7
Straw, 15% moisture	5,400	12.5
Wood, air dry	5,400-6,800	12.5-15.8

Hogged Fuel A fuel consisting of the waste wood from lumber or wood pulp operations, including bark, which might be dried.

Holding See *Equalization; Resting; Sweating; Tempering.*

Honeycombing The formation of splits and voids in a wet material due to uneven shrinkage during drying.

Hop Drying Hop or hops, the ripe cone-like flowers of a plant *Humulus lupulus,* containing an aromatic oil used in beer manufacture, are often dried, particularly for commercial use. After picking and sorting, hops are placed into a bulk holder and into pokes. The pokes are carried or moved by trolley to the green stage room and placed adjacent to the kilns for loading. Each poke is then carried into the kiln and placed on lifter cloths on the floor. After the kiln is loaded sulfur is burned in pans in the plenum chamber. The hops are dried in a batch unit and then tipped into piles from the lifter cloths. After conditioning for several hours the hops are made into bales. The hops may be dried in bulk boxes about 0.6 to 1 m (2 to 3 ft) deep, through which heated air is moved. Also see *Green Stack Floor; Oast; Pocket; Poke.*

Horizontal Agitated Batch Vacuum Dryer A horizontal cylinder which has radial paddles, which is heated, in which the material to be dried is stirred or agitated in contact with the heated metal wall (Fig. H-8). Units can be operated under vacuum or at atmospheric pressure. A ribbon or scroll agitator blade may also be used.

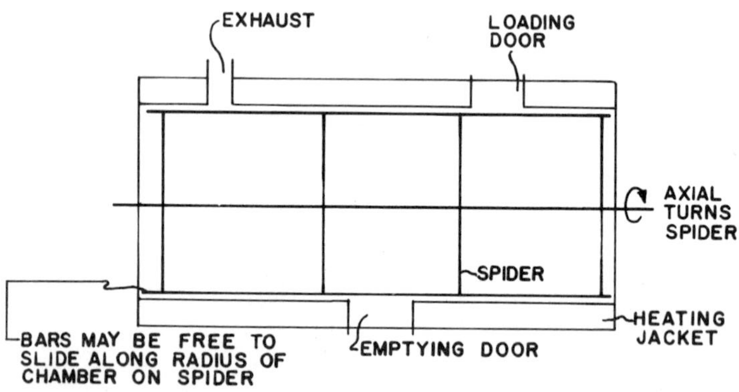

Figure H-8. Horizontal agitated batch vacuum dryer. *From Nonhebel, G., and Moss, A. A. H., Drying of Solids in the Chemical Industry (1971). Used with permission of Butterworth Publishers, London.*

Horizontal Belt Dryer An apparatus which carries the materials to be dried on a moving horizontal belt through drying air (Fig. B-4).

A cooling section may follow the drying section. The air may be forced through the product on a perforated, slotted, or screened belt or over a shallow layer of product on the belt. Also see *Belt Dryer*.

Horizontal Chamber The part of a spray dryer into which atomized products are forced in a horizontal direction, through which air is moved for moisture removal, and from which the dried products are removed (Fig. H-9). Horizontal chambers may be in the shape of a box, horizontal cylinder, or other shape in which the major axis of the chamber is in a horizontal direction.

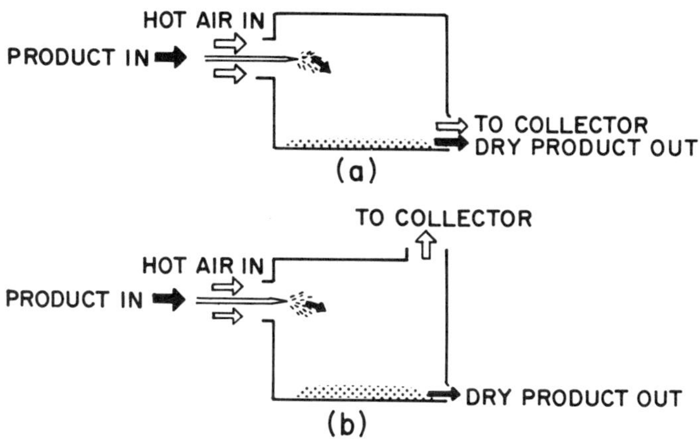

Figure H-9. Horizontal chamber, spray dryer. *From Hall, C. W., and Hedrick, T. I., Drying of Milk and Milk Products, 2nd edition (1971). Used with permission of Avi Publishing Co., Westport, CT.*

Horizontal Tube Evaporator An evaporator for removing water from a liquid, usually under vacuum, in which heat is transferred to the product in the evaporator through horizontal tubes (Fig. H-10).

Horsepower A measure of the rate of doing work, which is 33,000 ft lb/min for 1 horsepower, or 746 watts. The useful work which can be performed by a power device is called brake horsepower. Also see *Air Horsepower; Fan; Fluid Horsepower*.

Horseradish Drying The roots of horseradish may be dried and used as a flavoring, seasoning, or condiment. After slicing the roots may be dried in a two-stage tunnel dryer, with inlet temperatures of 68°C (150°F) and 57.2°C (135°F) for the first and second stages,

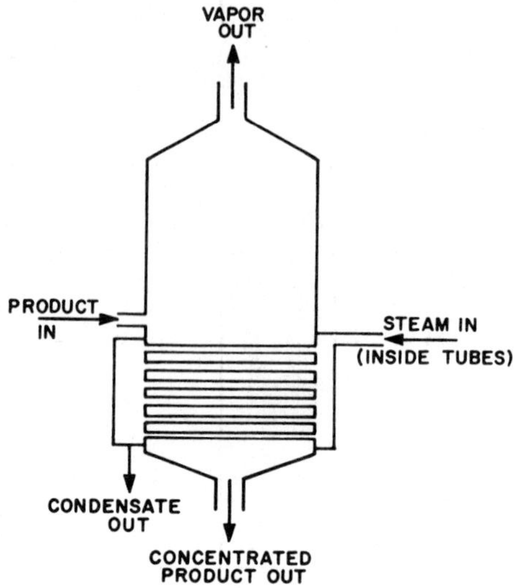

Figure H-10. Horizontal tube evaporator.
From Hall, C. W., and Hedrick, T. I., Drying of Milk and Milk Products, 2nd edition (1971). Used with permission of Avi Publishing Co., Westport, CT.

respectively, to 8% moisture content. The product may then be dried in a bin dryer to 5%. Powder or granules are usually made from the dried slices.

Hot Wire Anemometer An instrument consisting of a resistance wire heated electrically for measuring the velocity of a fluid, usually a gas. The unit operates on the principle that a small diameter electric conductor gives off heat in relation to the velocity of air passing over the wire. A constant current is maintained through the wire providing a fixed heat output and the observed temperature of the wire indicates the air flow, as obtained from an appropriate calibration chart.

HUF See *Heat Utilization Factor*.

Hukill's Analysis A method formerly used for analyzing drying of deep layers in which the moisture content of the product of different depths is obtained or estimated for the time of drying. To obtain the graphical plot, moisture content ratio is represented on the ordinate, $M - M_e/M - M_o$, and the time unit, which is the time of one-half response of drying fully exposed grain, is represented on the abscissa. The depth factor is the depth of grain or product which contains enough product so that if all the theoretical available heat were used the product would dry to equilibrium in a period of time equal to the time of one-half response. For a particular drying condition, the moisture content ratio is converted to moisture content, percent; time units are converted to hours; and the depth factor is converted to feet (Fig. D-4). Also see *Depth Factor*.

Humectants A material (chemical) which is a wetting or moistening agent which is used to maintain the moisture content of products such as baking products, glues, inks, crops, foods, and textiles. When added to food, humectants help to bind water, such as for sugar, honey, salt, glycerol, and sorbitol. Humectants are the opposite of desiccants.

Humid Heat The quantity of heat required to raise the temperature by one degree of one unit mass of dry air plus the water vapor which accompanies it. Expressed in terms of kJ/kg K, Btu/lb dry air-°F or kilo cal/kg dry air-°C.

Humid Volume The volume occupied by one unit mass of dry air plus the water vapor it contains, expressed as ft^3/lb dry air or m^3/kg dry air.

Humidification The process of adding moisture to air resulting in an increase in humidity. By circulating air of a higher humidity, the moisture content of a hygroscopic product can often be increased, a process known as conditioning, such as is done for some grains for milling and processing.

Humidity The weight of water vapor in a unit volume of atmosphere, expressed in terms of grams per cubic meter, grains per cubic foot, pounds per cubic foot, kilograms per cubic meter, etc. Often expressed in terms of relative amount of moisture which could be held by the air, known as relative humidity. Also see *Absolute Humidity; Relative Humidity*.

Hydrate The combinations of certain chemical compounds with water (Fig. H-11). The water molecules (H_2O) are attached to the chemical. As an example, $CaSO_4 \cdot 5H_2O$, upon drying, goes to $CaSO_4 \cdot 3H_2O$ and $CuSO_4 \cdot H_2O$ to $CuSO_4$ when dried at 25°C (77°F). Also see *Moisture Representation*.

Hydrated The addition of water to a chemical to form a hydrate. In the process of adding water, heat of hydration is released. The equipment for carrying out this process is known as an hydrator.

Hydrator See *Hydrated*.

Hydraulic Radius (H_R or R_H) The cross section or the area through which a fluid flows divided by the wetted perimeter; that is, the

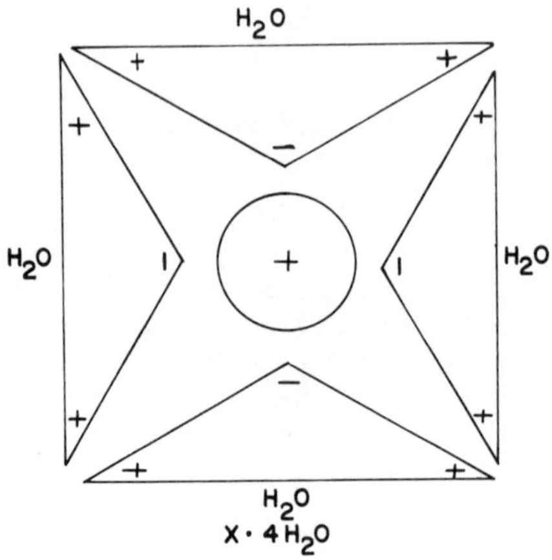

Figure H-11. Hydrate.

perimeter touched by the fluid. For a gas flowing in a round tube or for a liquid fully filling the tube,

$$H_R = \frac{A}{P_W} = \frac{\pi R^2}{\pi D} = \frac{\pi R^2}{2\pi R} = \frac{R}{2}$$

where

A = cross sectional area
p_W = wetted perimeter
R = radius, $2R$ = diameter, D

For a partially filled pipe or tube or a channel the wetted perimeter is the distance touched by the fluid.

Hydro Prefix meaning water, such as hydroelectric generator, that is, water powered generator; also hydrogen or water production, as a compound of hydrogen.

Hydrodynamic Parameter A dimensionless number, Ho, which represents or describes the rate of change of fluid flow with time:

$$Ho = \frac{vt}{\ell}$$

where

v = velocity
t = time
ℓ = length

Hydro-Extractor   A centrifugal drying machine with the basket specially designed to handle textile fabrics and fibrous materials. Also see *Whizzer*.

Hydrolytic Rancidity   See *Rancidity*.

Hydrometer   An instrument specially designed and calibrated to measure the density of a liquid in which it is immersed. The hydrometer is used to determine the density of a solution being condensed by an evaporator.

Hydrophilic   A substance having an affinity for water. The substance can be used to remove moisture from the environment and for drying solids. Also see *Adsorption; Air Drying; Deliquescent; Hydrophilic Fiber*.

Hydrophilic Fiber   In textiles the ability of fibers to absorb or hold moisture. Under normal conditions 20°C (68°F) and 75% relative humidity, the following hydrophilic fiber moisture contents are maintained: acetate, 6.5%; cotton, 7%; linen, 8%; worsted (long stapled wool), 10%; rayon and silk, 11%; and wool, 15-16%. Also see *Absorbent Cotton; Equilibrium Moisture; Moisture Regain*.

Hydrophobic   A substance having no or very little affinity for water.

Hydrous   A substance which contains water of hydration or water of crystallization.

Hygro   A prefix representing humidity and moisture.

Hygroscopic Equilibrium   See *Equilibrium Moisture*.

Hygrodeik   An instrument with a wet bulb and dry bulb used for determining relative humidity which is determined by reading from a chart.

Hygrometer   An instrument for measuring the humidity in the atmosphere and air. The humidity may be determined from a hygrometer which measures the temperature at which the moisture condenses from vapor to water, known as the dew point. Also see *Dew Point; Psychrometer*.

Hygroscopic   Describes a material which gains or loses moisture depending on the product and the humidity and temperature of the surrounding environment. Biological products are hygroscopic, whereas inert materials are not hygroscopic. Also see *Absorption; Adsorption; Deliquescent; Equilibrium Moisture*.

Hygroscopic Equilibrium   See *Equilibrium Moisture*.

Hypotonic Refers to the exertion of less osmotic (pressure is negative) pull than the medium on the opposite side of a semi-permeable membrane. Thus the material has a lower concentration and loses water during osmosis.

Hysteresis The lagging of an effect from the expected value due to resistance to movement, such as friction in an instrument or controller, resistance to flow of moisture in a product, or trace of an electrical response in circuits containing magnetic elements (Fig. H-12).

A hygroscopic product which is subjected to a wetting cycle, followed by a drying cycle, will exhibit an equilibrium moisture content of the product which follows the trend but lags the wetting cycle. The effect is called "hysteresis" because of the "lagging."

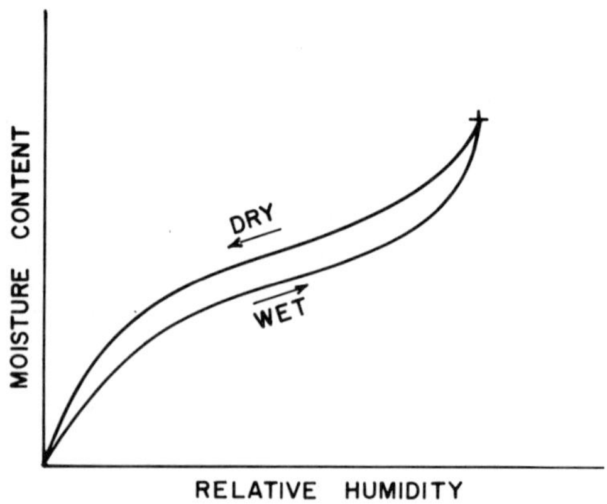

Figure H-12. Hysteresis curve.

I

Ice Cream Mix, Dry Ice cream mix consists of the ingredients which make up the ice cream before it is frozen and before air is incorporated. The mix consists of milk fat, milk solids not fat, egg yolk, sugar, stabilizer, flavors, fruits, nuts, and water. Ice cream and ice cream mix contain from 35 to 40% total solids, with the remaining 60 to 65% water. The water can be removed from ice cream mix by drying to less than 3% in a spray dryer. The dry mix is reconstituted by adding 1.65 parts of water to 1 part of mix, following which it can be frozen.

Ideal Fluid See *Fluid, Ideal.*

Ideal Gas A low density gas consisting of elastic molecules in which the effective volume occupied by the molecules is zero and which obeys the ideal gas laws (PV = MRT). Also called perfect gas. Also see *Boyle's Law; Charles Law; Gas Constant.*

Ideal Gas Law See *Boyle's Law; Gas Constant.*

Ignition The process of causing a substance to burn. A phenomenon occurring when the rate of gain of heat of a substance due to oxidation is greater than the rate of heat loss. The temperature of ignition is dependent on the properties of the product and the environment. Ignition temperatures are associated with the fuels used for drying (Table I-1). Also see *Combustion; Flash Point.*

Immersion Length The immersion length of a thermometer extends from the free end of the well, bulb, or element to the point of immersion in the medium being measured for temperature.

Imperial Gallon See *Gallon; Quart.*

Table I-1
Ignition Temperatures in Air at 1 Atmosphere, Approximate

	Temperature	
	°F	°C
Acetylene	580- 825	305-441
Gasoline (regular-high test)	500- 800	260-427
Hydrogen	1065-1095	574-590
Kerosene	490- 560	255-294
Methane	1170-1380	630-750
Propane	900-1000	480-540
Wood	Approx. 525	Approx. 275
Ethylene	1000-1015	540-545

Inactivation of Enzyme See *Enzyme*.

Incidental Water The amount of water in percent weight which a product needs to maintain its state.

Incineration The burning of materials, usually for the purpose of disposal. The heat could be used for heating and drying.

Inclined Tube Manometer See *Draft Gage*.

Incombustible Not capable of burning. Also see *Combustible; Flammable; Inflammable*.

Incompressible Fluid A gas or liquid which maintains essentially the same density under the pressure and temperature conditions involved. A fluid flowing under these conditions is called incompressible fluid flow. Incompressible flow is often assumed to represent the conditions of flow even when some change of density occurs particularly if the results are not affected by the assumption.

Indexes of Quality for Dehydration Many of the steps in the preparation of dehydrated foods have a marked effect on the survival of microorganisms in the finished products. The most significant steps include:

Method of treatment
Handling before dehydration
Time required for dehydration
Amount of moisture removed
Temperature reached by product at end of process

These parameters may be related to the quality based on microorganisms present after drying.

Indirect Dryer See *Indirect-Fired Dryer; Indirect Heater*.

Indirect-Fired Dryer A type of dryer in which the products of combustion do not come in contact with the product being dried. Sometimes called an indirect dryer. Heat exchangers are often used in which a wall separates the products of combustion and drying air.

Indirect Heater A heat exchanger used for heating a fluid, which for drying is usually air, in which the products of combustion and air heat a wall through which heat is transferred to heat air on the opposite side, without mixing with the air being heated (Fig. I-1). Steam heated and hot water heat exchangers for dryers are indirect heaters.

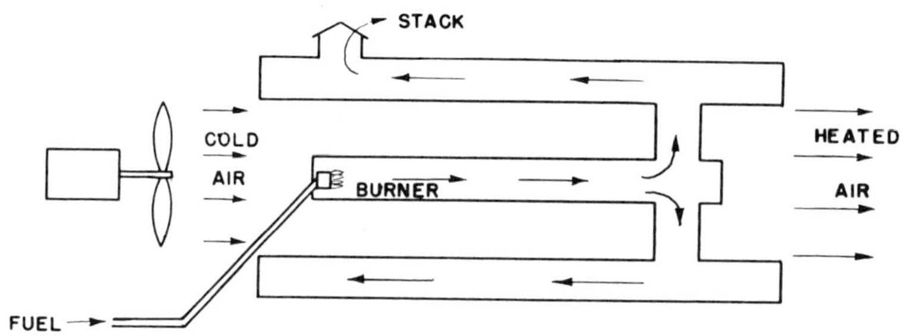

Figure I-1. Indirect heater.

Induction Heating A method of heating in which a material which conducts electricity is heated by electric currents induced in the material by an alternating magnetic field. Electric heating is accomplished with an electric alternating current of high value and voltage of low magnitude. The frequency of the induction coil varies over a large range depending on the method of obtaining the effect. Uniform heating is obtained in the normal frequency range: standard, 480, 960, 3000, and 9600 cycles/sec; with a spark gap, 20,000 to 40,000 cycles/sec; and electronic, 450,000 cycles/sec. High frequency induction gives surface heating at 100,000 to 500,000 cycles/sec. Induction occurs as a result of current flow induced.

Infestation The presence and/or multiplication of living organisms usually of unwanted varieties in products, before or after drying.

Inflammable Combustible or ability to support combustion. In contrast the word incombustible means lack of ability to burn. Also see *Combustible; Flammable; Limits of Inflammability*.

Infrared Dryer A device for removing moisture from products in which the energy for moisture removal is supplied primarily from an infrared energy source (Fig. I-2). The design must carefully consider the surface area to be radiated, method of exposing the surface area, and removal of moist air. The infrared energy does not penetrate the solid product, can rapidly heat and possibly scorch the product, and can be more rapidly absorbed by the air between the product and heat source if the air is nearly saturated with water (Table I-2). Also see *Infrared Drying*.

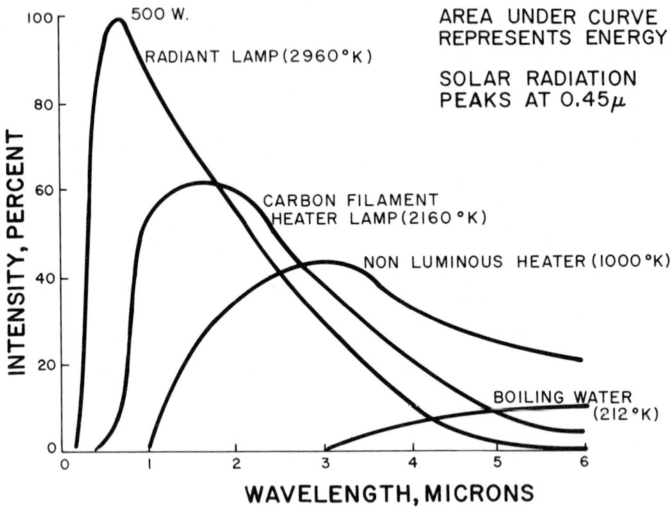

Figure I-2. Infrared radiation energy from various sources.

Table I-2
Infrared (Dryer) Absorption Bands

Carbon dioxide
2.36- 3.02μ
4.00- 4.80μ
12.5 -16.5μ
Water vapor
2.3- 3.3μ
4.8- 6.5μ
12.0-25μ

Infrared Drying Moisture removal from materials or products using electromagnetic radiation in the wavelength 0.76 μm to 400 μm is known as infrared radiation, IR, or radiant heating. The peak emission at 190°C (374°F) is 6.1 μm; 880°C (1616°F) (gas), 2.5 μm; and at 2200°C (3992°F), 1.1 μm (electric). Infrared energy is given off by electric or gas heated units. The transmission of the energy from the source to the product depends on surface characteristics of the surfaces and absorption characteristics of the gases between the two, particularly water vapor and carbon dioxide. Infrared drying is used for sheets of materials, surface finishes, and surfaces of materials, usually in thin layers, or for agitated materials in the radiation field. Also see *Infrared Dryer*.

Inherent Moisture The chemically combined and absorbed moisture of a product, such as in coal, as separate from the surface or free moisture. In bituminous coal the inherent moisture is 10 to 30% and lignites, 30 to 45%.

Initial Moisture Distribution See *Parabolic Initial Moisture Content*.

Ink Drying Ink dries by oxidation, penetration, absorption, precipitation, evaporation, or a combination of these. Oxidation drying is accomplished with linseed oils and driers, possibly using hard resins to improve setting. Resin or gilsonite is used for penetration drying mixtures. Absorption drying is accomplished by the liquid portion of the ink moving into the pores of the paper. Precipitation drying is based on upsetting the structure by introducing certain glycol or hot wax. Evaporation drying at high temperature is assisted by dissolving hydrocarbon solvents which have a narrow range high boiling point. Also see *Separation Drying*.

In-Package Desiccation Nearly dry products with less than 5% moisture can be dried further with drying agents (desiccants) placed in the package or container with the product. Desiccants are used which have a substantial moisture absorption ability at low relative humidities (below the relative humidity in equilibrium with a moisture content below 5%). Desiccants with absorption/adsorption characteristics similar to calcium oxide are used. The desiccant is placed in a container (usually a small package) which will prevent the desiccant from contaminating the product and which will permit water vapor to pass into the desiccant. The rate of desiccation is related to (1) adsorption property of desiccant, (2) size of package of adsorbant, (3) size of particle and porosity of product, (4) size and shape of product container, and (5) location of desiccant in product package. An inert gas, such as nitrogen, may also be a part of the in-package environment.

Input The quantity of material, energy, resource, or power directed into a system, such as a drying system, either as a batch or quantity per unit time. The input is one of the parameters in determining efficiency. Also see *Efficiency*.

Insertion Tube Dryer Tubes made of perforated metal sheet and rounded metal points which are pushed into open sacks of grain or bins of grain to remove moisture by forcing air through the tube and product. Heated air passes through the tubes from flexible piping into which air is forced by a small fan and heater. The unit is portable and is used for small-scale drying or cooling. It is particularly applicable to drying small seeds.

Insolation The solar radiation which reaches the earth or a body on the earth, including both direct and sky radiation. For drying in the U.S., the insolation is approximately 408 Langley/day (1500 Btu/ft^2 day).

Insoluble A substance which will not dissolve, at least in more than traces, in water or another identified solution.

Inspissation The process of a liquid substance gaining consistency as a result of evaporating or boiling.

Installation Fuel supplies and electrical connections, up to the point of connection with the dryer, should comply with the following standards, or their successors, of the National Fire Protection Association:

NFPA No. 31, Standard for the Installation of Oil Burning Equipment.
NFPA No. 54, Standard for the Installation of Gas Appliances and Gas Piping.
NFPA No. 58, Standard for the Storage and Handling of Liquefied Petroleum Gases.
NFPA No. 70, National Electrical Code.

The fuel line from the natural gas meter to the fuel train inlet of the dryer shall be sized and installed to meet American National Standard, 283.1, Installation of Gas Piping and Gas Equipment on Industrial Premises and Certain Other Premises. Also see *Safety*.

Instantizing See *Agglomeration; Puff Drying*.

In-Storage Dryer A dryer in which the drying bin or compartment is also used to store the product after it is dried.

Insulation Material used to isolate or limit the effect or flow of an external phenomenon, such as vibration, sound, heat, voltage (electric), and moisture (vapor barrier). Also see *Thermal Conduction*.

Interfacial Tension The surface tension at the surface separating two non-miscible liquids.

Intermediate Moisture Foods Those foods which in general have a moisture content between 10 to 35%, wet basis, and a water activity of 0.60 to 0.85. As the moisture content of a product increases so does the water activity. Some food products are not dried below the intermediate moisture level because of cost and/or effect of drying on the product. Also see *Water Activity*.

Intermittent Drying Moisture removal by periodic operation of the fan and/or application of heat, so that the greater concentration of moisture in the center of the product may diffuse toward the surface during "resting" periods. If the total time for drying is not critical from the standpoint of product quality, the energy for drying by fan and heater operation can be reduced by intermittent operation. Also see *Dryeration*.

Intermittent Fan Operation A procedure for periodically operating the air moving device (AMD) for a drying or aeration system. In drying some products, particularly those with a short or non-existent constant rate period of drying, the air flow can quickly carry away the moisture on the surface of the product, and without changing the air flow, the air carries less and less moisture with the passage of time. By stopping the fan for a time the moisture distributes more uniformly in the product, after which the fan is again operated. For aeration systems, the fan may be stopped when the outside humidity is high or when the cold grain would cause condensation with the flow of ambient air. When these conditions cease the fan is operated again. Also see *Intermittent Drying; Tempering*.

Internal Diffusion The movement of moisture, either as a vapor or liquid, without change of state, through a material, usually a solid. The movement of moisture obeys the basic laws of diffusion.

Internal Energy That energy which a body contains which depends on the internal state of the system, represented by U. For an ideal gas, $\Delta U = m\ c_V\ \Delta T$. Internal energy is associated with the molecular level with translocational energy of molecules and energy levels within the molecules. U is represented by thermodynamic rather than mechanical parameters. Also see *Enthalpy; Free Energy; Gibb Function; Helmholtz Function*.

International System of Units See *SI Units*.

Interstitial The space between products, cells, etc., through which moisture or air may pass.

Intrinsic Energy Same as *Internal Energy*.

Invisible Loss See *Invisible Shrinkage*.

IR See *Infrared*.

Invisible Shrinkage That loss or reduction in weight which occurs in excess of the water removed, which varies from 0.1 to 1.0% of the weight. Commercial concerns use values from 0.1 to 0.8% loss to represent invisible shrinkage. Also called invisible loss. These losses occur as a result of respiration, handling, product particles, dust, etc.

Ionizing Radiation The use of subatomic particles in which beta or gamma rays are used to treat a product so that chemical change through ionization of microorganisms or of the substance occurs to enhance keeping quality. If vital organs of the organism are ionized the capacity to grow, reproduce, or live is affected. Various levels of treatment provide different effects or results (Table I-3).

Table I-3
Approximate Effect of Different Energy Levels of Ionizing Radiation

Product	Energy, rad	Effect
Enzymes	10^7	to destroy
Virus	$2\text{-}3 \times 10^6$	to destroy
Spore-forming bacteria	10^6	to sterilize
Non-spore-forming bacteria	10^5	to pasteurize
Insects	$2\text{-}5 \times 10^4$	to kill
Insects	10^4	to sterilize
Plants	10^3	to induce mutations
Man	5×10^2	to kill

Rad is a minute quantity of energy of 100 ergs per gram of irradiated material.

Irreversible Thermodynamics That branch of thermodynamics which deals with non-equilibrium processes. All real thermodynamic processes are irreversible due to friction, heat transfer, unrestrained expansion, and mixing. However, most processes can be approximated by a series of equilibrium states and studied with the equations of classical thermodynamics. Irreversible thermodynamics deals with processes which cannot be approximated by equilibrium states.

Isentropic A thermodynamic system in which the entropy remains constant in a reversible adiabatic process. Also see *Polytropic*.

Isentropic Compressibility See *Compressibility*.

Isobar Refers to conditions of equal barometric pressure. Isobars may be shown by lines representing points of equal barometric pressure, such as on a weather map.

Isobaric A constant pressure process.

Isochore A plot or graph representing the state of a system of two variables, such as pressure and temperature, when the volume remains constant. Also see *Isometric*.

Isometric A constant volume process. Also see *Isochore*.

Isopieste A curve or line representing equal moisture contents, or other appropriate parameters, plotted for a constant pressure.

Isostere An equation or curve representing moisture content relationships plotted for a constant moisture content. Also see *Isopieste; Isotherm*.

Isotonic The exertion of the same osmotic pull by one medium as the other medium on the opposite side of a semi-permeable membrane. Thus, the state where the same concentration of particles exists on both sides of the semi-permeable membrane between two fluids.

Isotherm The relation between the moisture content of a substance and the water vapor (relative humidity) in the air surrounding the substance, at constant temperature, expressed as an equation or plot. Also see *Adsorption Isotherm; Freundlich Isotherm; Harkins-Jura; Henderson Equation; Sigmoid; Smith Equation*.

Isothermal A constant temperature process. Isothermal lines connect points at the same temperature.

Isothermal Compressibility See *Compressibility*.

Isotropic The condition in which the same properties exist in all directions from a point or location.

i-x Diagram A form of the Mollier diagram in which the axes are the enthalpy and the absolute humidity of the air with other lines or curves representing the pressure and temperature. The i-x axes are not necessarily at right angles. In German literature, i, enthalpy, kcal/kg, is plotted on one axis versus x, absolute humidity, on the other axis. Also see *T-s Diagram*.

J

Janssen Equation A relationship which represents the lateral pressure of material in deep bins or silos (Fig. J-1):

$$L = \frac{\rho R}{\mu}\left(1 - e^{\frac{-K\mu h}{R}}\right)$$

where

L = lateral pressure of grain, lb/ft^2
ρ = weight of grain, lb/ft^3 (40 to 50)
μ = coefficient of friction of grain on walls (0.30 to 0.48)
R = hydraulic radius of bin
h = distance from top of product, ft
K = ratio of lateral to vertical pressure, L/Y (commonly 0.60 for grain)
Y = vertical pressure of the grain, lb/ft^2

Also see *Hydraulic Radius; Lateral Pressure; Rankine Equation.*

Jet Dryer An apparatus for moisture removal in which high velocity air is impinged on the product. For drying veneer, microwave heating to 93.3 to 148.9°C (200 to 300°F) using air at 15.3 to 22.6 m/s (3000 to 8000 fpm), with 0.12 to 0.25 (1/20 to 1/10 in) veneer, the drying occurs approximately twice as rapidly in a jet dryer as in the conventional dryer. A yellow popular 0.25 cm (1/10 in) thick material veneer dries in 15 to 20 min in a conventional dryer and 7 min in a jet dryer.

Joule The amount of work done in 1 sec by a current of 1 amp flowing through a resistance of 1 ohm. One joule equals 1 x 10^7 ergs or 0.7375 ft lb or 0.7390 gram calories. Also see *Erg.*

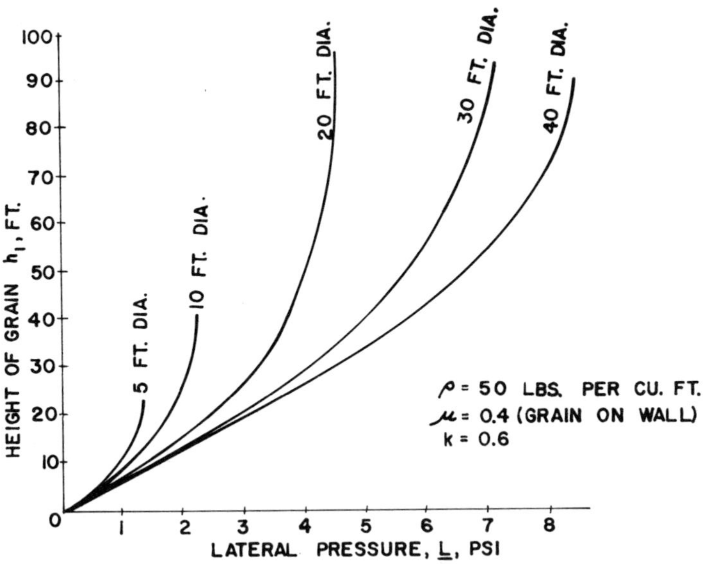

Figure J-1. Janssen lateral pressure relationships.

Joule Law An equation which gives the heat generated in calories, H, when a current passes through a conductor:

$$H = I^2Rt$$

where

I = electric current, amperes
R = resistance, ohms
t = time, sec.

Joule-Thomson Effect The cooling which results when a highly compressed gas is permitted to expand adiabatically such that no work is done.

Jute Drying Jute is a textile fiber, made from the stems of a woody plant, used for making burlap and twine. The stems are soaked, the fibers retted, and separated from the other tissues. After mechanical working and pounding, the jute fibers are dried.

Juice See *Fruit Juice Drying; Apple; Cherry; Grape; Strawberry.*

K

k A letter commonly used to represent the thermal conductivity of a material in heat transfer calculations or the constant for an ideal gas or mixture of gases, sometimes called adiabatic exponent, also represented by γ. In the thermal conduction equation, where q is Btu/hr, Cal/sec, watts (W), or J/s:

$$q = \frac{k}{x} A \, \Delta t$$

where

A = area
Δt = temperature difference
x = thickness

In the foot-pound-hour system, k is in units of Btu per hr per ft^2 per ft °F (sometimes expressed in terms of per inch of thickness instead of foot); in c-g-s system, Calories per sec per cm^2 per cm per °C; in the SI units, W/mK.

$$1 \; \frac{\text{Btu - ft}}{\text{hr ft}^2 \; {}^\circ\text{F}} = .004134 \; \frac{\text{Cal - cm}}{\text{sec cm}^2 \; {}^\circ\text{C}} = 1.731 \; \frac{\text{Wm}}{\text{m}^2\text{K}} \text{ or } 1.731 \; \frac{\text{W}}{\text{mK}}$$

For the ideal gas at any definite temperature, specific heat ratio, k = pV or $k = c_p/c_v$ where c_p is the specific heat at constant pressure and c_v the specific heat at constant volume. Also see *Boltzmann Constant; Thermal Conductivity.*

K, Kelvin An abbreviation and scale of absolute temperature on the Celsius scale for which 273°K is equivalent to 0°C and 32°F. Absolute zero is -273°C; and 373°K is equivalent to 100°C and 212°F. In SI nomenclature, K represents kelvin temperature without the degree symbol. Also see *Pressure Parameter; Saturation Temperature; SI Units; Temperature, Absolute.*

Karman See *Von Karman.*

KD A standard lumber abbreviation in common use in contracts and other documents involving transactions for purchase and sale of lumber, meaning kiln dried. Also see *AD.*

Kerosene A liquid used as a fuel which is obtained from petroleum in the fractional distillation of oil after the gasoline has been removed. It is a light oily liquid with a net heating value of approximately 43,270 kJ/kg (18,600 Btu/lb).

Kiln An oven which is heated for curing, baking, dehydrating, and/or drying products such as sand, tile, wood, brick, and some food products. The products may be treated in a batch or moved continuously through a stationary kiln or a rotating cylinder through which heated air is forced and the product is moved (Fig. K-1). Also see *Compartment Kiln; Oast (Ost); Tunnel Kiln.*

Kiln-Brown Stain A chemical stain that may develop during kiln drying of lumber as a result of chemical changes in the materials in the lumina of xylary cells. Also see *Yard-Brown Stain.*

Kiln-Dried Lumber Lumber dried to a specified moisture content, usually between 6 to 19% moisture content, using heated air under controlled conditions.

Kilning To process a material in a kiln.

Kilo Calorie See *Calorie.*

Kilogram A unit of weight in the metric system equal to 1000 grams or 2.2046 lb.

Kilogram-Calorie See *Calorie.*

Kilometer A unit of measure of distance in the metric system which is 1000 meters or 0.62137 miles.

Kilowatt A unit of power with one kilowatt equal to one thousand watts or 10^{10} dyne-cm/sec; 746 watts is equal to 1 horsepower.

Kilowatt-Hour Represented by kwh, kwhr, kwHr, or kWh, a unit of energy consisting of power used over a period of time in hours used in commercial applications; 1000 watt-hr is equal to 1 kilowatt-hour. 1 kWh = 1.340 hp.

Kindling Temperature The temperature at which rapid oxidation and burning of a material occurs. Also see *Combustion; Ignition.*

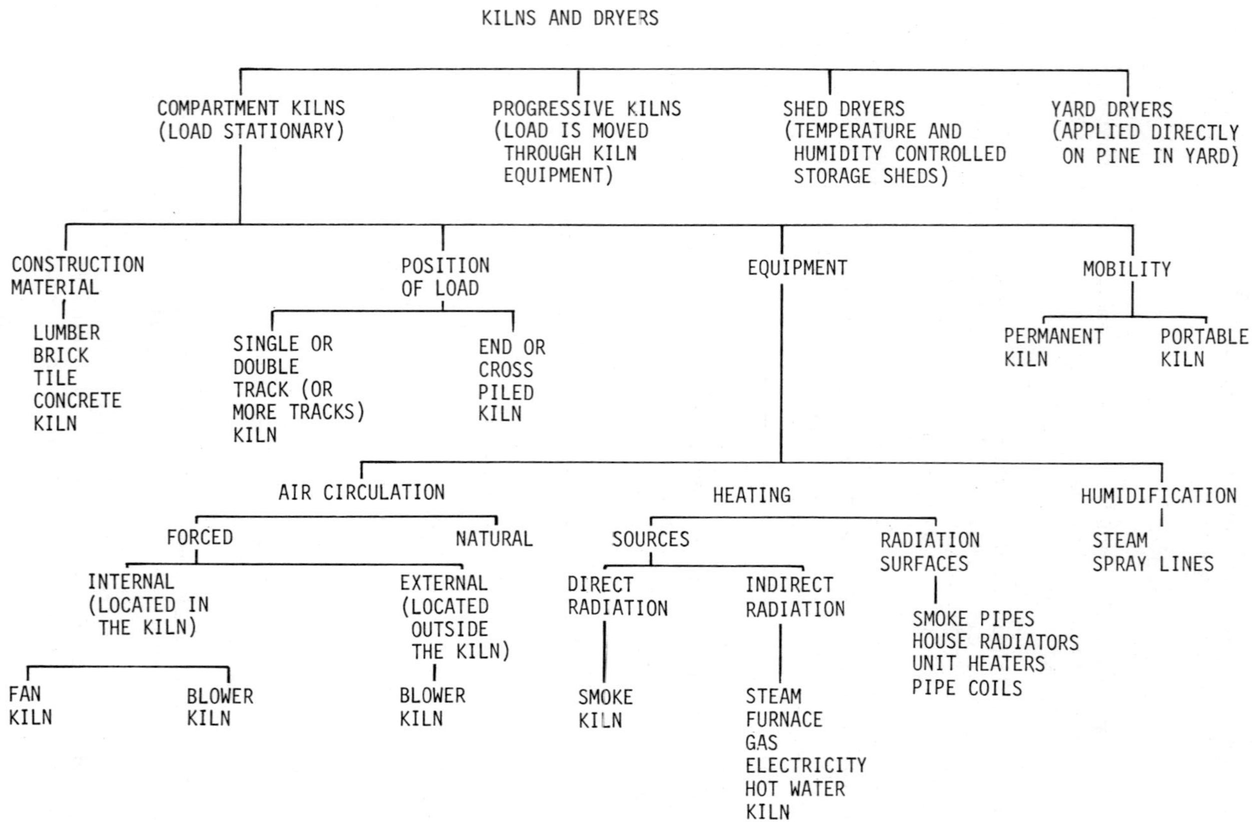

Figure K-1. Kilns, classification.
From Kollman, F. P., and Cote, W. A., Principles of Wood Science (1968). Used with permission of Springer-Verlag, New York.

Kinematic Viscosity The ratio of the absolute viscosity to the density of a fluid in the English system of units. The ratio of absolute viscosity to specific gravity or density in the metric system. English units are ft^2/sec, c-g-s unit is the stoke, cm^2/sec, and the SI units are m^2/s. Also see *Absolute Viscosity; Conversion Factors; Viscosity.*

Kinetic Energy The energy possessed by a body by virtue of its motion, K.E. = $1/2\ mv^2$, where m is the mass of the body and v the velocity. Units are in $m\ell^2 t^{-2}$. Also see *Energy.*

kWh, kwh, KwHr See *Kilowatt-Hour.*

Kirpichev Numbers Dimensionless numbers represented by Ki_h and Ki_m, which are used for heat and mass transfer used to relate drying conditions, defined respectively as

$$Ki_h = \frac{q^* L}{k \Delta T}$$

where

q^* = heat flux density
L = value of coordinates on body surface
k = thermal conductivity
ΔT = temperature difference, $T_c - T_o$

$$Ki_m = \frac{GL}{D\rho n}$$

where

G = mass velocity
L = value of coordinates on body surface
D = diffusivity
ρ = density
n = specific mass content

Also see *General Equation.*

Knudsen Diffusion; Knudsen Flow At low pressure for a gas or vapor the mode of motion in which the individual molecular free path motion is the principal mechanism, as exemplified by flow of a gas through a long tube where the mean free path is greater than the radius of the tube. Also applies to movement of gas through a small hole where the mean free path in the gas is at least ten times the diameter of the hole. The relationship is particularly applicable to freeze drying applications. Knudsen flow is represented by:

$$Q = \frac{(p_1^2 - p_2^2)\ \pi\ a^4}{16\eta\ell} \cdot \frac{M}{RT}\ \text{g/sec}$$

where

p_1, p_2 = inlet and outlet pressure
M = molecular weight
a = radius of tube
ℓ = length of tube
η = dynamic viscosity

Also see *Poiseville Equation*.

Kossovich Number A dimensionless number, Ko, equal to the ratio of the heat required for evaporating the moisture in a material to the heat required for heating the material from 0° to T_c°. Ko = $L\bar{u}/c\ T_c$, where

L = latent heat of vaporization
$\bar{u}$ = moisture, decimal
c = specific heat of moist material
T_c = temperature difference, $T_c - T_o$

L

Lacerator A machine for breaking or shredding forage plants at or slightly after the time of harvest in the field to hasten drying. Flails are commonly used.

Laminar Flow A type of flow of a fluid in which the velocity at every point does not change with time. The Reynolds number is generally less than 2000 for laminar flow. Laminar flow is also called streamline flow or viscous flow. Also see *Critical Velocity; Reynolds Number; Turbulent Flow.*

Lanolin A natural oil in wool which can be removed during a washing process, followed by drying. The lanolin suspended in water is passed through a centrifuge. The concentrated lanolin is then neutralized, bleached, washed, and run through a smaller centrifuge to remove the water. The lanolin-water mixture is dried to 0.25% water content in a dryer. Lanolin is used in pharmaceuticals and cosmetics.

Latent Heat The quantity of heat required to change a substance from a solid to a liquid at constant temperature is the latent heat of fusion. To change a substance from a liquid to a vapor at constant temperature requires the addition of heat, called the latent heat of vaporization. The reverse process is accompanied by the release of the same quantity of heat (Table L-1).

Lateral Ducts Tubes, channels, or pipes extending, usually horizontally, from a main duct located at one side or center of a drying chamber, which distribute air through the product being dried under a pressure system, or collect air from the product for a "suction" system. The spacing and size of the lateral ducts depends on quantity of air flow, product, and depth of product. The cross-section of the duct is selected to provide a velocity not in excess of 5.8 to 7.6 m/s (1000 to 1500 ft/min). Openings are provided so that the air velocity from the duct is about 0.25 m/s (50 ft/min). Also see *Duct; Grain Drying; Hay Drying*.

Table L-1
Latent Heat of Water

For Water	Metric Units	English Units	SI Units
Latent heat of fusion	79.1 cal/g (at 0°C)	144 Btu/lb (at 32°F)	334.9 kJ/kg (at 0°C)
Latent heat of vaporization (or evaporation)	539.55 cal/g (at 100°C)	970.4 Btu/lb (at 212°F)	2257.0 kJ/kg (at 100°C)

Lateral Pressure in a Bin Products in a container exert a sideways or lateral pressure which the container must be designed to withstand. Products in a shallow bin under 6.25 m (20 ft) depth exert a lateral pressure calculated from the Rankine equation. Products in a deep storage exert a lateral pressure that can be estimated from the Janssen equation. Also see *Janssen Equation; Rankine Equation*.

Layer of Charge The quantity of product placed on the bed or at one loading of a storage or dryer. An additional charge or charges may be placed over another partially dried or previously dried layer.

Leaf That part of a plant which has a large surface area, receives energy from the sun, builds up sugar for the plant, and plays an important role in the photosynthesis and respiration of the plant. The surface of the leaf is covered with a waxy cuticle which hinders evaporation. Stomata are provided through which air may enter, moisture may flow, and carbon dioxide or oxygen are given off. Leaves may be crushed to assist moisture movement for drying. During harvesting and drying special efforts are made to avoid leaf loss. Leaf loss increases as the moisture content decreases. Thus, harvesting wet, followed by drying, reduces losses.

Least Squares A statistical technique for fitting an equation to a straight-line relationship, or an assumed straight-line relationship, in which the distance of data points from the straight line is minimized. Also see *Standard Error*.

Leather Drying Leather is raw hide converted by tanning, with or without the hair, which involves moisture removal. Tanning is preceded by soaking the hide in soft water for 24 hours followed by drying. Leather drying is done in open sheds, closed rooms, or in drying tunnels with humidity and temperature controls, with the first stage of drying using a low temperature and high humidity to avoid too rapid drying. Leather may be pasted on glass plates and dried to give a smooth surface. Also see *Hide or Skin Drying; Tanning.*

Leather Hard Moisture Content The moisture content of clay at which the major shrinkage as a result of drying ceases, usually at 9 to 10%, dry basis. Also see *Clay Drying.*

Lebedov Number A dimensionless group, Le, involved in drying porous products relating the molar expansion flux to the molar vapor transfer flux:

$$Le = \frac{e\ b_T\ (t_s - t_o)}{c_b\ p\ \rho_s}$$

where

e = voids
b_T = vapor expansion in body capillaries
t_s = surrounding medium temperature
t_o = initial temperature
c_b = specific vapor capacity
p = pressure
ρ_s = density of solid

Leek Drying Leek, a plant of sturdy stem and large drooping leaves, is used primarily for soups. The leek plant is diced and dried in a continuous dryer on a belt, at 4.88 kg/m^2 (1 lb/ft^2) or less and sulfited. Temperatures of air from 82 to 54°C (180 to 130°F) are used as the leek moves through the dryer.

Lewis Number A dimensionless number, Le, which is the ratio of the thermal and mass diffusivity of a material. In the Russian literature the above relationship is called the Lykov-Lewis number.

$$Le = \frac{\alpha}{D}$$

where

α = thermal diffusivity, $k/c\rho$
D = mass diffusivity, or coefficient of diffusivity, L^2/T

Also see *Lykov Number.*

Lignin An organic substance found in the woody parts of plants, associated with cellulose. Lignin cells are usually easy to dry because plants with lignified cells are generally dead.

Limit In instrumentation and control, a value beyond which a controlled variable is not permitted to go. A limit switch is used to control a variable, such as maximum temperature in a dryer, maximum pressure in a boiler, or minimum-maximum relative humidity for an aeration system. In mathematics, the value which a function approaches, but does not exceed. As an example:

$$\lim_{x \to a} f(x) \to k$$

If f(x) is 1/x and a is infinity, k is zero. Thus, the limit of f(x) approaches zero as a approaches infinity.

Limit of Inflammability The high temperature below which and the low temperature above which (borderline values) a mixture of combustible gas and air sustain combustion for a particular fuel or product (Table L-2). Also see *Flammability*.

Table L-2
Limit of Inflammability, Approximate Values at Atmospheric Pressure

Fuel	Volume of Gas in Air, %, Range Lower - Upper
Hydrogen, H_2	6.2 - 71.4
Carbon monoxide, CO	16.3 - 71.2
Methane, CH_4	5.8 - 13.3
Ethane, C_2H_6	3.3 - 10.6
Ethylene, C_2H_4	3.4 - 14.1
Propane, C_3H_8	2.4 - 7.3
Butane, C_4H_{10}	1.9 - 6.5

Limit Switch A switch which is operated to control electricity to an operating unit to provide limits or boundary conditions on the unit, such as a switch which will turn off the burner for a heater if the exhaust system is not functioning, or will close the fuel valve if ignition to the burner does not occur. A malfunctioning of one unit limits the action of another unit in the system.

Line Pressure The pressure of the fuel in the supply line to the dryer.

Liquid A state of matter between a gas and a solid in which the molecules are free to move with respect to each other while being held together by cohesive forces such that the liquid maintains a fixed volume. By contrast, in a gas the molecules move to uniformly occupy the volume of the container and in a solid the molecules maintain a fixed position.

Liquid Diffusion Diffusion of liquid in capillary-porous solids at high moisture contents such as for clay, flour, wood, paper, textiles. Also applies to the diffusion of liquid in solids which are soluble in the liquid, such as soap, gelatin, and glue.

Liter (or Litre) A unit of volumetric measure in the metric system of 1 kilogram of water at 4°C (39.2°F) and 760 mm pressure. One liter has a volume of 0.001 m^3, 1.000028 cubic decimeter (dm^3), 61.026 in^3 or 1.0567 liquid quarts. Also see *Conversion Units; Gallons; Quarts*.

Logarithmic Average or Logarithmic Mean A single number representing two numbers as used in heat transfer, determined by dividing the difference of two numbers by the logarithm of the ratio of the two numbers. The logarithmic mean is used to represent the area of a cylinder for heat transfer equations,

$$A_m = \frac{A_o - A_i}{\ln(A_o/A_i)}$$

This relationship is generally used to represent the logarithmic mean area, A_m, if the outside area, A_o, is at least twice the inside area, A_i.

Long Ton See *Conversion Units; Ton*.

Long Tube Vertical Evaporator (LTV) A device for removing water in the form of a vapor from a liquid solution, in which the heat from the steam is transferred through vertical tubes of considerable length to the product being evaporated. The long tube vertical (LTV) evaporator may have either a falling film or rising film (Fig. L-1).

Loop Dryer An apparatus for drying fabric without tension in which the cloth passes over a roller in a chamber with warm circulating air. Also see *Festoon Dryer; Tenter Dryer*.

Loss The amount or portion of energy, force, power, pressure, or product lost during an operation or process such as drying, storage, handling, manufacturing, packaging, etc. Usually, an increase in loss causes a reduction of efficiency. Also see *Invisible Shrinkage*.

Louvered Column A vertical chamber, usually from 0.1 to 0.6 m (4 to 24 in) wide, in a dryer, with louvered openings for moving heated air for drying and unheated air for cooling under pressure (Fig. L-2). Also see *Column Dryer*.

Low Heating Value The energy output of the fuel used in a dryer, or other heating device, is expressed in heating values, Btu per unit weight or volume (Table L-3). The low heating value is usually more useful than the high heating value because the exhaust temperature of the heater is above the original temperature of the fuel, and the heat in the moisture in the products of combustion is not utilized. The low heating value, Btu/lb of fuel, based on the components is

$$\text{Low heating value} = 14{,}400\ C + 52{,}000\ (H - \frac{O}{8}) + 4{,}500\ S + 1{,}100\ w$$

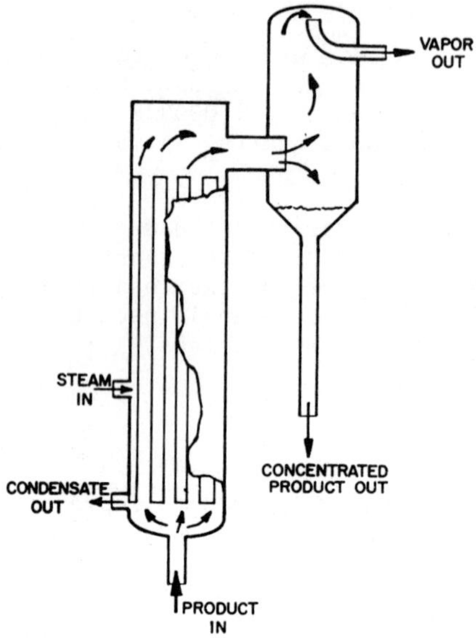

Figure L-1. Long tube vertical evaporator.
From Hall, C. W., and Hedrick, T. I., Drying of Milk and Milk Products, 2nd edition (1971). Used with permission of Avi Publishing Co., Westport, CT.

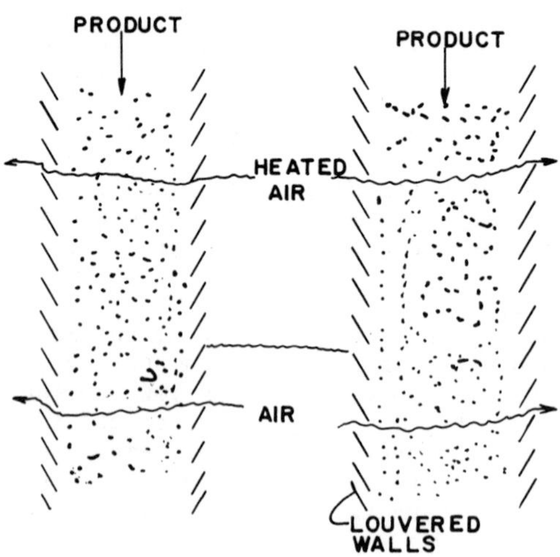

Figure L-2. Louvered column.

Table L-3
Low Heating Value of Various Fuels (Approximate)

	Density	Low Heating Btu/lb	Value kJ/kg
Coal			
Anthracite		12,800	29,772
Bituminous		13,500	31,400
Semibituminous		14,600	33,960
Manufactured briquets		13,050	30,355
Fuel Oil	lb/gal		
No. 1	6.79	18,690	43,470
No. 2	7.21	18,310	42,590
No. 4	7.78	17,740	41,260
No. 5	7.83	17,640	41,030
No. 6	8.05	17,180	39,960
Gasoline	6.15	18,910	43,960
Kerosene	6.82	18,660	43,400
Gas (C_nH_{2n+2})	lb/cu ft		
Natural (CH_4) methane	0.046	22,190	50,010
Ethane (C_2H_6)	0.085	20,416	47,480
Manufactured	0.048	10,580	24,610
Propane (C_3H_8)	0.12	19,944	46,390
Butane (C_4H_{10})	0.15	19,680	45,775
Wood, dry		8,000	18,600

where

C = the weight of carbon/lb of fuel
H = weight of hydrogen/lb of fuel
O = weight of oxygen/lb of fuel
S = weight of sulfur/lb of fuel
w = weight of moisture/lb of fuel

For conversion to SI units, 1 Btu/lb = 2.324 kJ/kg. The low heating value is also known as the net heating value. Also see *High Heating Value*.

Low Temperature Grain Drying The drying of grains at moisture contents not exceeding 25%, such as shelled corn and soybeans with a limited amount of heat ($\Delta T < 6°C$ or 10°F), the temperature increase of heated air less than 28°C (50°F) supplied by electricity or fossil fuel, in a storage not exceeding 4.6 to 6.1 m (15 to 20 ft) depth, circulating air through a perforated floor and grain with a fan with a flow of 0.005 to 0.015 m^3/m^2s (1 to 3 cfm/bu). Typically, a 10 hp motor-fan would be used with a 10,000 bu bin. With electric heat, a 10 to 15 kW heater would be used to heat the air providing a temperature rise of at least or approximately 27.8°C (50°F). From 1/3 to 1/2 kWh was used to remove each 1% (point) of moisture per bushel. The fan should normally be operated continuously. A humidistat may be used to control the fan, turning it off during periods of high humidity. Also see *Supplemental Heat.*

Low Temperature Green Crop Dryer See *Green Crop Dryer.*

LP Gas LP represents liquid petroleum, a mixture of gaseous petroleum products normally stored and transported as liquid under pressure. The principal constituents are propane and butane. LP gas has an ignition temperature of approximately 450°C (810°F) and a net heating value of 45,300 kJ/kg (19,500 Btu/lb). Also see *Natural Gas.*

LTV See *Long Tube Evaporator.*

Lukomskiy Number A dimensionless number, Lu, used in combined heat and mass transfer, which is the ratio of the thermal diffusivity to the mass transfer of the product, seldom used outside of USSR, sometimes written as reciprocal of the following:

$$Lu = \frac{\alpha}{a_m}$$

where

α = thermal diffusivity, L^2/T
a_m = potential conductivity of mass transfer, L^2/T

Also see *Lykov Number.*

Lumber Boards or timber cut from logs. Boards usually have a nominal thickness of 5 cm (2 in) or less in the USA. Timber is a piece of wood or lumber at least 12.7 cm (5 in) thick and 15.24 cm (6 in) wide which may have a square or rectangular cross-section. Intermittent steaming (2 to 4 hr) is used to prevent case hardening during kiln drying. Also see *Bark Drying; Dry Kiln Automation; Particle Board; Vapor Drying; Wood Drying.*

Lumber Drying The process whereby the moisture content of lumber is adjusted from a high to a low predetermined level by techniques which result in maximum product value. Kiln drying is most commonly accomplished by circulation of temperature and humidity adjusted air over the surfaces of the lumber which has been stacked in stickered piles. The critical economic problem related to drying is product degrading due to warping or structural failure such as internal or external checking or splitting. Degrade is the result of stresses which result from shrinkage during removal of bound water from the wood. Drying schedules are designed to reduce the severity of the effects of shrinkage stresses by inducing flow and other stress redistribution reactions in the wood. Variables involved in drying schedules include temperature, humidity, quantity and direction of air flow. Most kiln schedules are arbitrary. The most damaging degrade occurs during the early drying stages when surfaces begin to shrink but inner regions do not. It is common practice, for many lumber items, to gently air dry the lumber to approximately the fiber saturation point or below before subjecting to the more drastic conditions of a kiln. Ideally, lumber is kiln-dried to a moisture content at which it will shrink or swell the least when placed into the atmosphere at which it is to be used. In order to provide maximum dimensional stability, equalization and conditioning steps are applied to the kiln drying. These processes involve exposure of the lumber to a predetermined humidity atmosphere until all of the boards are at the desired moisture content throughout.

There is considerable variation in the moisture content of green wood (Table L-4). Finished grade lumber is usually kiln-dried to values of 6 to 12%. As compared to air-drying, the advantages of kiln-drying include: reduction of weight accompanied by lower shipping costs, shorter drying time, controlled lower moisture content, setting of resins in some woods, killing of microorganisms that might be in the wood, and less degrade under proper drying.

Conventional drying temperatures range from 49 to 82°C (120 to 180°F), usually increasing the temperature from beginning to end over several days (5 to 30). Steam heating, under appropriate circumstances, can reduce drying time to one-third that of conventional drying.

Table L-4
Lumber Heartwood Average Moisture Content

Species	Moisture, %, d.b.
Fir	30 - 100
Maple	55 - 70
Oak	60 - 85
Pine	30 - 100
Spruce	35 - 55

Source: USDA. 1955. Wood Handbook. Handbook No. 72. Forest Products Lab.

Lykov Number (Luikov Number) A dimensionless number, Lu, used in combined heat and mass transfer, representing ratio of mass diffusivity and thermal diffusivity,

$$Lu = \frac{D}{\alpha}$$

where

D = mass diffusivity, L^2/T
α = thermal diffusivity, k/cp, L^2/T

The symbol Lu is also used to represent the Lukomskiy Number in Russian literature translated to English. Also see *Lewis Number; Lukomskiy Number*.

Lyophilization The process of moisture removal in which the biological cultures and other biological materials are preserved by rapidly chilling to -78.9°C (-110°F) followed by drying in a vacuum, often with the addition of some heat to aid in vaporization of water. Also see *Freeze Drying*.

M

Macrocapillary Those capillaries which have radii greater than 10^{-5} cm and in which the radii exceed the mean free path of a molecule. In the macrocapillary the saturated vapor pressure is equal to the vapor pressure over a flat surface. Bodies with macrocapillaries do not absorb moisture from moist air. Such bodies will give up moisture in an atmosphere saturated with water vapor. Also see *Microcapillary; Mean Free Path.*

Maillard Reaction The deteriorating interaction of protein (amino acids) and reducing sugars (aldehyde or ketone), causing a brown color which may occur as a result of heating during drying of fruits. The browning may be retarded by sulfuring. The reaction may occur in dried milk and some fruits after extended storage. Also known as Browning reaction or non-enzymatic browning, although browning discoloration may be caused by several other phenomena. Also see *Sulfuring.*

Malt A food or flavoring product formed by germinating barley in a moist atmosphere. The product is then dried and cured. Curing is done by slowly increasing the temperature over several days. Malt is used in beer, food, and medical products. Kilning or drying of malt can be carried out in two steps: (1) drying the green malt from 45% moisture using air at 21.1 to 48.9°C (70 to 120°F); (2) heating after the malt is down to 10% moisture, using air at 48.9 to 65.6°C (120 to 150°F). Direct fired gas heated kilns are commonly used. The temperature and method of drying have a great influence on quality of product. Also see *Brewers Malt Drying.*

Manifold A pipe or chamber which directs air to a series of gas burners; a pipe from which air under pressure is directed into a series of aeration or drying bins. Also see *Mixing Chamber.*

Manometer An instrument used to measure the pressure of fluids. The difference in elevation of the liquid in a tube can be measured to determine the pressure developed on one end of the manometer. A convenient material to use for the liquid in the manometer is water, although liquids of lower density might be used for greater accuracy. When material of lower density is used the magnitude of change of liquid level is greater for a given pressure, thus providing an opportunity for greater accuracy in reading the value. The manometer is commonly used to measure the static pressure of air in various parts of the drying system (Fig. A-1). Also see *Draft Gage*.

Manure Drying See *Paunch Drying; Waste Drying*.

Marsh Gas See *Methane*.

Mass Transfer The movement of matter from one place to another, usually considered with reference to a defined boundary, as in the transfer of water within or from a wet product during drying. Mass transfer coefficients, k_ga_m, are measures of moisture flow during drying (Fig. M-1). Also see *Diffusion*.

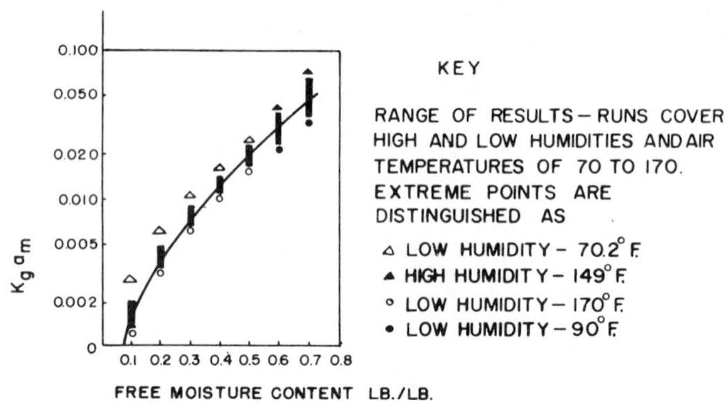

Figure M-1. Mass transfer coefficients for drying.

Mass Velocity A measure of the mass of a fluid passing through a unit cross-sectional area of a body in unit time.

Materials Categories for Drying The following categories of materials may be used to classify products for moisture removal. These categories may be helpful in identifying the method of drying to be used:

1. Crystals, usually > 1 mm
2. Porous solids

3. Granules
4. Pastes and sludges
5. Powders, usually < 200 microns
6. Slurries
7. Liquids, extracts, solutions, and emulsions

Maximum Average Concentration of Contaminants The maximum average concentration of contaminants to which workers may be exposed in an 8-hr day without injury to health. Examples in millions of particles per ft^3 of air are talc, 20; cement, 50; and dust, 50. Tolerance levels are adjusted as new information is obtained. Designers should obtain the latest requirements from EPA, OSHA, and other appropriate state and federal agencies.

Maximum Permissible Time See *Drying Time*.

Maximum Rate Drying In deep layer drying or bed drying, the maximum speed of moisture removal. Also see *Decreasing Rate of Drying; Hukill's Analysis*.

Mean Calorie See *Calorie*.

Mean Free Path The mean distance between successive collisions of molecules in a gas. Also see *Macrocapillary*.

Mean Moisture Ratio A method of representing the moisture ratio in a deep bed

$$\overline{MR} = \frac{1}{D} \ln\left[\frac{e^{D} + e^{\theta} - 1}{e^{\theta}}\right]$$

where D is a dimensionless depth variable

$$D = \left[\frac{h \, \rho \, L \, (M_o - M_e)}{Q \, c_a \, (T_o - T_e)}\right] Y$$

where θ is dimensionless time variable

$$\theta = ht$$

and h is the drying constant

$$h = b \, P_s^m V^n$$

ρ = product bulk density, kg/m^3
L = latent heat of moisture evaporation, kcal/kg
M_o = initial moisture content, decimal, dry basis
M_e = equilibrium moisture content, decimal, dry basis
Q = mass flow rate of drying air, $kg/hr\ m^2$
c_a = specific heat of dry air, kcal/kg°C
T_o = air temperature, averaged over time, °C

T_e = air temperature, outlet, equilibrium with grain, °C
t = drying time, hr
b,m,n = constants
P_s = vapor pressure at saturation, g/cm^2
V = superficial air velocity, based on empty containers, m/s

Mechanical Drying The removal of water or moisture from solids or semi-solids by mechanical means, e.g., filter press, rollers, and hydro-extractors, etc. Mechanical means, such as rollers, may be used to force moisture to the surface of the product, such as with forages and paper, which facilitates subsequent moisture removal. Also see *Filter; Hydro-Extractor; Press; Whizzer.*

Mechanical Efficiency of a Fan A measure of the power output of a fan in relation to the power input:

$$\text{Mechanical Efficiency} = \frac{(\text{air flow, cfm})(\text{total pressure, inches } H_2O)}{(6356)(\text{BHP})}$$

where BHP is the brake horsepower input at the shaft of the fan. Also see *Air Horsepower; Fan; Horsepower.*

Mesh A means of designating the number, not the size of openings in a screen or sieve, where the number of openings per linear inch is the mesh number. Ten openings per inch (100 openings per square inch) represents a mesh of ten. Standards have been established for wire size making up the screen sieve so that the area of openings is standardized.

Meter (or Metre) A unit of length measurement in the metric system. The basic unit of length for the SI units is defined as 1,553,164.13 wavelengths of the red line of cadmium in air at 760 mm pressure and 0°C. A yard is 3600/3937 meters. A meter is 39.37 in.

Methane A colorless gas consisting of carbon and hydrogen, CH_4, which burns with a pale flame, with a low heating value of 50,000 kJ/kg (21,500 Btu/lb). The weight relative to air is 0.555. Natural gas contains at least 80% methane. Marsh gas is predominantly methane.

Metric System A system of measurement based on the meter for length and gram or kilogram for weight. The meter and gram are subdivided or multiplied in decimal form. The time is in seconds. The metric system is represented by the CGS (or cgs), centimeter-gram-second, or MKS (or mks), meter-kilogram-second, designations. Recently the SI (International System) has been adopted by many groups. Also see *Conversion Factors; SI Units.*

Microbial Index The choice of microbial index depends on public health recommendations, product, and the use of product. All low-acid dehydrated products should be checked for sanitary quality. A low incidence of heat-resistance spore-forming bacteria is necessary if the dehydrated product is to be an ingredient in other foods. Checks for occurrence of mold spores and specific groups of bacteria must be made. Yeasts and molds are used as indexes of quality for dried fruits. Higher moisture products (not completely dehydrated) need to be protected from spoilage organisms by pasteurization, fumigation, or other chemical environments.

Microcapillary Capillaries in a solid with a radius less than 10^{-5} cm in which the saturated vapor pressure of moisture depends on the curvature of the meniscus. The rules governing transfer are molecular in nature and molecules move under thermal motion. Neither laminar nor turbulent flow develops. Micropores in grain, such as corn, have a surface area of 26.3 km^2 (6.5 acres) for 0.1 kg dry grain. Also see *Macrocapillary*.

Microprocessor A subclass of digital minicomputers, or a microcomputer dedicated to a single task which is usually not reprogrammed; the unit has a large-scale integrator and provides a general-purpose data source for connection with external devices.

Microwave Drying The process of removing moisture from a product in which the product is heated with microwave energy. Microwave heating is with energy at a frequency in the range of 900 to 18,000 megacycles/sec (wavelengths in the range of 13 to 0.7 in). As compared to drying with electric resistance heating, a vacuum microwave dryer is about 50% more efficient. Microwave drying can be used for a wide range of products, from fine to large particles, and can be carried out at low temperatures. Also see *Dielectric Heating*.

Milk Drying Milk and milk products are dried to provide a product of less volume for handling and storage and a product which will keep longer. The most common dairy products dried are non-fat dry milk (NDM), whole milk, whey, cream, and buttermilk, plus several manufactured products such as cheese, ice cream mix, malted milk, and other beverages.

Drying is usually preceded by evaporation of 2:1 to 3:1 (40-48% T.S.). Drying is done in a spray dryer or on a drum dryer. The spray dryer is most commonly used for food products because the flavor of product is superior for reconstitution as compared to the drum-dried product. With the spray dryer there is generally less heat effect, better stability, and dispersibility of product. Foam spray drying produces a product of low density and improved dispersibility during reconstitution. Instantizing of dried milk products is of particular value for beverage products. The product is dried to 3 to 5%. Inlet air is heated to 121.1 to 260°C (250 to 500°F) which leaves the dryer at 79.4 to 98.9°C (175 to 210°F). The dry product should be cooled quickly to 32.2 to 43.3°C (90 to 110°F) (Fig. M-2).

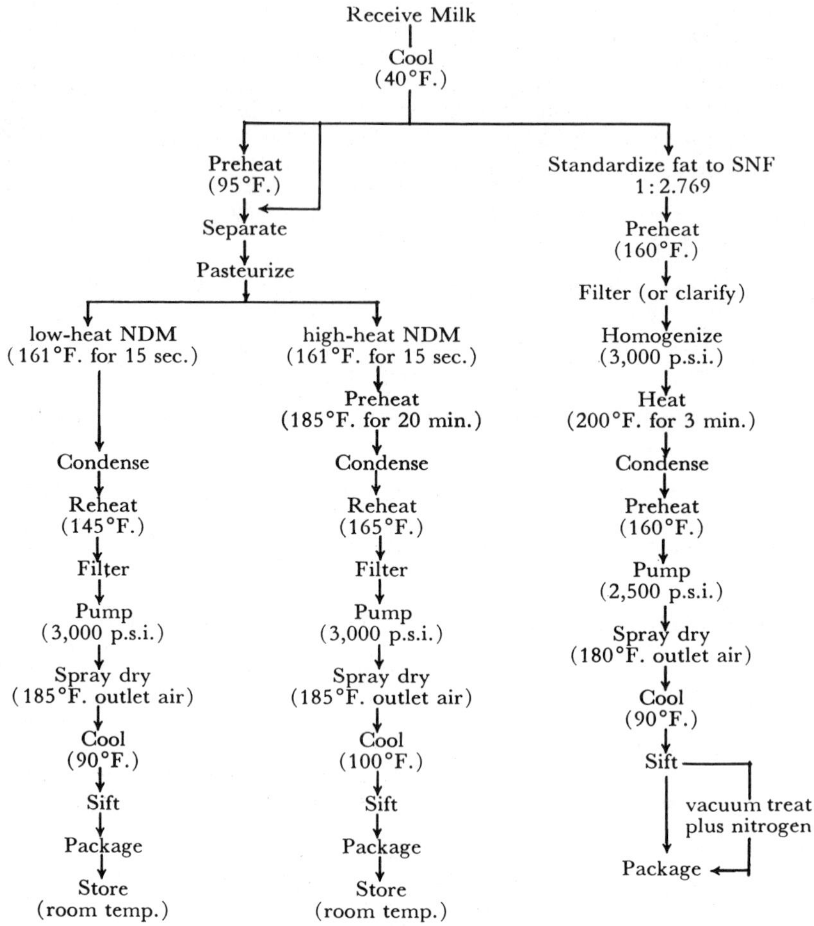

Figure M-2. Milk drying flow chart for the spray process. *From Hall, C. W., and Hedrick, T. I., Drying of Milk and Milk Products, 2nd edition (1971). Used with permission of Avi Publishing Co., Westport, CT.*

Dry whole milk can be made from whole milk with either a spray or drum dryer using a procedure similar to the manufacture of nonfat dry milk. The products produced, however, are generally not as acceptable for food products as those made with NDM by adding fat and homogenizing (Table M-1).

Elimination of oxygen and addition of nitrogen are used to improve stability of packaged product, particularly dry whole milk. Also see *Condensed Milk; Instantizing; Spray Drying*.

Table M-1
Approximate Moisture Content of Various Milk Products Involved in Condensing and Drying

	%		%
Skim milk	90.5	Whole milk	85.0
Condensed skim milk	71.0	Whey	93.2
Evaporated milk	73.0	Dried whole milk	2.0
Plain condensed milk	70.0	Non-fat dry milk	3.25
Condensed buttermilk	72.0	Dry buttermilk	3.9
Sweetened condensed skim milk	29.0	Dry whey	6.1
		Dry malted milk	3.3
Sweetened condensed whole milk	27.5	Sodium caseinate	4.0
Condensed whey	48.5	Sweetened condensed whey	28.5

Source: From Arbuckle, W. S., Ice Cream (1972). Used with permission of Avi Publishing Co., Westport, CT.

Mill Drying The process of adding heat to the air circuit of the grinding mill to reduce the moisture content of the product. The procedure is also known as attritor drying and grinder drying. Products may be dried before, during, or after milling. Wet coal, for example, is now primarily dried during the milling process. Hot air or flue gas may be introduced into the grinding circuit and the drying may be continued in the storage following pulverization. Coal may be dried 1 to 30% depending on the intended use, type of coal, and degree of pulverization. The grinder-dryer may also be used for chalk, limestone, clay, and filter cakes. Explosions and fire hazard are safety aspects which must be given special consideration in design and operation. Also see *Hammer Mill.*

Miniovich Number A dimensionless group, Mn, used in drying, relating pore size and porosity of product:

$$Mn = \frac{S\,R}{e}$$

where

R = pore radius
S = particle area/particle volume
e = porosity

Mixing Chamber A compartment preceding a burner in which gas and air are mixed to provide correct ratio for burning; a compartment preceding an aeration or ventilation system where gaseous chemicals and air are mixed, for insect control. Also see *Manifold.*

Mixing Cup Temperature See *Bulk Fluid Temperature.*

Modified Drum Dryer See *Doctor Blade; Drum Dryer.*

Modulate In a dryer, automatically governing the rate of fuel flow by a control which is temperature-sensitive in order to maintain a constant temperature at the location of the sensing device.

Moist The representation that a substance contains water. Also see *Moisture; Moisture Content.*

Moistless A substance which is dry or lacking in moisture.

Moisture Liquid, especially water, diffused or condensed in relatively small quantities, dispersed through a gas as a vapor or in a solid body as a liquid or solid or in pores of a solid as vapor; the water component of a body.

Moisture Content (a) Several classifications or descriptions of moisture content are used, such as bound moisture, unbound moisture, equilibrium moisture, free moisture, capillary moisture, gravitational moisture, and osmotically held moisture. (b) The relationship of weight of moisture in a product to the weight of product, expressed as a percent. Several terms are used to describe moisture content (Fig. M-3). Expressed as dry basis:

$$\text{M.C., \%d.b.} = \frac{\text{weight of water in product}}{\text{weight of dry matter}} \times 100\%$$

Expressed as a wet basis,

$$\text{M.C., \%w.b.} = \frac{\text{weight of water in product}}{\text{weight of dry matter + weight of water}} \times 100\%$$

The M.C., %w.b. and %d.b. are related as,

$$\text{M.C., \%d.b.} = \frac{\text{moisture content, \%w.b.}}{100 - \text{moisture content, \%w.b.}}$$

The wet basis of moisture content representation is used for the commercial trade for grain, hay, food, organic and inorganic aggregates. The dry basis of moisture content representation is used for research results, lumber, wool, and textiles. Also see *Dry Basis; Final Moisture Content; Hydrophilic Fiber; Incidental Moisture; Inherent Moisture; Moisture Regain.*

Moisture Content of Air See *Humidity.*

Moisture Content Representation See *Moisture Content.*

Moisture Determination Several standards have been established by various organizations for determining the moisture content of various products. For details on the methods consult publications of:

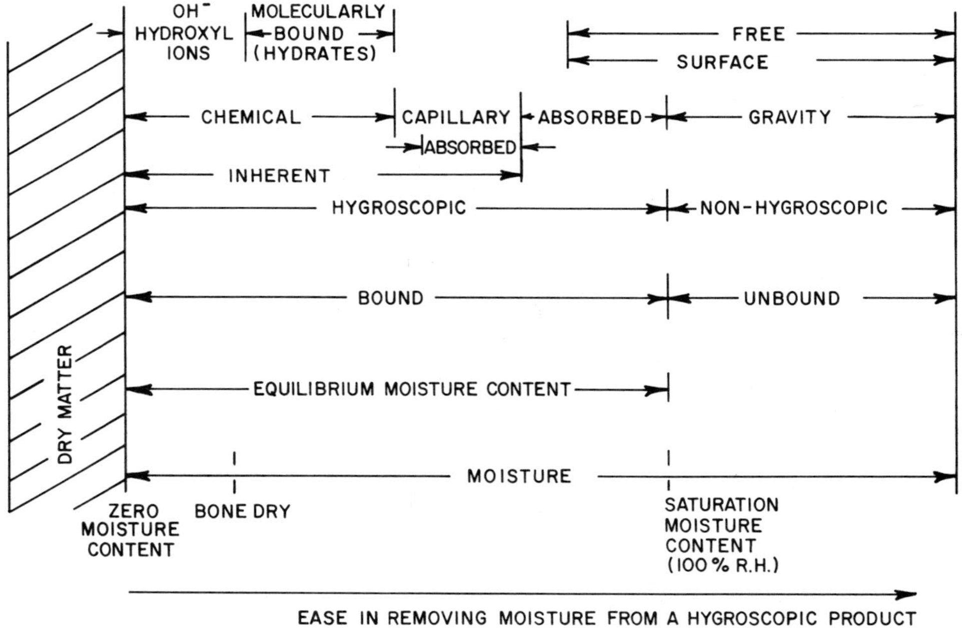

Figure M-3. Moisture content representation.

American Association of Cereal Chemists (AOAC)
American Institute of Chemical Engineers (AIChE)
American Society of Agricultural Engineers (ASAE)
American Society for Testing Materials (ASTM)

Also see *Dielectric; Electric Resistance; Exhaust Oven; Fischer Method; Moisture Meter; Neutron Scattering; Nuclear Resonance; Oven Method; Toluene.*

Moisture Determination, Wood For wood the moisture content is usually expressed on a dry weight basis, or d.b. Oven drying, distillation, titration, electrical properties, and hygroscopic methods are used to determine the moisture content. The most exact method for much research is the oven drying method, using a temperature of 100 to 103°C (212 to 215°F) with small samples of 15 to 22 grams until a constant weight is obtained, which for 20 grams is 5 to 20 hr for a single piece or 4 to 6 hr for chips.

Moisture Distribution The distribution of moisture throughout a wet material will be more uniform for slow drying or where the vapor pressure difference of the product and the surrounding atmosphere is small. For a product dried rapidly with heated air by conduction, the variation of moisture content throughout the product may be large (Fig. M-4). A parabolic initial moisture distribution may exist after the free water is removed. The vapor pressure of the moisture near the external edge of the product might increase upon heating, causing some moisture to flow toward the interior, temporarily increasing the interior moisture content. These changes may cause some products to change dimensions such that undesirable checking or cracking may develop. Also see *Intermittent Drying; Parabolic Initial Moisture.*

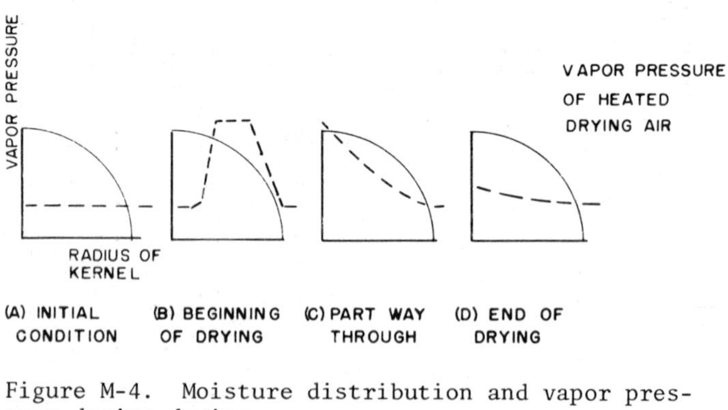

Figure M-4. Moisture distribution and vapor pressure during drying.

Moisture, Equilibrium See *Equilibrium Moisture.*

Moisture Equivalent The water content expressed as a percentage of the dry weight that soil can retain against a centrifugal force of 1000 x force of gravity.

Moisture Gradient The variation of moisture content in a solid material or drying bed at a given time in the drying process. The moisture gradient may be expressed in percent moisture per unit distance.

Moisture Meter An instrument or device for determining the moisture content, usually in percent, of a material, such as grain, hay, food, lumber, soil, tobacco, cotton, and flour. Also see *Capacitance; Dielectric; Direct; Electric Resistance; Indirect; Resistance*.

Moisture Migration The movement of moisture in stored products as a result of moisture and temperature changes. Moisture migration may occur with a product generally considered safe for storage. Moisture may accumulate on cold surfaces, usually at the top of stored products. The moisture content of the top layer may increase from 12 to 30% due to moisture migration. Greater moisture accumulation, and therefore greater losses, occur as a result of (1) a high moisture content at the beginning of storage of materials, (2) a tall structure, (3) a large difference in atmospheric and grain temperatures, and (4) when produce is placed in storage during warm weather followed by cold weather (Fig. M-5).

Moisture Ratio; Moisture Content Ratio A dimensionless ratio,

$$MR = \frac{M - M_e}{M_o - M_e}$$

where

M = moisture content at a particular time
M_e = equilibrium moisture content, usually on a dry basis
M_o = original moisture content, usually on a dry basis

MR or MCR is used in equations and in graphical representations, usually on the ordinate. Also see *Hukill's Analysis*.

Moisture Regain A term used in the textile industry to represent the percentage by weight of moisture in a textile material in relation to the dry weight.

$$\text{Moisture regain, \%} = \frac{\text{weight of moisture}}{\text{dry weight}} \times 100\%$$

Also see *Hydrophilic Fiber*.

Moisture Representation See *Moisture Content*.

Moisture Tension A numerical index of the energy with which water is held in soil expressed on a log scale, called pF.

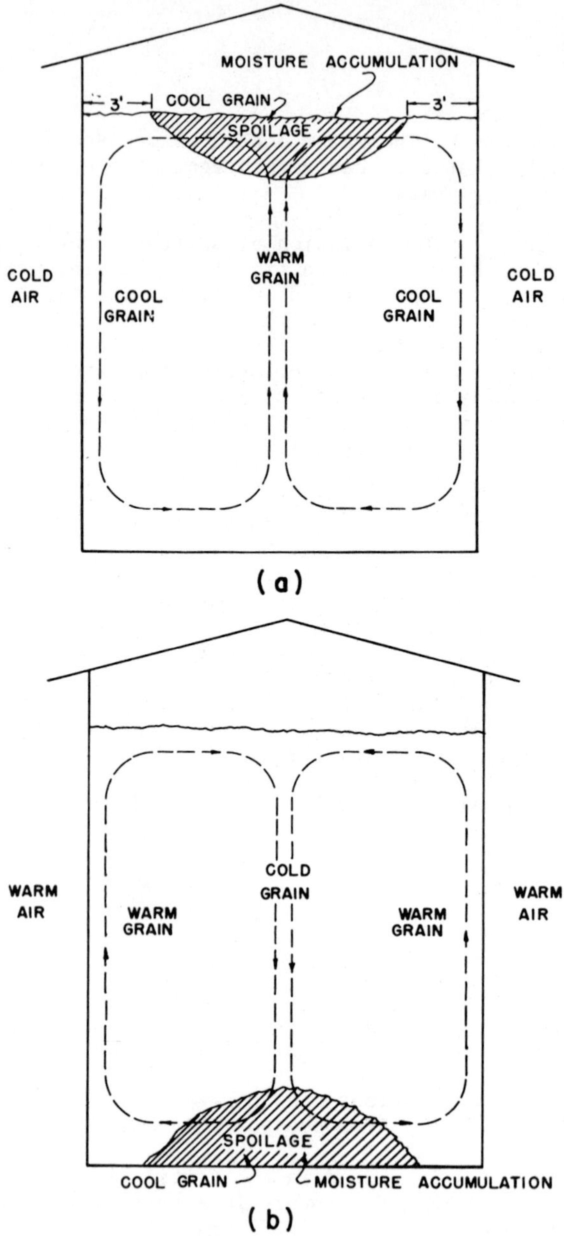

Figure M-5. Moisture migration and accumulation. (a) convection air currents with warm grain in bin with colder surrounding air. (b) convection air currents with cold grain in bin with warmer surrounding air.

Mold Growth See *Organic Acid*.

Molecular Diameter A diameter of molecules, usually given in angstrom units (Å) for molecules of elements and compounds. The molecular diameter is the same as the thickness of each molecular layer or for a monomolecular layer (Table M-2).

Table M-2
Molecular Diameter at 15°C, Å

Water	2.9
Carbon dioxide (CO_2)	4.59
Oxygen (O_2)	3.61
Air	3.72
Methane (CH_4)	4.14
Water vapor	4.6

1 cm = 10^8 Å

Molecular Diffusion See *Diffusion*.

Molecular Sieve Drying A method of moisture removal, usually from gases but also from organic liquids, using a class of adsorbents known as aluminosilicates or zeolite. Previously molecular sieves were made with naturally occurring zeolites. There is now an increasing use of synthetic crystalline zeolites. The adsorbent has a high porosity of uniform size of pores which have a diameter of 4 to 5 Å thus providing a large surface area. The adsorbents have a high adsorptivity at low water vapor pressures (Fig. M-6). The zeolite pellets used for drying are dried in a rotary dryer or tunnel dryer.

Mollier Chart The name applied to a method of representing graphically the thermodynamic properties of a pure substance with the enthalpy, h, on the ordinate and the entropy, s, on the abscissa. Other properties represented on the graph include pressure, superheat, moisture, and temperature. Also see *i-x Diagram; Psychrometric Chart*.

Motor Nameplate data should appear for each electric motor used on drying and aeration equipment. These data may be included with the compliance certificate data, or may be attached separately to each motor in a location clearly visible to the operator. Manufacturers whose equipment bears a compliance certificate shall provide printed instructions for installation and operation of units.

Mow Drying See *Barn Drying; Green Crop Drying; Hay Drying*.

Multiple Effect A type of evaporator used for a liquid where moisture is evaporated from more than one evaporator in series with another evaporator(s) (Fig. M-7). The moisture removed as vapor from the first evaporator (or effect) is used as steam to heat the second

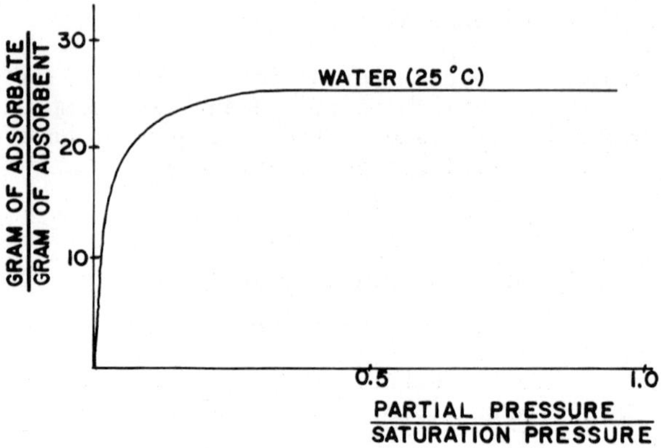

Figure M-6. Molecular sieve. *From Encyclopedia of Science and Technology, Vol. 8. Copyright 1977. Used with permission of McGraw-Hill Book Company.*

evaporator (effect) at a lower pressure (effect). As the number of effects is increased, the quantity of steam input per pound of water evaporated is decreased, with a greater cost for equipment (Table M-3).

Table M-3
Multiple Effect Evaporator, Approximate Quantity of Steam (Lb/Lb) Required to Vaporize 1 kg of Water, kg

	Range	Average
Single effect	1.33-1.00	1.17
Double effect	0.63-0.50	0.57
Triple effect	0.40-0.34	0.37
Quadruple effect	0.30-0.26	0.28
Quintuple effect	0.24-0.22	0.23
Sextuple effect	0.20-0.18	0.19
Septuple effect	0.18-0.16	0.17

Multiple Tube Dryer A type of dryer in which several tubes are placed in a circular configuration, mounted at an angle so the product will move by gravity from one end of the tubes, which all rotate as a unit, and in which the tubes are heated externally with hot air, which also circulates through the wet incoming product (Fig. M-8).

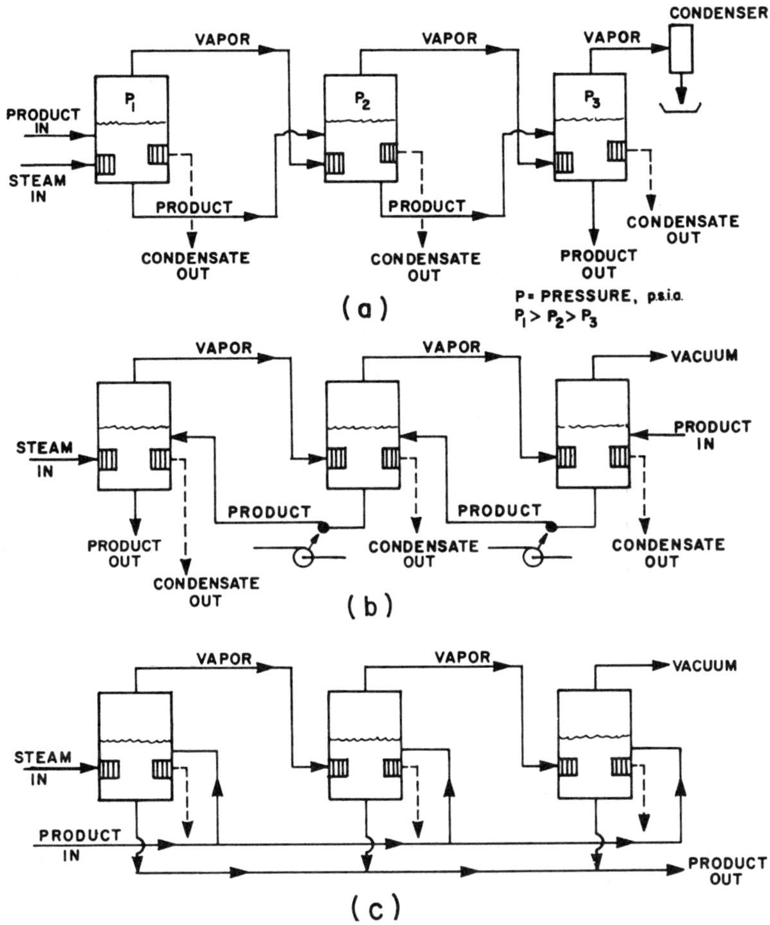

Figure M-7. Multiple effect evaporator. (a) forward feed. (b) backward feed. (c) parallel feed.
From Hall, C. W., and Hedrick, T. I., Drying of Milk and Milk Products, 2nd edition (1971). Used with permission of Avi Publishing Co., Westport, CT.

Multi-Stage Drying The procedure of removing moisture from products in more than one step. Accomplished by either moving the product from one drying chamber to another, perhaps with different drying conditions, or by changing the drying conditions of product being dried in a particular chamber. The purpose of multi-stage drying is to avoid damage to the product (such as cracking of rice) to maintain desirable product conditions during moisture transfer, and to provide more economical drying.

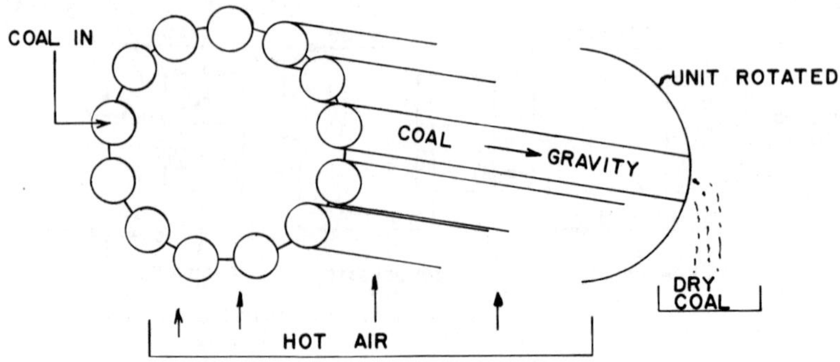

Figure M-8. Multiple tube dryer.

Multi-Stage Flash Operation Used for making salt from brine or in making fresh water, similar to multiple-effect forced-circulation evaporator. The liquid is flashed down from high to low temperatures, successively. The temperature difference available for heat transfer in a flash evaporator is:

$$\Delta t = \frac{\Delta T}{N \ln \left[\frac{1 - (\Delta T/2000) - (BR/\Delta T)}{1 - (\Delta T/2000) - (BR/\Delta T) - (R/N)}\right]}$$

where

ΔT = total flashing range, °F
N = number of flash stages
B = sum of average boiling point elevation and average short-circulating loss
R = steam economy desired in lb evaporation/1000 Btu extracted from the steam

Mushroom Drying The mushroom, a fungi devoid of chlorophyll, whole or cut into pieces, is blanched then dried. When mushrooms are air dried, an air temperature not in excess of 65.6°C (150°F) is used to dry to 5% moisture content. Freeze drying is often used in which mushrooms are dried to 3%.

Mycotoxin A general term applied to those chemical substances produced by some molds which are poisonous or produce toxic substances when eaten by man or animals. Mycotoxins remain in food after the mold that produced them has died. Grain that has become moldy might produce a mycotoxicosis causing an illness. Among the most prominent fungi which may cause mycotoxin are *Aspergillus flavus, Penicillium rubrum, Aspergillus ochracens, Fusarium tricinctum,* and *Fusarium graminearum.* Also see *Aflotoxin; Mold.*

N

Natural Convection   See *Convection*.

Natural Gas   A gaseous hydrocarbon, odorless and flammable, found in its natural state in particular geologic formation as a product of decomposition of organic matter. The composition is chiefly of the methane series with varying amounts of other components often present, such as carbon dioxide, hydrogen, and helium. Also see *LP Gas; Methane*.

Natural Ventilation   The use of wind or change in temperature of the air to obtain air movement to ventilate buildings or to remove moisture from grain. Natural ventilation can be accomplished by: (1) air movement through the storage, as with ear corncribs, (2) a down-draft or pressure ventilator, (3) an up-draft or suction ventilator, (4) an A-type frame or vertical duct which provides an air movement through the center of the bin or crib, and (5) tubes made of steel spring or screened channels placed horizontally in the bin. Also see *Cowls; Wind Pressure*.

Negative Radiation   See *Back Radiation*.

Negative Pressure   As applied to gage pressure scale when the pressure considered is below atmospheric, a differential in negative direction, more commonly designated as vacuum. At sea level, the lowest possible pressure is -760 mm, -14.7 psig, 0 psia, or 1 bar. Does not apply to the absolute pressure scale where the lowest possible pressure is 0 psia.

Net Heating Value   See *Low Heating Value*.

Neutron Scattering   Neutrons emitted from a radioactive source are directed into the product being tested to determine the moisture content. In passing through the material, the atomic nuclei of the material are bombarded by the neutrons. When a neutron collides with a heavy nuclei of a solid, less energy is transferred from the neutron than if it hits a light nuclei as that of the hydrogen in water. When more hydrogen nuclei are present in the material, a greater number of the slow neutrons will be scattered back to the vicinity of the source which can be measured by a sensitive radiation counter. An accuracy within 1% has been obtained in measuring moisture content of soils.

Newman Equation See *Parabolic Initial Moisture Distribution*.

Newton Law of Cooling An empirical law which states that the rate of cooling is proportional to the temperature difference between the cooling body and the surrounding medium. The relationship is applicable for small temperature differences.

$$t = t_e + (t_o - t_e)\, e^{-A\theta}$$

$$\frac{t - t_e}{t_o - t_e} = e^{-A\theta}$$

where

t = temperature at time, θ
t_e = temperature of air
t_o = temperature when $\theta = 0$
A = constant of cooling

The relationship is also used as a basis of moisture content change,

$$\frac{M - M_e}{M_o - M_e} = e^{-k\theta}$$

where k is a drying constant.

Nip The name given to the space above two drums or rolls of a double-drum dryer, which holds and feeds a slurry to the rolls for drying (Fig. N-1). Thus, a nip feed dryer. Also see *Atmospheric Double-Drum Dryer; Drum Dryer*.

Nonhygroscopic Product A product which cannot contain bound moisture, such as sand. The drying of nonhygroscopic products is a surface phenomena. Also see *Hygroscopic*.

Normal Atmosphere The ambient atmospheric conditions at 1 atmosphere pressure (760 mm), 0°C (32°F), and at a latitude of 45°. Also referred to as NTP, normal temperature and pressure. Also see *NTP*.

Normal Distribution A Gaussian distribution, normal curve, which

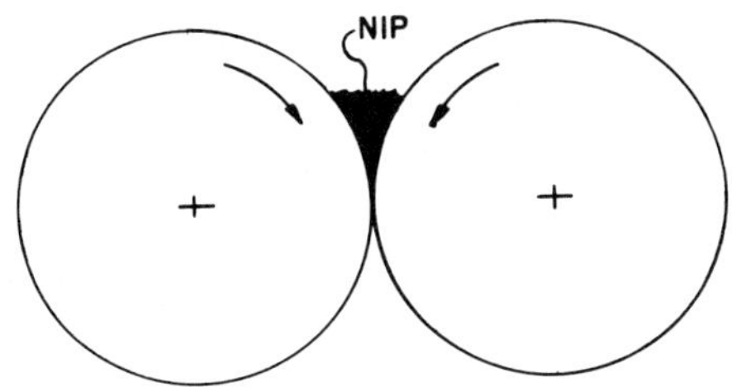

Figure N-1. Nip on a double-drum dryer.

is used in statistical analysis in which the data are symmetrical with a mean, μ, and standard deviation, σ, and probability density function, Pr(x):

$$Pr(x) = \frac{1}{\sigma\sqrt{2\pi}} \exp[-(x - \mu)^2/2\sigma^2]$$

The distribution closely approximates variables in the population.

Nozzle A device for converting pressure to velocity in a fluid. In spray drying systems an atomizing nozzle is used to produce uniform drops while creating a large surface area. Also see *Atomization*.

NTP or STP Normal temperature and pressure or standard temperature and pressure of a gas for standard conditions of calculations and comparisons, usually taken at 0°C (32°F) or 20°C (68°F) or sometimes 25°C (77°F), and a pressure of 1 atmosphere (760 mm).

Nuclear Resonance-Absorption A method for determining moisture content of products using a double electrical field: a magnetic field and a radio-frequency field. The hydrogen nuclei absorb energy from the radio-frequency field. The maximum absorption of energy occurs when resonance occurs which is shown by an increased flow of current through the coil producing the field. The more water present, the greater is the energy absorption from the field.

Number Prefixes Names placed before words to represent large to small quantities. The names representing large numbers, 10^6 and above, have as a symbol capital letters, such as G for giga. The names representing numbers of 10^3 and less are represented by small letters, such as k for kilo (Table N-1).

Nusselt Number A dimensionless parameter, Nu, of heat transfer which characterizes the relation between the heat transfer of the boundary layer of the fluid and thermal conductivity of the fluid:

$$Nu = \frac{h \cdot d}{k}$$

Table N-1
Number Prefixes for SI and Metric Units

Multiples and Submultiples	Prefixes	Symbols
10^{18}	exa*	E
10^{15}	peta	P
10^{12}	tera	T
10^{9}	giga**	G
10^{6}	mega	M
10^{3}	kilo	k
10^{2}	hecto	h
10	deka	da
10^{-1}	deci	d
10^{-2}	centi	c
10^{-3}	milli	m
10^{-6}	micro	μ
10^{-9}	nano	n
10^{-12}	pico***	p
10^{-15}	femto	f
10^{-18}	atto	a

*Also giga giga, GG
**Also bera, B
***Also micromicro, μμ

where

h = film coefficient
d = thickness of fluid film
k = thermal conductivity of the film

The number is used to relate data on free and forced convection from solids to fluids. Also see *Biot Number*.

Nut Drying Nuts dry from 30 to 40% on the tree to 5% for normal storage. Nuts will dry on the tree in appropriate weather conditions. Otherwise, the nuts may be dried in a bin with forced air. Drying is necessary to prevent molding, discoloration, and undesirable chemical changes of the nut meat, and to provide a product which will shell easily. Some nuts may crack if dried too rapidly (Table N-2).

Dryers may be used to dry nuts which have been rewet. Forced natural or slightly heated air can be used for drying rewet nuts. To avoid the undesirable effect of heated air, the air may be dehumidified over an evaporator plate before forcing over wet nut products.

The nut meats may be 6 to 10% moisture upon shelling. The moisture content must be reduced to the range of 3 to 5% to provide a product with safe-keeping quality. Forced heated air is used for drying, but the temperature must be kept below that which would

Table N-2
Nut Drying

Nut	Tree Moisture Content When Harvested, %	Nut Meat Desirable Moisture Content for Storage, %	Maximum Kernel Temperature for Drying
Pecans	30	4.5	48.9°C (120°F)
Pistachio	up to 45	4.0	35-43°C (95-100°F)
Walnuts, English	35	4.0	~ 62.8°C (145°F)
Hickory	10	5.0	Atmospheric

otherwise cause migration of the oil to the surface, cause off-flavors, or undesirable taste and appearance, usually below 48.9°C (120°F). Nut meats may be dried with an internal rotating drum dryer, continuous belt dryer, or tray dryer. Also see *Bin Dryer; Peanut Drying; Wagon Dryer; Tunnel Dryer.*

Nutritive Value of Dehydrated Food In drying, moisture is removed from a food product, which results in increasing the concentration of nutrients in the remaining mass. Proteins, fats, and carbohydrates are present in larger percentages in dried foods than in the fresh product from which the dry product is made (Table N-3).

Table N-3
Nutritive Content of Fresh and Dried Beef

Beef, Meat Components	Composition, %	
	Fresh	Dried
Protein	20	55
Fat	10	30
Carbohydrates	1	1
Moisture	68	10
Ash	1	4

O

Oast; Ost   A kiln used to dry products such as hops, malt, or tobacco.

Oats Drying   See *Cereal Drying*.

Oil, Non-Drying   An oil whose film does not become hard or gummy when exposed to air as, for example, olive oil.

Oil, Semi-Drying   Oil which absorbs some oxygen and becomes gummy, but not hard as, for example, cottonseed oil, when exposed to air.

Onion Drying   (a) The common practice at harvest has been to place onions in crates and cure them in the field before moving to storage. With mechanical harvesting, however, onions can be moved in bulk or in pallets to storages and cured in bins with forced ventilation. The onions may be topped in the field before storage or may be stored with the tops on. Excess moisture is removed from the skin of the onions and tops during the curing operation. The use of heated air above 35°C (95°F) has not been accomplished without excessive shrinkage and is not generally recommended. Most successful results have been obtained with heated air by limiting the temperature rise to 11°C (20°F), with an air flow of 0.01 to 0.015 m^3/m^2s (20 to 30 cfm/ft^2). The onions should be cooled after curing to about 10°C (50°F) by November 1 and 0 to 3.3°C (32 to 38°F) by December 1 in northern areas of the U.S. The optimum storage condition for onions is 0°C (32°F) without warming periods during storage. The dormancy of onions is easily broken if they are warmed after being cooled, and sprouting will result. Forced ventilation of .013 m^3/m^2s (15 cfm/ft^2) of floor area, assuming a depth of 3 m (10 ft), is satisfactory for cooling the onions and maintaining a uniform temperature throughout the mass. To avoid bruising, the onions should not be stored more than 3.75 m (12 ft) deep.

(b) Onions are prepared for tunnel drying by cutting into 1/6 in. slices. A two-stage tunnel dryer, with the first stage at 71 to 87.8°C (160 to 190°F) and the second stage at 54.4 to 60°C (130 to 140°F), is used to dry 90% product to 5 to 7% in 10 to 15 hr. Bin drying to 4% with 48.9°C (120°F) air follows the tunnel dryer. More rapid drying is accomplished on a continuous stainless steel perforated belt, using air forced through the product with drying completed in 6 hr. Dried onion is sold as sliced, chopped, or powdered material. Also see *Garlic Drying; Gladiola Drying*.

Optimum Moisture See *Final Moisture*.

Order of Magnitude A designation relating to the large difference between two numbers, with one number approximately ten times another. Thus, the difference between 15 and 140 to 160 is an order of magnitude. Also see *Scale-Up Dryer*.

Ordered See *Bright-Leaf Tobacco*.

Organic Acid for Grain Storage Certain organic acids, compounds made up of CHO, are used for treating high moisture grains to prevent mold growth and deterioration. The organic acids used separately or in combination are propionic (CH_3CH_2COOH), acetic (CH_3COOH), and formic (HCO_2H) acids. Organic acids are sprayed on the wet grains, reduce the pH to 4, kill the mold organisms, and destroy the germinating capacity of the embryo. Organic acids are being used to control mold growth in cut forage in bales, wafers, and pellets. Application rates vary from 0.5 to 2.5 kg of organic acid per 100 kg of product, depending on the moisture content, product, and chemical used.

Orifice An opening with a smaller diameter than the tube or passage for the purpose of metering or regulating flow of fluid (Fig. O-1). A manometer may be used to measure upstream and downstream pressure of fluid from an orifice from which the flow can be determined. The opening through which gas is admitted to the burner is called an orifice.

Orsat Apparatus An instrument used for determining the quantity by percent volume of oxygen, carbon dioxide, and carbon monoxide in a mixture of gases, usually in the exhaust of a combustion device, such as an engine, furnace, heater, boiler, or in a controlled atmosphere storage.

Osmosis The movement of a liquid from a weaker to a more concentrated solution through a semi-permeable membrane. The movement is in a direction as to equalize concentrations. An osmotic pressure of 5 to 25 atmospheres is commonly exerted by sap in cells of living plants.

Osmotically Bound Water Water which is held to a substance by the measure of entropy, in contrast to water bound by chemicals and molecular forces.

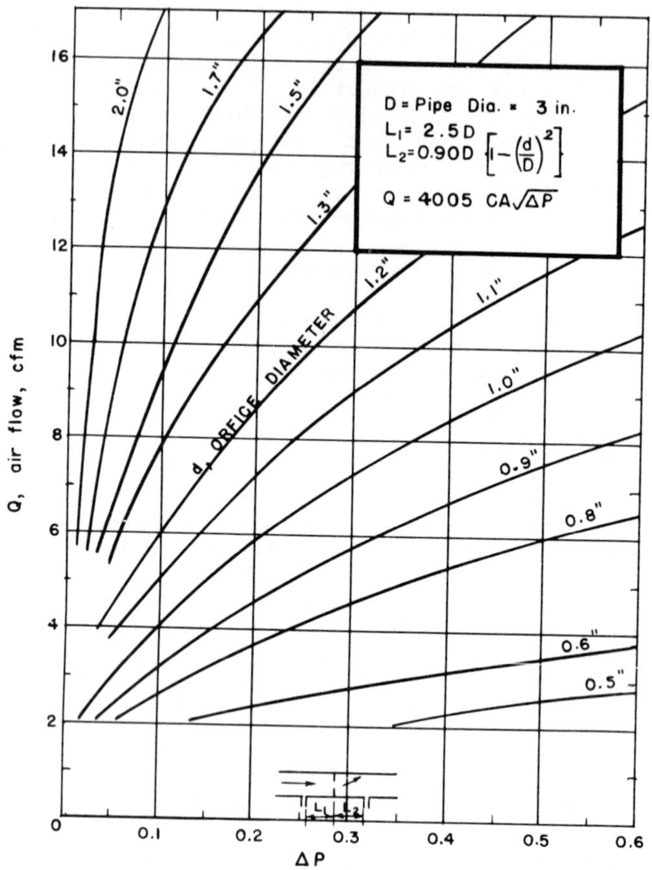

Figure O-1. Orifice, pressure drop through 3 inch diameter orifice for measuring air flow.

Ost See *Oast*.

Othmer Plot A plot of vapor pressure and equilibrium moisture content which can be used for determining latent heat of vaporization of products with low amounts of water:

$$\frac{L}{L'} = \frac{\log P_2 - \log P_1}{\log P_2' - \log P_1'}$$

where L and P represent the latent heat and vapor pressure of the product; L′ and P′ represent the latent heat and vapor pressure of free water (Fig. O-2). The ratio of the numerator to denominator is the ratio of latent heat of the product and free water. (L/L' plots as straight lines on logarithm paper.) If L/L' = 1.1, the latent heat of vaporization of water from the product is 1.1 x the latent heat of free water at the temperature. Also see *Clapeyron Equation*.

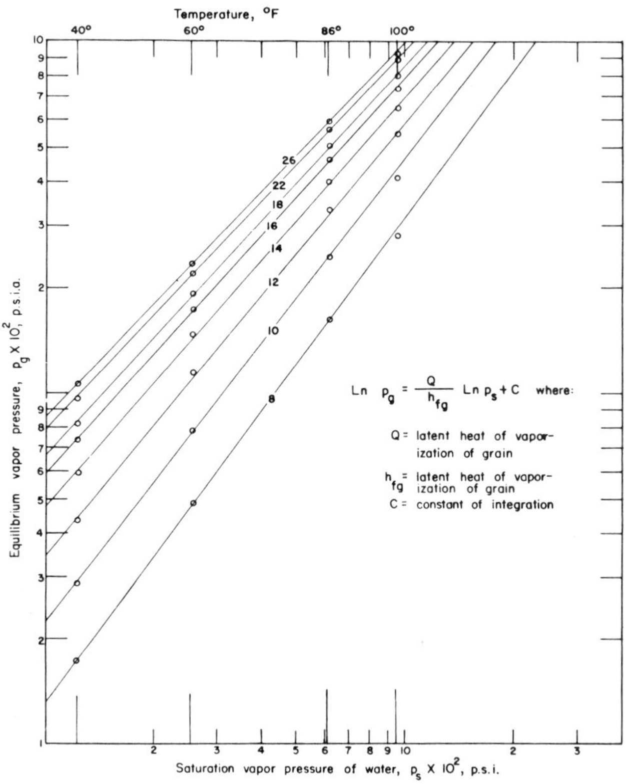

Figure O-2. Othmer plot.

Oven Method of Moisture Determination The usual oven methods for moisture determination of grain are:

(a) Air-Oven Method, 130°C ± 1°C (266°F). (1) One-stage, for grain under 13% moisture. Grind duplicate samples of 2 to 3 grams each. Heat 1 hr at 130°C (266°F). Place in desiccator; then weigh. Samples should check within 0.1% moisture. (2) Two-stage, for grain above 13% moisture. Remove moisture until a 25 to 30 gram sample is below 13% (usually about 14 to 16 hr are required). Continue as discussed under (1).

(b) Water-Oven or Air-Oven Method, 100°C (212°F). Place

duplicate samples of 15 grams or more for grain and 25 grams for forage in oven, heated to 100°C or higher, for a specified length of time as established by standards (Table O-1). Place in desiccator for cooling to room temperature; then weigh to the nearest 0.01 gram. Samples should check within 0.2% moisture. For products over 25% moisture, use 100 gram sample.

The AOAC has set up slightly different requirements which give a slightly higher value for moisture determination of grain and stock feeds. The material is ground and heated at 135°C (275°F) for 2 hr. The recommended procedure for determining moisture content of different products should be consulted. Specific procedures for moisture determination are given by ASAE Standard S352 for grains and seeds; ASAE Standard S353 for meat and meat products; ASAE Standard S358 for forages. Also see *Exhaust Oven; Vacuum Oven.*

Table O-1
Oven Temperature and Heating Period for Moisture Content Determinations

Seed	Oven Temperature, ± 1°C	Heating Time hr	min
Alfalfa	130	2	30
Barley	130	20	0
Beans, edible	103	72	0
Bluegrass	130	1	0
Bromegrass	130	0	50
Cabbage	130	4	0
Carrot	100	1	40
Clover	130	2	30
Corn	103	72	0
Fescue	130	3	0
Flax	103	4	0
Oats	130	22	0
Onion	130	0	50
Orchardgrass	130	1	0
Radish	130	1	10
Rye	130	16	0
Ryegrass	130	3	0
Safflower	130	3	0
Sorghum	130	18	0
Sunflower	130	1	0
Timothy	130	1	40
Turnip	130	4	0
Wheat	130	19	0

Source: ASAE Standard S-352, 1972.

Overall Heat Transfer Coefficient The drying during the constant rate period is supplied the heat represented by the heat transfer, q, Btu/hr (Table O-2) for U = values and various units (Table O-3).

$$q = UA\ (t_1 - t_2)$$

where

U = overall heat transfer coefficient
A = area of heat transfer
$t_1 - t_2$ = temperature difference between the body of the heated air and the body of the product being dried

Table O-2
Overall and Convection Heat Transfer Coefficient, U and h, Conversion Factors

1 Btu/hr ft^2 °F	= 0.1355 x 10^{-3} gcal/s cm^2 °C
	= 3.94 x 10^{-3} hp/ft^2 °F
	= 4.886 kcal/hr m^2 °C
	= 5.6783 x 10^{-3} kW/m^{2}K
	= 1 pcu/hr ft^2 °C
	= 0.5678 x 10^{-3} W/cm^2 K
	= 2.04 x 10^{-3} W/in^2 °F
	= 5.678 W/m^{2}K

Table O-3
Overall Heat Transfer Coefficients, U, Approximate Range of Values

General	Btu/hr ft^2 °F	W/m^{2}K
Freon condenser, water as a coolant	50-150	250- 800
Water-to-water heat exchanger	150-300	800-1650
Water-to-oil heat exchanger	20- 60	100- 350
Steam-to-water	200-700	1000-3900
Steam-to-gas	5- 50	25- 275
Types of Dryers		
Vacuum shelf	1	5- 6
Indirect rotary	2- 10	10- 55
Jacketed trough	2- 15	10- 85
Rotary vacuum	5- 50	25- 275
Agitated pan	5- 60	25- 325
Drum	200-300	1150-1700

Conversion factor: 1 Btu/hr ft^2 °F = 5.678 W/m^{2}K

Overall Shrinkage Ratio The number of pounds of fresh, untrimmed vegetable, as harvested, required to produce one pound of acceptable dry product at a specified moisture content. The overall yield ratio is the reciprocal of shrinkage ratio (Fig. O-3). Also see *Drying Ratio; Overall Yield; Yield.*

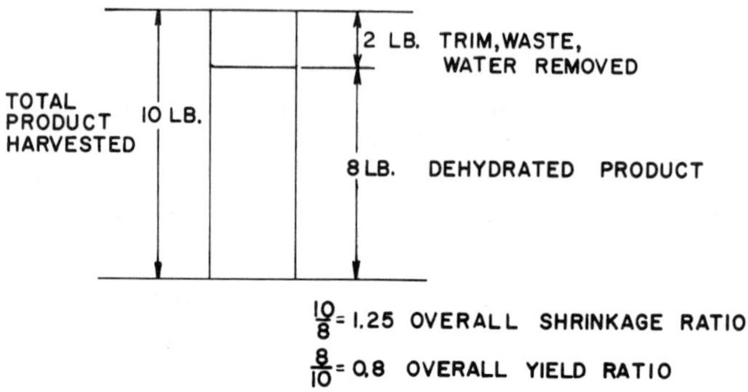

Figure O-3. Overall shrinkage ratio.

Overall Thermal Efficiency The overall thermal efficiency is the heat utilized for drying divided by heat input. The overall thermal efficiency of heated air dryers will be greater for: (1) direct-type heaters, (2) low air flows, (3) high initial moisture contents, (4) high atmospheric temperatures, (5) thick columns or deep beds of grain, and (6) counterflow dryers. The average overall thermal efficiencies for an oil-fired direct heater is 35%, for an oil-fired indirect heater is 30%, and for a coal-fired indirect heater is 20% (Table O-4). The overall thermal efficiency,

$$\text{percent} = \frac{(\text{water evaporated, lb})(\text{latent heat, Btu/lb})(100\%)}{(\text{fuel used, lb})(\text{net heating value of fuel, Btu/lb})}$$

The efficiency of units in a drying system is in Table O-5. Also see *Dryer Efficiency.*

Overall Yield The units of finished, acceptable product produced from 100 units (kg, lb, grain) of raw material, expressed in percent. The overall shrinkage ratio is 100 times the reciprocal of the overall yield. Also see *Overall Shrinkage Ratio; Yield.*

Overhead-Fired A direct fired dryer in which the oil or gas burner is installed above the drying tunnel usually done to save floor space.

Ozone An unstable gas, O_3, bluish in color with a distinctive odor, formed by an electric discharge in oxygen. Formed naturally in the atmosphere by lightning. Ozone is used for purification of air and water and as a bleaching agent. The use of ozone has been suggested for maintaining white color of seeds and for reducing growth of associated microorganisms by incorporating ozone in drying or aerating systems, such as for white pea beans.

Table O-4
Overall Thermal Efficiency of Units, Percent

Steam engine, thermal to mechanical	15
Internal combustion engine, gasoline	15-25
Diesel engine	27-34
Counterflow conveyor dryer	60-65
Rotary high temperature dryer	65-70
Boiler, steam	70-85
Heat exchanger, air to air	70-85
steam to fluid	70-80
water to water	80-90
Electric heating elements to liquids	over 90

Table O-5
Overall Thermal Efficiency, Percent, Approximate

Drying System	
Direct-fired heater	
Summer	60
Fall	50
Winter	36
Indirect-fired heater	
Summer	45
Fall	36
Winter	28
Heater-Burner for Drying	
Direct-fired heater	
Summer	85
Fall	70
Winter	50
Indirect-fired heater	
Summer	60
Fall	50
Winter	35

P

Packing   As materials are placed into a bin or storage the weight of material becomes compacted, and fines accumulate under conveyor discharges, which increases the static pressure drop for air flow. The increase in static pressure as a result of packing may be as much as 100%, depending on size, depth, moisture content, and fines. The density is related to the amount of packing. Also see *Compaction*.

Paint Drier   A chemical added to paint or other coatings to hasten drying. Some dryers are oxides of metals. The term may also apply to an apparatus for heat drying of paint, using electric or gas infrared heaters or heated air convection dryers. Also see *Driers, Paint; Oil, Semi-Drying*.

Pallet Drying   See *Potato Drying*.

Pan Dryer   See *Charge Pan Drying Machine; Vertical Agitated Batch*.

Parabolic Initial Moisture Distribution   In drying situations where drying has proceeded to the point that diffusion equations apply and the initial moisture content of the body is not uniform, the distribution of moisture through the product will approximate a parabolic function. The moisture content will be less near the surface than at the center and at various depths. In many cases the parabolic moisture will exist before drying starts with lower moisture content at the surface increasing to the center. Also see *Moisture Distribution*.

Parallel Flow Dryer   See *Concurrent Flow Dryer*.

Parsley Drying   A green plant which is used as an herb and is harvested like hay and often dried. Drying to 4 to 5% is done in 30 min in a continuous belt three-stage drier. After drying the leaves are separated from the stems by air. The leaves are merchandised as flakes or granules. The leaves and stems are ground for powder.

Partial Pressure In a mixture of vapors or gases, the pressure exerted by each gas component. According to Dalton's ideal gas law, the total pressure of a mixture of gases is the sum of the pressures that each gas exerts if present by itself.

Particleboard Furnish Drying Wood chips, flakes, curls, slivers, and shavings are dried to provide a desired uniform moisture content for manufacturing wood products such as particleboard and chipboard. Furnish may be dried on a screen or open belt which carries the product through a heated air chamber or tunnel. Much particle drying is done in tubes in which the particles are suspended in and transported by the heated drying medium. A more uniform product moisture content is secured in a pneumatic or fluidized bed dryer in which high velocity air moves over the furnish. A rotary dryer, in which the product is tumbled inside the rotating cylinder and heated air is blown through the falling product, provides a more uniform moisture content than when the product is on a belt, tray, or screen. Two- or three-stage dryers are used where considerable moisture is to be removed, and in which a wide variety of products, size, shape, and moisture content are included. The initial moisture content is from 50 to 200% (d.b.) (more commonly 60 to 100%). Jet dryers are also used where the heated air sweeps around a cylindrical drying chamber, drying the furnish and carrying out the dried product. Direct-fired heaters are commonly used for wood dryers, with inlet temperatures of about 260 to 371°C (500 to 700°F). One three-pass rotary dryer has an inlet temperature of 650°C (1200°F) entering the first pass. Indirect heated units usually have lower temperatures, 176.6 to 260°C (350 to 500°F). The final moisture content is 3 to 5%.

Particulates Small diameter liquid or solid particles in air, often emitted from a process, which may include dust, dirt, smoke, soot, mist, mold, fog, spray, fumes, and pollen. Also see *Air Pollution Count; Dust*.

Pascal A derived SI unit, Pa, for pressure in newtons per square meter. One lb/in^2 equals 6895 Pa, 6.9 kPa. Also see *Conversion Factors*.

Paste Dryer See *Slurry Dryer*.

Pasteurization The heat treatment of a liquid sufficient to kill all pathogenic and nearly all non-pathogenic organisms present and to inactivate certain enzymes which would otherwise decrease product storage life. Pasteurization is usually accomplished by heat treatment at a specified temperature for a controlled time. Cold pasteurization may be accomplished by radiation and/or chemical methods. Pasteurization precedes the drying of many liquid food products. Also see *Thermization*.

Paunch Drying Paunch is the partially digested material contained in the rumen of cattle, which is normally a waste product at time of slaughtering. Cattle paunch products may be dried using a heated air rotary dryer. The dried paunch at 5 to 7% moisture may be pelletized and used for livestock or fish feed. A beef animal has 50 to 60 lb of paunch with approximately 80% water.

Pea Drying Peas that are dried may have the skin slit (1/8 in. in each pea) to provide more rapid drying and wetting, following a blanching operation. Slit peas may be treated by spraying or dipping in a sulfite solution. The slit peas may be dried in belt, cabinet, or tray dryers, using 88 to 94°C (190 to 200°F) in initial stages of drying. The peas are dried to about 8%, then placed in a bin dryer with drying air at 49°C (120°F).

Dehydrofreezing, in which 70% of the water is removed from the raw pea, is also a procedure used to provide a product with good quality which is easily rehydrated. The moisture is removed on a belt trough dryer and then frozen. Also see *Beans; Dehydrofreezing; Green Pea Drying*.

Peach Drying Peaches are halved, pitted, peeled (lye), and sulfured before drying. A countercurrent tunnel dryer, using incoming heated air at 68.3°C (155°F), requiring 25 to 30 hr for 13 kg/m^2 (3 lb/ft^2), drying to 25% moisture content. The drying time may be reduced 30% if preceded by blanching.

Peanut Drying The moisture content of peanuts at the time of digging is 36 to 60%. Peanuts can be dried to about 8% in field stacks in 4 to 8 weeks and to 9% in the windrow in 1 to 2 weeks. In an atmosphere of 64.4% relative humidity (approximately room temperature) the peanut kernels will be at 7% and the shells at 12.5% moisture. The major problem in drying peanuts is to prevent the splitting of the skin caused by rapid drying. Heated air, if properly managed, can be used for drying of peanuts. The air should not be heated above 43.3°C (110°F) for peanuts which are to be used for table stock and temperatures up to 51.7°C (125°F) can be used if the nuts are to be used for oil. When drying peanuts in piles, 3 sacks deep, 12 hr were required to dry peanuts in piles 3 sacks deep from 18 to 8% with 48.9°C (120°F) air. With heated air the shells lose more moisture than the nuts and during storage gain moisture from the nuts. It is advisable to stop drying when the peanuts are at 9% and the shells will pick up enough moisture to bring the peanuts down to 7%. On a continuous dryer with the peanuts on the vines placed 0.6 m (2 ft) deep the best drying and quality were obtained by using an air velocity of 0.3048 m/s (60 ft/min) at 29.4 to 32.2°C (85 to 90°F). The static pressure per foot of depth was 0.05 in water with an air flow of 0.03 m^3/m^2s (60 cfm/ft^2) for unshelled Virginia runner peanuts.

Peanut Flakes A product made from peanuts to be used for a high protein food, similar to high protein soybean products. Peanut flakes are made from the whole peanut from which the shell and red skin have been removed. After processing of the peanut, the product is dried on a rotating drum dryer and removed in a thin sheet.

Pear Drying Pears are prepared for drying by halving, peeling, slicing, and sulfuring. Drying is done on trays in a countercurrent flow dryer with heated air in at 65.5°C (150°F).

pcu The representation for the pound centigrade unit, which is the amount of heat to raise the temperature of one pound of water 1 degree centigrade, and equals 1.8 Btu. The pcu is equal to the pound calorie or centigrade heat unit, Chu.

Peat Drying Peat from an undrained bog has over 90% moisture which can be used for a fuel if dried. Peat is commonly dried by natural means by exposure to sun and air flow. One method of drying is to cut the peat into brick shapes, stack, and air dry. Peat may also be milled and dried and compressed into pellets or briquettes for heating. Dry peat has a heating value up to 18,608 kJ/kg (8,000 Btu/lb). Peat with a moisture content of 50% can be burned, which has a heating value of about 7,675 kJ/kg (3,300 Btu/lb).

Peclet Number A dimensionless number, Pe, for thermal similarity, which is a ratio of molecular and convective heat transfer in a stream:

$$\text{Pe} = \frac{v\ell}{\alpha} = (\text{Reynolds Number})\cdot(\text{Prandtl Number})$$

where

v = velocity of flow
ℓ = distance
α = thermal diffusivity

Pelican Sampler A bucket which is periodically inserted into a falling stream of grain to obtain a representative sample for evaluation, testing, or quality control.

Pendular The condition in drying a porous solid when a continuous film of liquid between discrete particles ceases and capillary flow ceases. Pendular flow follows the funicular condition of drying. Also see *Funicular*.

Pepper Drying Peppers which are used for spice and color are dried with the sun and/or heated air drying. Whole or sliced pods are dried with heated air at 65.6 to 79.4°C (150 to 175°F), requiring 6 hr for slices and 12 hr for whole pods to dry to 7 to 8%. Drying may be with single or two-stage dryers. The final product may be whole, sliced, chunks, or powdered.

Perfect Gas See *Gas Constant; Ideal Gas*.

Performance Ratio In a heated air dryer, the performance ratio, PR, is the actual temperature decrease divided by the wet-bulb depression during drying:

$$PR = \frac{(t_1 - t_2)}{(t_1 - t')} = \frac{\text{temperature decrease}}{\text{wet-bulb depression}}$$

where

t_1 = entering air temperature
t_2 = leaving air temperature
t' = wet bulb temperature of air

The PR decreases with drying and varies from 0.75 to 0.25. Also see *Effective Heat Utilization; Heat Utilization Factor.*

Period A quantity which is a function of time, which repeats itself after constant time intervals, T_p, is the period of the function, in sec, hr, days, weeks, etc.

Periodic Temperature Change The temperature of a product, particularly in storage, varies greatly from the atmospheric temperature, except on the surface. The atmospheric temperature is periodic and can be represented as a sine wave for daily and yearly variations. The equation representing periodic temperature:

$$t = t_o e^{-x\sqrt{\frac{\pi}{\alpha P}}} \left[\sin \left(\frac{2\pi\theta}{P} - x\sqrt{\frac{\pi}{\alpha P}}\right)\right] + t_m$$

where

t = temperature at any depth x, °F
t_o = amplitude (half range) of external fluctuations, °F
x = depth, ft
P = period, hr
α = diffusivity, $ft^2/hr = k/c\rho$
t_m = mean of external fluctuations, °F
θ = time, hr

The damping of temperature fluctuations is represented by equation:

$$\frac{t_a}{t_o} = e^{-x\sqrt{\frac{\pi}{\alpha P}}}$$

where t_a is the amplitude at depth, x (Fig. P-1). Also see *Transient Heat Transfer.*

Period of Decay See *Decay Coefficient.*

Permeability The ability of a body to permit passage of a gas or liquid through it.

Pervaporation An evaporation process from a colloidal suspension utilizing a semipermeable bag suspended in air. Crystals, if present,

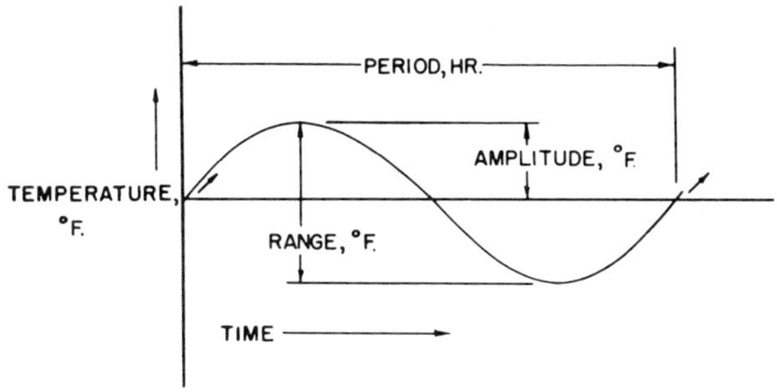

Figure P-1. Periodic temperature representation.

pass through the membrane and are deposited on the outside of the bag.

pH A measure of the degree of alkalinity or acidity based on hydrogen ion concentration. The pH of various products is in Table P-1. pH is the logarithm of the reciprocal of the hydrogen ion concentration:

$$pH = \log_{10} \frac{1}{[H^+]}$$

A neutral substance has 10^{-7} hydrogen ions and 10^{-7} [OH] ions, which is pH = 7.

	H^+	OH^-
pH 1 acid	10^{-1}	10^{-13}
pH 7 neutral	10^{-7}	10^{-7}
pH 13 base	10^{-13}	10^{-1}

Water is neutral and slightly dissociates $H_2O \rightleftarrows H^+ + OH^-$.

Table P-1
pH of Some Materials Dried

Sea water	7.7-8.25
Blood	7.4-7.5
Animal tissues and fluids	~7.0-7.5
Milk	6.5-7.0
Plant tissues and fluids	~5.2
Water, pure	7.0
Tomato juice	4.2-4.4

pF The measure of energy with which moisture is held in the soil which is expressed as the common logarithm of the head necessary to produce the suction corresponding to the capillary potential, in cm H_2O.

Piezoelectric The phenomenon exhibited by some solids whereby an imposed loading resulting in stress-strain causes a voltage, and vice versa. Some biological materials, such as wood, may exhibit piezoelectric properties. These properties may be used as a basis of quality control measurement.

Piezometer An instrument for determining the pressure in the fluid. The device is used to obtain the static pressure at a point, or at several locations, to obtain the average static pressure of a cross-section. Design and use must be such that the velocity head is not included if the fluid is moving. Also see *Pitot Tube*.

Pilot Device on the burner of a dryer for igniting the fuel for heating. The following pilots are available:

(1) Pilot-continuous--The pilot remains lit during the time power is supplied to the burner.
(2) Pilot-interruptible--The pilot is lit during the time combustion is being established. When combustion is established the pilot is de-energized.
(3) Pilot-standing--The pilot is manually lit and remains lit until manually shut off.

Pilot Plant A small unit, larger than laboratory apparatus and smaller than the operating plant, of a drying, processing, or manufacturing activity. The design of a pilot plant is based on the best information available from bench, laboratory, or research results. Information from operating a pilot plant is used to evaluate the potential of a full-size operation and to provide basic information for design. Also see *Scale-Up Dryer*.

Pipe Rack Dryer A device for drying, particularly for brick or tile, made up of a series of steam pipes or hot water pipes on which the product is placed, usually in pallets, for heating, over which air is moved to carry away heat.

Pitot Tube An airflow measuring device consisting of a tube directed towards a flowing fluid which receives the impact of the fluid, from which the velocity produces a pressure head, $v^2/2g$. A piezometer is connected to the pitot tube to measure the static pressure which can be distinguished from the total pressure (Fig. P-2). A manometer can be connected to a pitot tube and piezometer tube so that velocity head is measured, from which the velocity can be determined, $v = c\sqrt{2gh}$, with c obtained by calibration.

Plasma A state of matter resulting from ionization of gases in

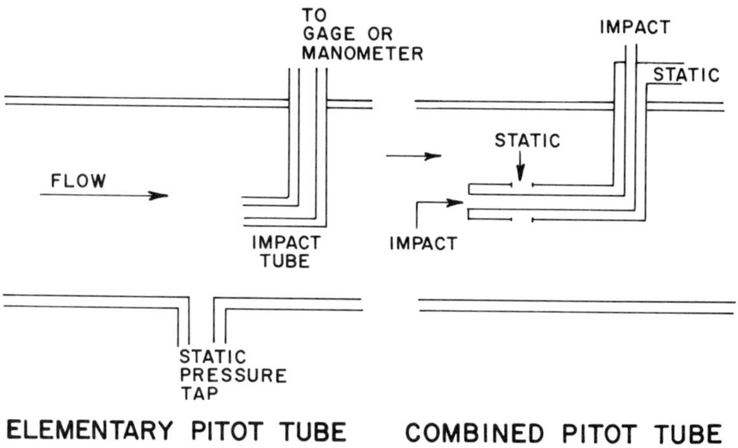

Figure P-2. Pitot tube.

which the number of positive ions and negative ions is approximately equal. The exposure of biological materials to plasma radiation may affect (1) the permeability of the material and subsequent diffusion and drying or wetting, and (2) the physical properties of fiber and the germinating ability of seeds and spores.

Plasticity   The physical property of a material such that the deformation from a particular loading is retained after removal of the load. A material such as clay will exhibit certain plastic properties depending on the moisture content. Drying may remove the plasticity of a material.

Platform Dryer   A device for removal of moisture on which crops are placed over openings through which air is forced. The openings are provided in the top of a level platform. Products may be in bulk or placed in sacks or bags. Also see *Bag Dryer*.

Plenum   An air chamber maintained under pressure, either negative or positive, usually connected to one or more distributing ducts or false floor in a drying or aeration system. The term is also used to designate the air chamber under the perforated floor in a grain bin and the pressure chamber between grain columns in some types of batch or continuous dryers.

Pneumatic Dryer   A device for removing moisture from products in which small particles, chopped, ground, or pulverized products, are conveyed by the drying heated air with temperatures up to 750°C (1382°F) (Fig. P-3). Also called pneumatic conveying dryer. Used for oxide by-products, dyestuffs, fine chemical powders, starch, fruit pulp, distilling wastes, and crops.

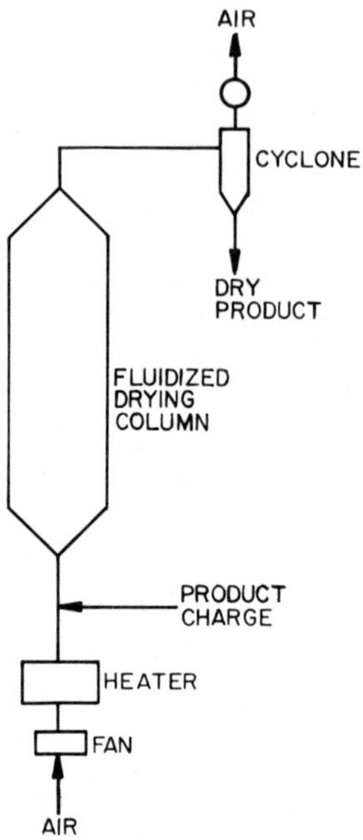

Figure P-3. Pneumatic dryer.

Pneumatic Ring Dryer An apparatus for moisture removal in which the product is propelled by hot air through a duct in the shape of a ring, as a circle or oval. Humid air is continuously exhausted. The dried product is removed at the center. The variables involved in operation are temperature of the air, rate of airflow, rate of product feed, and drying time. The dryer is used for granular materials such as starch, flour, powdered soup, cereals, and coffee. Also see *Pneumatic Dryer*.

Pocket A strong jute sack used for holding dried hops for sale. A pocket is 2 m (6 ft) high with a diameter of 0.6 m (2 ft) with a capacity of about 68 kg (1.5 cwt) of dry hops.

Poise A measure of absolute viscosity in the cgs system of units, which is one dyne-second/cm^2. One kg/m sec = 10 dyne sec/cm^2 (or poise) = 1 Ns/m^2. Centipoise or 0.01 poise is commonly used as a measure of viscosity. Also see *Rhe; Viscosity*.

Poiseuille Equation A mathematical relationship for the quantity per unit time, Q, of laminar flow through a cylindrical tube:

$$Q = \frac{\pi p a^4}{8 \ell \eta}$$

where

p = pressure difference between two ends of tube
ℓ = length of tube
a = radius of tube
η = dynamic viscosity

Also see *Viscosity*.

Poisson Ratio The lateral strain (decrease or increase in diameter) divided by the longitudinal strain (decrease or increase in length) when a body or a mass of bodies is appropriately loaded within the elastic limit (Table P-2). For values of the Poisson ratio up to 0.50, elongation results in an increase in volume of the material. If the material is not isotropic there will be more than one value of Poisson ratio.

Table P-2
Poisson Ratio, Selected Values

Wood	0.3 -0.5
Apple	0.21-0.34
Potato	0.45-0.49
Cork	0
Cereal grain	0.32

Source: Mohsenin, N. N., 1970. Physical Properties of Plant and Animal Materials, Gordon and Breach, New York, NY.

Polymer Curing and Drying See *Ultraviolet Curing*.

Polytropic Gas The name given to changes represented by pv^n = constant, applied to compression and expansion of air in a compressor. Special cases are:

$n = 1$	isothermal	pv = constant
$n = k$	isentropic	pv^k = constant
$n = 0$	pressure constant	p = constant
$n = \infty$	volume constant	v = constant

Porcelain Drying Hard porcelain (china), which has little shrinkage during drying, is fired at 800 to 1200°C (1472 to 2192°F).

Porcelain Pottery Drying Porcelain pottery is made from a slurry composed of white clay and water. The slurry can be poured into a plaster of paris mold, which absorbs water from the slurry. After approximately 24 hr a thin solid layer forms on the mold. The remaining slurry is poured out of the shell so formed. The shell is placed in a kiln, heated to 1900°C (3452°F), with the entire kiln process taking another 24 hr.

Porosity Refers to the proportion, either in decimal or percentage, often represented by ε, of the total volume of a granular substance which is in pore space not occupied by the substance. Also see *Voids*.

Porous Body See *Capillary-Porous Body*.

Portable Dryer An apparatus for moisture removal which can be readily moved on wheels or skids or can be easily carried. Commercial portable dryers may be moved from one location to another for use but are not commonly used for drying except in a stationary position (Fig. H-5). A hand-held hair dryer is portable but products are not moved through the dryer.

Pot Burner A type of apparatus for supplying heat by combustion or oil fuel, in which the fuel is fed into and burned from a gravity-fed sump or pot. Not recommended for direct-fired heaters for dryers.

Potato Drying The handling and storing of potatoes may involve: (1) curing period, including drying, (2) cooling period, (3) warming period, and (4) the conditioning period. During the curing period the potatoes heal after being injured or bruised during harvest. Potatoes are maintained at about 15.6°C (60°F) with air at 90 to 95% relative humidity during the curing period with an air circulation of 0.001 m^3/m^3s (1/3 cfm/bu) for about 2 weeks. The heat of respiration from the potatoes will provide the necessary heat for maintaining the desirable temperature in the fall. A temperature of approximately 4.4°C (40°F) and relative humidity of 95% are desirable for storage during the holding period. Excessive shrinkage and early sprouting occur depending on the variety if held at temperatures above 10°C (50°F). At temperatures below 4.4°C (40°F) the sugar content may increase and the sprouting is retarded. The potatoes can be cooled to 4.4°C (40°F) by ventilation within one or two months after harvesting with an air movement of 0.002 m^3/m^3s (1/6 cfm/bu) (Fig. P-4). If adequate insulation is used it is not necessary to supply artificial heat to maintain a temperature of 4.4°C (40°F) in northern climates. If potatoes are graded and handled at 4.4°C (40°F) some injury will result. Before handling after storage it is desirable to warm the potatoes to about 10°C (50°F) to reduce injury. Potatoes which are to be used for chipping are conditioned by heating to 15.6 to 21.1°C (60 to 70°F) and to prevent shrinkage a humidity of about 85% is maintained.

Either a gravity or forced air circulation system can be used. Air is distributed by shell- or through-circulation. Air is moved by gravity or is forced through slotted walls of the bins, which are about 4 m (12 ft) wide with shell-circulation. The shell system is satisfactory in the northern climates. With through-circulation, air is forced around each potato by a duct system or slatted floor arrangement such as is used for grain and hay drying. A slatted floor or sidewall ports can be used to direct air through pallet boxes.

Potatoes can be dried for animal feed with equipment used for dehydration of hay in which an entering air temperature of 650°C (1200°F) are used. The steps in drying are (1) washing to remove stones and dirt, (2) grinding in a hammer mill with 1 cm (3/8 in.) diameter holes in the screen, (3) mixing some of the dried material with the entering wet material while keeping the entering moisture content to 40% to prevent formation of balls, and (4) drying. Also see *Sweet Potato Drying; Sweet Potato Flake Drying*.

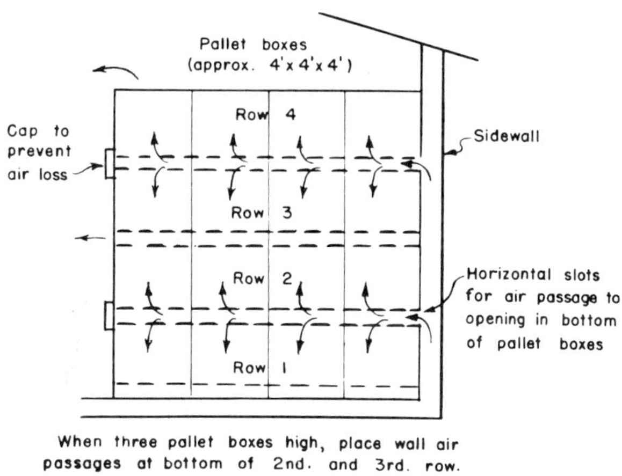

Figure P-4. Potato drying in pallets.

Potential, Drying — See *Drying Potential*.

Potential Gradient — A difference in driving force, such as for temperature or vapor pressure, from one point in the body or system to another point.

Potentiometer — An instrument for determining the voltage or voltage difference of a circuit or sensor. A simple potentiometer is a slide wire device in which the variable resistance is used to balance an internal circuit. A galvanometer indicates the voltage and is used to measure temperature or pressure.

Pottery Dryer — Pottery, wares of porcelain, electrical ware, sanitary, stoneware, etc., are dried using many different methods depending on the specific product, shape, size, etc. Small batch units for the individual artist to large continuous units have a variety of methods of heating with electricity, gas, steam, hot water, in the floor, walls, or pipes. Also see *Earthenware Drying; Porcelain.*

Pound A unit of measure of mass, abbreviated lb. The most common is the avoirdupois pound of 16 ounces or 7000 grains or 0.45359 kg as contrasted to the Troy or apothecaries pound of 12 ounces or 5760 grains or 0.3732 kg.

Powder Drying Smokeless powder, pyrocellulose, used for manufacturing explosives, is first dehydrated, using a dehydrating press at 250 psi, and after several other processes, is then water-dried. For water-drying, water at 65°C (149°F) is circulated throughout the powder causing cracks permitting rapid removal of alcohol. The wet powder is dried to 0.9 to 1.5% moisture content in a continuous dryer at 55 to 65°C (131 to 149°F). Also see *Dehydrating Press.*

Powdered Milk A term, now archaic, formerly used for dry milk.

Power Series An infinite series in mathematics of the form

$$\sum_{n=0}^{\infty} a_n x^n = a_o + a_1 x + a_2 x^2 + \ldots + a_n x^n$$

PPB Represents parts per billion. The weight or volume of a substance per one billion units of weight or volume of the solution. Equivalent to 1 milligram/m^3 (approximately) at the same weight, or 1 unit in 10^9 units. Also see *PPM.*

PPM Represents parts per million. The weight or volume of a substance per one million units of weight or volume of the solution. 10,000 ppm = 1%. Equivalent to 1 mgm/kg or 1 unit in 10^6 units. Also see *PPB.*

Prandtl Number A dimensionless number, Pr, representing the ratio of momentum diffusivity and thermal diffusivity in fluid convection, a measure of the similarity of the temperature and velocity fields:

$$Pr = \frac{c_p \mu}{k} = \frac{Pe}{Re} = \frac{\text{Peclet Number}}{\text{Reynolds Number}}$$

where

c_p = specific heat
μ = kinematic viscosity
k = thermal conduction

Also see *Kinematic Viscosity; Peclet Number; Reynolds Number.*

Precision Refers to the quality of measurement as stated. The degree to which a measurement represents the true value of the measured. In drying, precision pertains to the degree to which measurements and controls represent and/or maintain the actual or desired conditions.

Preconcentration A process before drying or dehydration in which an appreciable amount of moisture is removed, such as with an evaporator or vacuum pan. Removal of water in an evaporator, where applicable, is less expensive than in a dryer.

Predvoditelev Number A dimensionless number, Pd, used in drying, which represents the rate of temperature change of the medium as compared to the rate of change of the material:

$$Pd = \frac{b\ R^2}{\alpha(t_o - t)} = \frac{bA}{\alpha\Delta t}$$

where

b = rate of temperature change, dT/dt, of the external medium
R^2 = A = surface area
α = thermal diffusivity
Δt = temperature difference of medium, where t_o is the original temperature

Prefixes See *Number Prefixes*.

Preformers Devices for placing products to be dried into uniform sizes of particles, aggregates, or granules, so that the products can be dried more or less uniformly. Paste must be partially dried before or as a part of preforming. Evaporation of moisture from liquid materials may be necessary before forming the paste. Various types of preformers include: grooved drum, which may be heated and accomplish some drying; band, in which material is fed out of a bottomless box onto a conveyor; mincing, pressing devices; reciprocating extruder, a trough-like die plate containing many holes over which a pair of heavy rubber rollers reciprocate; direct pressure extruders, with produce being forced out through a die with product under pressure behind the die; and pelleting. Rough grinding and screen classification of solids provide another means of getting aggregates.

Press Drying A method of drying by applying heat to opposite faces of a board using heated platens to remove moisture from the board. The heating is done by conduction heating provided by contact between heated platens and boards by using a platen pressure of 1.75 to 5.3 kg/cm^2 (25 to 75 psi) with temperatures in the range from 121 to 232°C (250 to 450°F). Ventilated openings are provided to carry the moisture away. Drying rates of 1.3 to 0.60 g H_2O/cm^3-min for temperatures of 104 to 137.8°C (220 to 280°F) are representative.

Pressure A property of a material, which is the force per unit area acting uniformly on a surface. The gauge pressure is measured from atmospheric pressure, usually as above atmospheric pressure, with vacuum as below atmospheric pressure. Absolute pressure is taken from zero pressure and is always positive (Table P-3). Also see *Absolute Pressure; Atmospheric Pressure; Conversion Factors; Critical Pressure; Gage Pressure; Negative Pressure; Partial Pressure; Pascal; Static Pressure; Total Pressure; Velocity Pressure; Water Column.*

Table P-3
Pressure Units and Conversion

1 lb/in^2	= 0.0703 kg/cm^2	= 6895 N/m^2	= 6.895 kPa
1 kg/cm^2		= 0.09807 MN/m^2	
		= 0.9807 bar	
1 kg/mm^2		= 9.807 MN/m^2	
1 lb/ft^2		= 47.88 N/m^2	= 47.88 Pa
1 ft H_2O	= 62.4 lb/ft^2	= 2989 N/m^2	
1 in Hg	= 70.73 lb/ft^2	= 3376 N/m^2	
1 mm Hg	= 1 torr	= 133.3 N/m^2	= 133.3 Pa
1 bar	= 14.5 lb/in^2	= 10^5 N/m^2	
1 atm	= 14.70 lb/in^2	= 1.013×10^5 N/m^2	

N = newton, kg (mass)/sec^2; Pa = N/m^2.

Pressure Drop The decrease in pressure exhibited by a fluid moving through a pipe, duct, valve, orifice, or a bed of material. Pressure drop may be identified by psi, in. Hg or in. H_2O in English units; or kg/m^2, mm Hg, torr, bar, cm H_2O, or Pascals in SI metric units in which a Pascal is a newton per square meter. The pressure drop increases with higher fluid (gas or liquid) flow, greater depth, smaller void space, and with a greater density of fluid (Fig. P-5). Also see *Bar; Conversion Factors; Darcy Equation; Duct; Pressure; SI.*

Pressure Energy See *Energy.*

Pressure-Freezing Air-Drying A procedure in which a product under pressure (up to 1000 psi) is frozen followed by air drying. The gas must be properly selected to provide the pressurized environment. Freezing is more rapid (up to four times) under gaseous pressure. Freon 13, methane, and nitrogen are gases used which provide a slight expansion to the product, lighter color, and a softer mouth feel. When the pressure-frozen product is substantially air-dried the resulting food is considerably larger in volume and rehydration faster than normal freeze-dried or atmospheric-dried.

Pressure Head The head or height of a column of fluid required to produce a given pressure. The fluid head is convertible to velocity head, $v^2/2g$. The pressure head, in feet, is p/w, p in lb/ft^2, divided by w, specific weight of fluid in lb/ft^3:

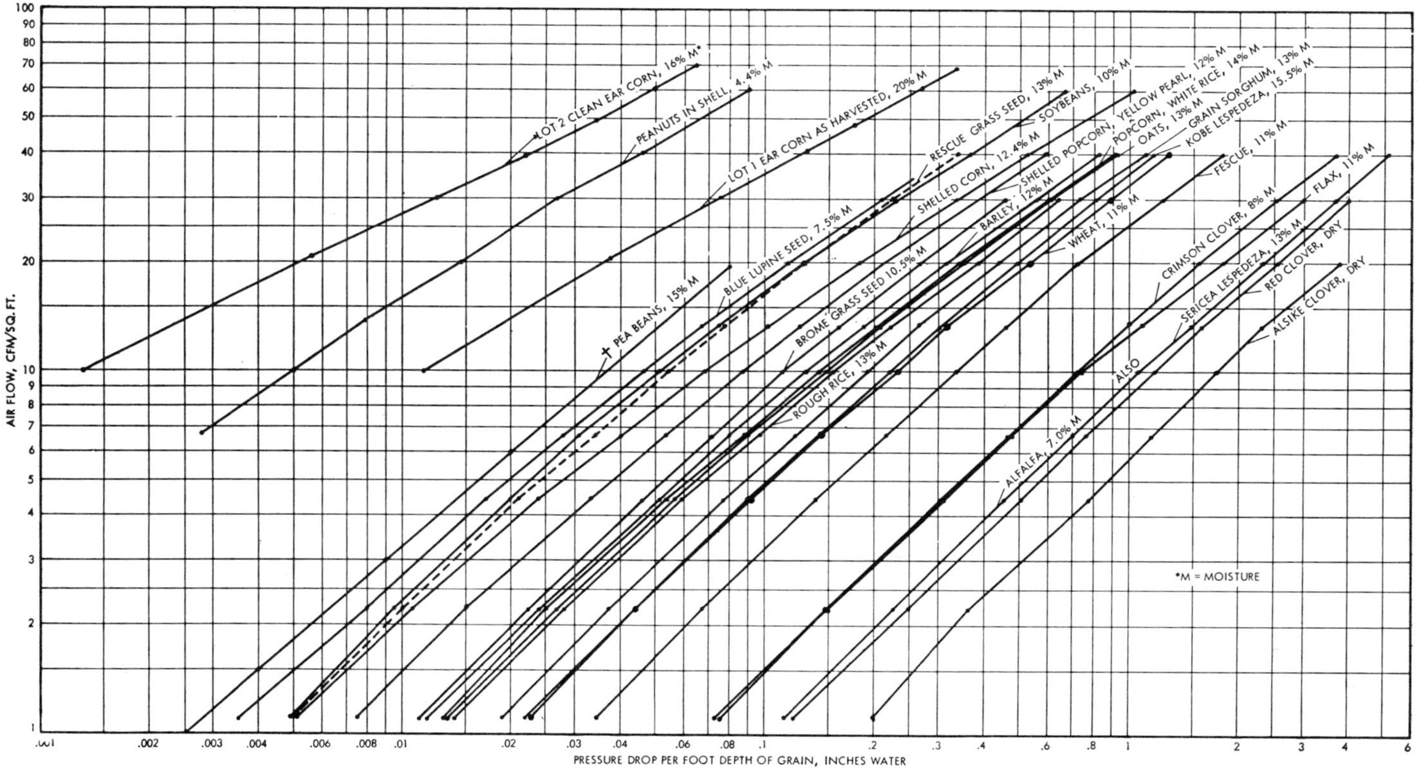

NOTES: This chart gives values for a loose fill (not packed) of clean, relatively dry grain.

Figure P-5. Pressure drop of air through grain.
From ASAE (1976). Agricultural Engineers Yearbook, St. Joseph, MI.

$$\frac{p}{w} + \frac{v^2}{2g} + z = \text{a constant (Bernoulli law)}$$

Also see *Bernoulli Theorem.*

Pressure Parameter   A dimensionless parameter, K_p, of pressure in heat transfer which characterizes the relation between the absolute pressure of a system and the pressure change across the phase boundary:

$$K_p = \frac{p}{\sqrt{\sigma(\Delta \text{sp.wt.})}} = \frac{\text{pressure}}{[(\text{surface tension})(\Delta \text{ specific weight})]^{1/2}}$$

Pressure Relief Valve   A valve designated as a safety device to open, and remain open, to discharge a fluid whenever the fluid pressure reaches a discharge setting of the valve. When the fluid pressure drops below this setting, the relief valve automatically closes.

Pressure System   An air or other gas or vapor system in which the ventilating, drying, or conditioning space is at a pressure above the atmosphere pressure to which the system is exhausted. Also called a pushing or forcing system of air movement, in contrast to a suction system.

Pre-Ventilation   As applied to crop dryers, refers to clearing or purging the plenum chamber or duct of any volatile gases prior to ignition of the burner. Usually accomplished by a device which insures that the fan must operate for a certain time before ignition is permitted.

Probable Error   In a random sampling of data from the population, those bounds which include 50% or one-half of the data, and which exclude 50% of the data. In a normal distribution the probable error is 0.6745 times the standard deviation. Also see *Error; Standard Deviation.*

Product Number   An empirical number obtained by pilot plant tests with a spray dryer representing the ease with which a product gives up moisture and the heat sensitivity of the product. A larger number represents a greater difficulty in drying and heat sensitivity. Examples of product numbers are: 5-11 for pigments; 17, milk; blood, 11-22; alumina, 11-15. *Reference:* Williams-Gardner, A. 1971. Leonard Hill, London.

Progressive Dryer   A tunnel dryer, operated periodically or intermittently, in which the product moves through the unit, used for clay, pottery, and lumber. The tunnel is filled with products, the heat and air are turned on, the temperature is held constant, and the air flow is uniform. The hot air at about 93.9°C (200°F) enters first, coming in contact with the nearly dry product leaving, and the air leaving the dryer comes in contact with green ware. The product then moves to the kiln for finishing. Also see *Progressive Kiln.*

Progressive Kiln A dry kiln in which the total charge of lumber is not dried as a single unit but as several units, such as kiln-truckloads that move progressively through the kiln and become drier as they progress. The temperature of the air is lower and its relative humidity is higher at the entering end compared to the discharge end for a counter current unit. Crossflow units are also used where equal or different air conditions prevail entering the tunnel. Also see *Compartment Kiln; Tunnel Kiln.*

Propane C_3H_8, a compound of carbon and hydrogen, a gas at normal temperature, used for a fuel. Propane is stored and transported in pressurized vessels. Also see *Fuel; High Heating Value; Low Heating Value.*

Propeller Fan A mechanical air moving device (AMD) which operates at low static pressures used for aeration, drying, heating, and cooling, consisting of blades, hub, shaft, and driving device. Most blades are at a fixed angle on a shaft, although some have an adjustable pitch. Fixed blade fans for drying are at an angle of 15 to 30° depending on pressure, speed, and width of blade (Table P-4). Also see *Air Moving Device; Axial Flow Fan.*

Table P-4
Propeller Fan Airflow and Power Requirements, Fan with Adjustable Blade

Blade Setting Numbers	Static Pressure, in Water						
	0.5	1.0	1.5	2.0	2.5	3.0	3.5
0				(15.5) 30,400	(16.5) 28,500	(18.5) 26,300	(20.0) 23,300
4		(7.0) 25,300	(8.0) 23,900	(9.2) 22,300	(10.5) 20,400	(11.5) 18,000	
8	(3.5) 18,400	(4.5) 17,400	(5.3) 16,200	(6.5) 14,600	(7.0) 12,400		
12	(2.0) 11,200	(3.0) 10,400	(3.5) 9,400	(4.2) 7,800			
16	(9.7) 5,200	(1.2) 4,700	(2.0) 3,800	(1.5) 2,000			

Air flow in cfm, horsepower in ( ) for blade setting; blade setting No. 16 corresponds to an angle of 30°.
Conversion factors: *1 cfm = 0.00047* m^3/s
1 in. water = 0.2488 kPa
1 hp = 0.746 kW
Source: Joy Manufacturing Co., Pittsburgh, PA.

Property A term used to describe or define a quality of a product, material, or substance, such as density, pressure, internal energy, kinetic energy, potential energy, enthalpy, entropy, electric charge, strength, elasticity, electric resistance, thermal, etc. (Work and heat are not properties but are measures of energy exchanged in a process.)

Protein The complex chemical nitrogenous compounds of plants and animals which are necessary for food for humans and animals (Table P-5). When heated in the presence of water, denaturization occurs, followed by coagulation, flocculation, and precipitation with continued heating. Enzymes are proteins, which are inactivated with appropriate heating. Overly heated protein which may occur during drying may be difficult to digest. Proteins are composed of amino acids, of which about twenty are known and eight are essential for the human diet. The eight essential amino acids are lysine, methionine, valine, tryptophan, threonine, leucine, isoleucine, and phenylalanine.

Table P-5
Protein Content of Selected Food Products

Product	Usual Range, %
Apples	0.4
Banana	1.5
Beans, dry	23
Beef	19
Corn	7-13
Eggs	15
Fish	14-22
Milk	3- 5
Oats	12-14
Peanuts	25-28
Peas	23-25
Potatoes	1.5-3.0
Rice	8- 9
Wheat	8-17

Prune Drying Prunes are dried on a tray (2 to 4 lb/ft^2) in a countercurrent tunnel dryer equipped to recirculate air, heated to 73.9 to 87.8°C (165 to 190°F). Prunes are dried to 16 to 19% moisture content.

PSI, PSIA, PSIG Psi (pounds per square inch): measure of pressure. Psia (pounds per square inch absolute): total pressure obtained by adding atmospheric pressure to psig. Psig (pounds per square inch gage): the pressure that a normal pressure gage measures. The difference between the pressure in the vessel, to which the gage is connected, and atmospheric pressure. When psi alone is used, it usually refers to psig. Also see *Bar; Pressure*.

Psychro A prefix meaning and indicating cold.

Psychrometer An instrument which uses a wet-bulb and a dry-bulb thermometer in an air stream with velocity of at least 4.5 m/s (15 ft/sec) used to measure the humidity of the air. The air can be forced past the thermometers or the thermometers can be moved through the air. A dewpoint hygrometer is often classified as a type of psychrometer. The psychrometric chart is used to determine the humidity of the air from the wet-bulb and dry-bulb readings or from the dew-point and dry-bulb readings.

Psychrometric Chart A diagram with the absolute humidity on one axis and temperature on the other axis, with curved lines representing the relative humidity, and other lines representing the density, wet bulb, etc., to determine the properties of vapor, usually at 1 atmosphere pressure (Figs. P-6a and P-6b). Also see *i-x Diagram; Mollier Chart.*

Psychrometry That body of knowledge and scientific practice dealing with mixtures of air and water. Also see *Psychrometric Chart.*

Puff Drying A method of drying generally used for removing moisture from a juice concentrate. The concentrate is expanded about 20 volumes in a vacuum of 2 to 6 mm Hg to incorporate vapor bubbles. The expanded material is dried in a vacuum heated 93.3 to 60°C (200 to 140°F) to remove water. Drying may be carried out in a vacuum shelf batch dryer or on a vacuum belt continuous dryer. Drying time varies from 1-1/2 to 5 hr. Juices are dried to 3% moisture. The products formed from puff drying are readily reconstituted.

Pulse Drying Pulse is a term applied collectively to peas, beans, and lentils. Pulse may refer to the edible pod produced by these plants. Drying is concerned primarily with drying of the pods, if wet, at time of harvest and of drying products made from the pods. Flour and soup are made from these leguminous pods. Procedures for handling pea beans are usually in such a manner to avoid cracking or checking. Peas, on the other hand, may be made into split pea soup. For soups the product(s) are processed before being either condensed and canned or dried and packaged. Condensed pea soup and condensed lentil soup are made the same way. Processing before canning consists of soaking followed by steam cooking for 30 min at 110°C (230°F) in a retort.

Pumpkin Drying Pumpkin or squash may be made into flour. The flesh is removed and cut into 1/4 inch pieces. After steaming on trays the product is air-dried to 6% moisture. The dry material is reduced in size with a hammer mill.

Pumpkin Flake Drying Pumpkin may be used for making instant flakes. Pieces of the pumpkin are softened by exposure to 104.4 to 121.1°C (220 to 250°F) for 15 to 45 min. The pumpkin is puréed in a pulper, resulting in a product with about 20% solids. The purée is dried on

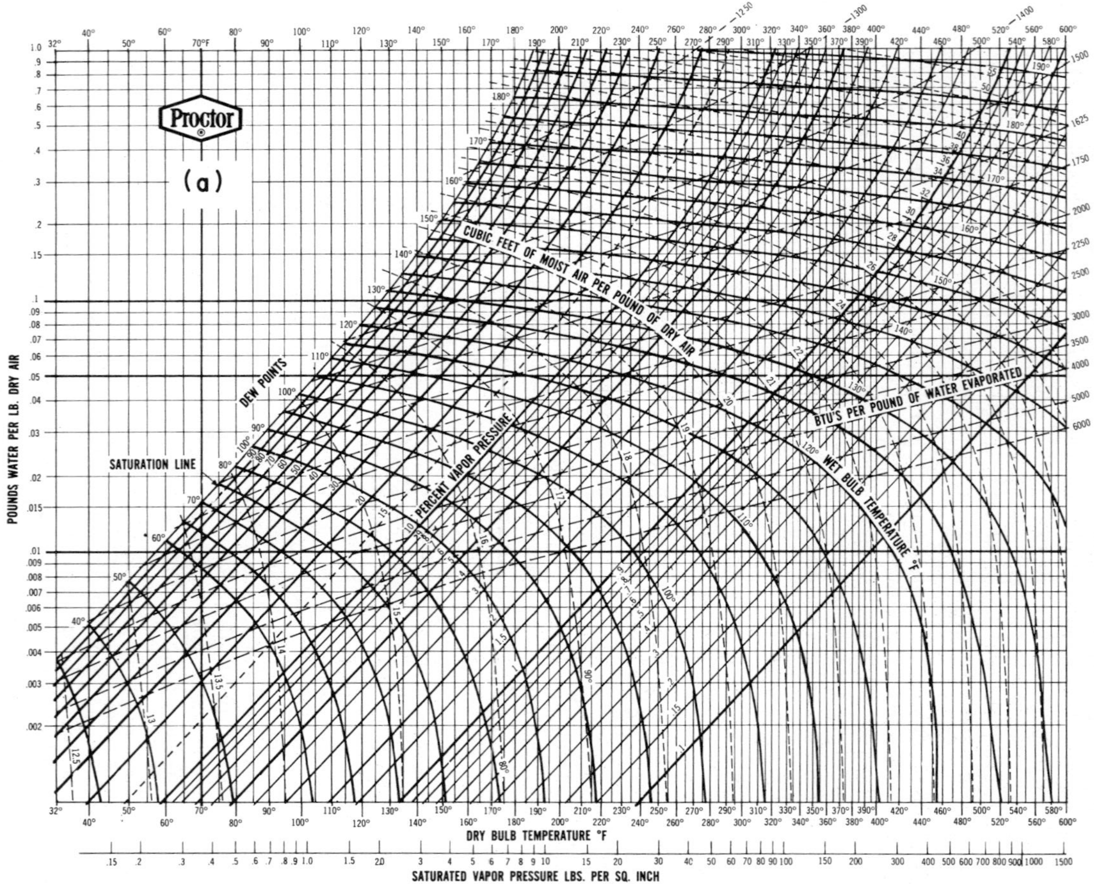

Figure P-6a. Psychrometric chart, English units.
Reprinted with permission of Proctor, Philadelphia, PA.

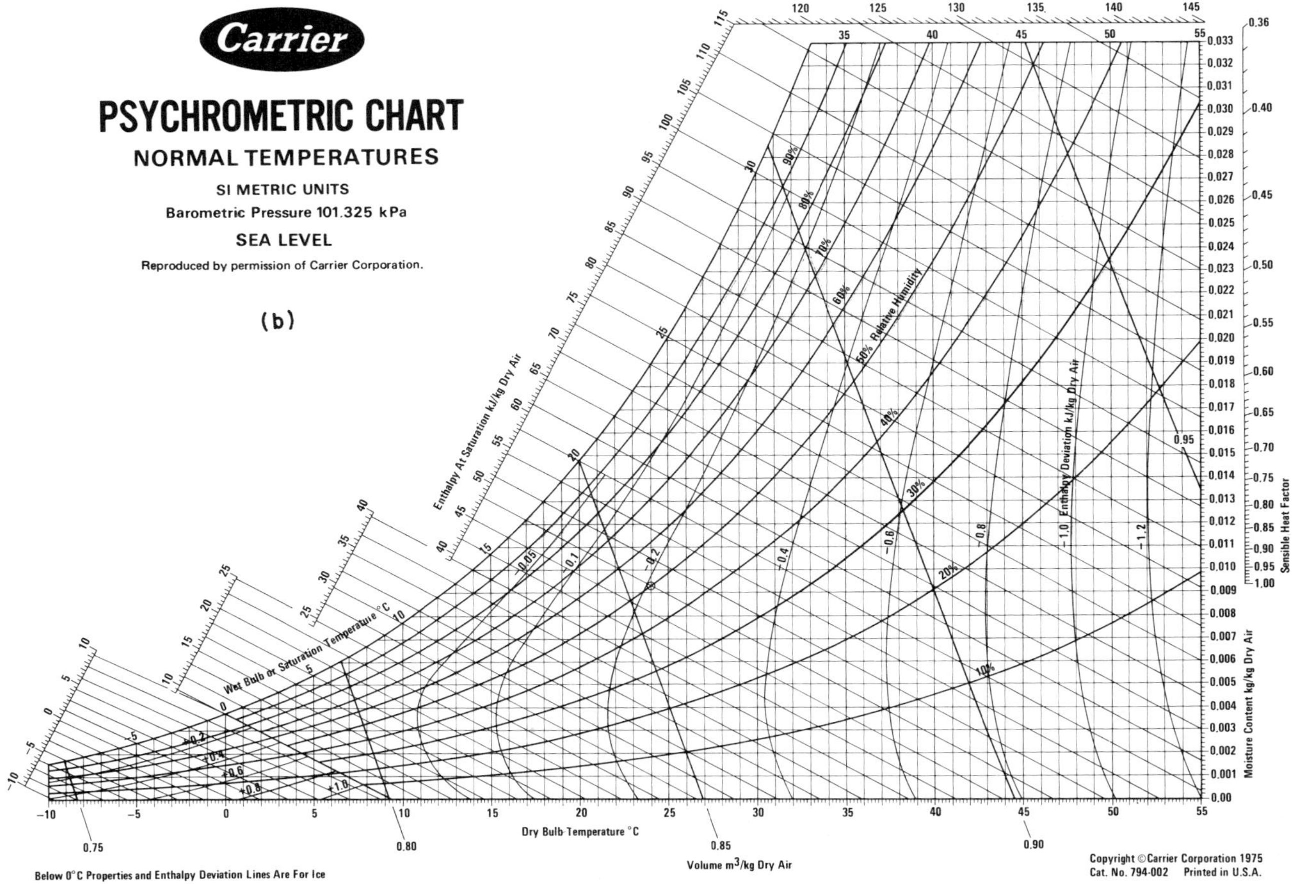

Figure P-6b. Psychrometric chart, SI metric units.
Reprinted with permission of Carrier, Division of Carrier Corporation (1976). Carrier, Syracuse, NY.

a double-drum dryer, heated with steam at 75 to 85 psi. Sheets are removed from the drums which are passed through a grinder to make flakes.

Pycnometer (Pyknometer, in English) An instrument for measuring the density of a fluid or of a solid which can be placed in a fluid. The density is determined on the basis of the weight of a known volume of contents. Some devices can be immersed in a temperature controlled water bath for determining the effect of temperature on density.

Pyro A prefix meaning fire, burning, heat, etc.

Pyrheliometer An instrument for measuring the intensity of solar radiation, such as in solar energy drying. Also see *Actinometer*.

Pyrometer An instrument for measuring high temperatures, as in the burner of a dryer. The major types of pyrometers are thermocouple, optical, radiation (mirror reflects energy onto thermocouple), and resistance. Radiation and optical pyrometers do not have a mechanical part in the high temperature field. Also see *Optical Thermometer*.

Q

Q A letter used to represent total heat transfer, charge of a capacitor, magnification factor, quintillion, and water volume. A quintillion, quint, Q, is 10^{18} Btu. In terms of energy, the Q was originally defined as the energy equivalent to 30 x 10^9 tons of charcoal.

$$Q \doteq 10^{18} \text{ Btu}$$
$$= 1.05 \times 10^{21} \text{ joule}$$

Recently, several authors have used Q for 10^{15} Btu. Also see *Quadrillion; Quintillion.*

q A letter of the alphabet which in heat transfer represents a rate of heat transfer, rate of water or air flow, etc. Also see *Q; Quadrillion.*

Quadrillion A number used to represent large values, particularly for energy. A quadrillion, quads, is used for 10^{15} units in the U.S.A. and 10^{24} in Great Britain. Also see *Q; Quintillion.*

Quadruple-Effect A multiple-effect unit, such as an evaporator, in which there are four stages. Each successive stage of an evaporator is at a greater vacuum. The heat in the vapor leaving the first effect is used for heating the second effect, etc. Also see *Multiple Effect.*

Quality The "goodness" of a product in comparison to a standard. The percent of saturation of a vapor. The ratio of vaporized portion to the total liquid and vapor, such that a fully saturated vapor is 100% quality, and if dry, that is, all moisture in liquid rather than vapor form, is 0% quality. 100% quality is dry saturated steam in which there is no liquid droplets and all the water is in vapor form.

Quality factors, such as color, shape, uniformity, conformity to a standard, appearance, etc., may be used as a basis of evaluation of effect of drying, handling, and storing of products. Also see *Saturated Vapor; Steam; Superheated Steam.*

Quick-Acting Valve A manually operated valve specially designed to accomplish rapid shut-off of flow of a fluid, such as flow of fuel to a dryer.

Quintal A unit of weight used in parts of Europe and in South America which varies considerably in exact weight, but is generally 100 lb in the English system of units or 100 kg for a metric quintal. Used to represent production of crops per hectare or per acre (Table Q-1).

Table Q-1
Weight of Product for Quintal in Various Countries

	Pounds	Kilograms
Argentina, Paraguay, Uruguay	101.3	45.94
Brazil	129.54	58.76
Colombia (100 libras)	110.23	50.0
Egypt	99.05	44.9
Great Britain, U.S.A.	100.0	45.36
	112,0	50.8
Metric Countries	220.46	100.0
Mexico	101.47	46.025
Portugal	129.53	58.752
Spain, Chile, Peru	101.43	46.01
Switzerland	110.43	50.0

Based on Webster's Intercollegiate Dictionary.

Quintillion A number, represented by Q, quints, for 10^{18} units in the U.S.A. and 10^{30} in Great Britain. Used to represent large quantities, such as world inventories, Btu of energy, etc. Also see *Q; Quadrillion.*

R

R, Gas Constant See *Gas Constant; van der Waals Equation.*

Radial Fan An externally driven air moving device (AMD) which takes in air near the axis of rotation and moves the air along the radius of the turning wheel, utilizing the centrifugal force of rotation. Radial fans can be used to develop higher static pressure than axial flow fans. Known also as centrifugal fan. Also see *AMD; Backward-Curved Fan; Forward-Curved Fan; Centrifugal Fan.*

Radiation A process of transferring heat from one body to another by electromagnetic wave motion through an intervening transparent medium. Part of the internal energy of the emitting body is converted into radiative energy which is propagated through an electromagnetic field and transformed back into energy of thermal motion of the elementary particles on reaching the body being heated. Also see *Electromagnetic; Infrared; Ionizing Radiation; Irradiation.*

Radiation Treatment The application of irradiation by subatomic particles or energy emitted from internal changes in atomic structure to control spoilage organisms in food, often associated with moisture control through dehydration or drying. The kinds of radiation based on effect on the product are identified as follows:

> Radicidation. A radiation treatment sufficient to reduce the non-spore forming pathogenic organisms below detection.
> Radurization. A radiation treatment sufficient to destroy the pathogenic organisms with the treatment similar to the effect of thermal pasteurization.
> Radappertization. A radiation treatment sufficient to provide commercial sterility of the product.

Also see *Thermization.*

Radical In chemistry, radical refers to a group of atoms which maintains identity in a chemical reaction; also, the group of atoms is associated with compounds. A radical is distinguished from an ion which has a charge, while a radical may or may not have a charge. Examples are: ammonia radical, NH_4, and water radical, H_2O. Associated with drying is the OH group which may be a radical, -OH, or an ion, OH^-. Also see *Moisture Content*.

Radio Frequency A signal in instrumentation or communications or power in the range of 10 to 30,000,000 kilocycles/sec (30×10^9 cycles) vibration. The abbreviation is r-f or RF. Also see *HFC; UHF; VHF*.

Radiosity The total radiant energy leaving a surface in terms of energy per unit area of surface, which includes reflected as well as emitted energy. The total radiosity of a surface includes all radiation leaving a surface, regardless of direction of radiation or spectrum involved.

Raisin Drying Grapes of certain varieties are processed for raisins followed by drying. Drying is preceded by dipping in lye and exposing to burning sulfur. Grapes are dried to 9 to 14% moisture content in 24 to 45 hr.

Rancidity The formation of flavor defects in fats giving a "fishy" or "tallowy" flavor by oxidation of fats, known as oxidative rancidity, designated by peroxide value determination. The formation of flavor defects giving a "soapy" flavor due to hydrolysis of fats into glycerol and fatty acids, known as hydrolytic rancidity, represented by acid value. Also see *Acid Number*.

Range The value of the complete swing of a cycle of events. For a sine curve the range is twice the amplitude. Also see *Amplitude; Periodic Temperature Change*.

Rankine Equation A relationship for calculating the approximate lateral pressure, L, psf, exerted by granular materials in a shallow bin, most appropriate for a depth under 20 ft (Fig. R-1),

$$L = \rho y \frac{1 - \sin\phi}{1 + \sin\phi}$$

where

ρ = density of material, lb/ft^3 (40 to 50)
y = distance from top to location in question, ft.
ϕ = angle of repose, deg (25 to 45)

Also see *Janssen Equation; Lateral Pressure*.

Rankine Scale °R, a scale for temperature based on the absolute zero of the Fahrenheit system, such that -459.7°F is equal to 0° Rankine. Also see *Celsius Scale*.

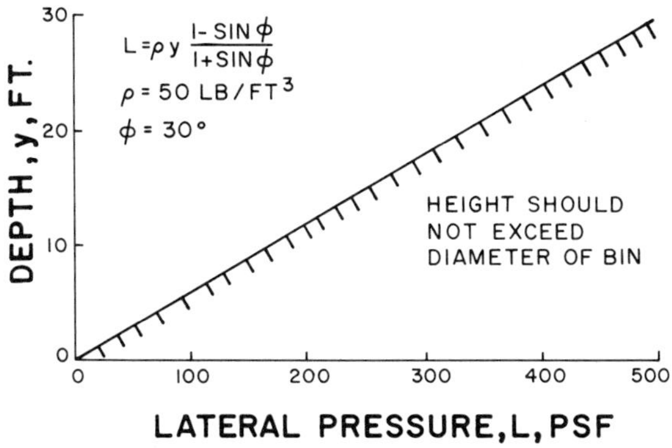

Figure R-1. Rankine relationship for lateral pressure.

Raoult's Law A relationship stating that when a solute is dissolved in a solution, not by dissociation, the ratio of the decrease in vapor pressure is proportional to the increase of molecules dissolved. Likewise, the decrease in freezing point and increase in boiling point are proportional to the molecules dissolved (or molecular weight of comparable solvents). Also see *Freezing Point*.

Raw Feed The product fed into the dryer which may be considered on the basis of the following physical and chemical properties: source of raw feed, number of products for same dryer, previous dewatering, raw feed storage, method of handling product to and from dryer, particle size distribution, physical characteristics for handleability, abrasiveness, odor, fire hazards, explosion hazards, temperature limitations, temperature effects, and corrosive effects.

Reaction The transformation of a substance or components into other substances, which may occur with the release of energy (exothermic) with inorganic compounds or combustion or with the consumption of energy (endothermic) as with organic compounds.

Reamur See *Réaumur,* which is a more common spelling.

Réaumur (Re) A scale of temperature divided into 80 divisions for which the freezing point is 0° and the boiling point is 80°:

$$\frac{Re}{4} = \frac{F - 32}{9} = \frac{C}{5}$$

where

F = Fahrenheit
C = Centigrade

Re = Réaumur

Also see Table R-1.

Table R-1
Réaumur Temperature Equivalents

°F	°C	°R	°F	°C	°R
32	0	0	131	55	44
41	5	4	140	60	48
50	10	8	149	65	52
59	15	12	158	70	56
68	20	16	167	75	60
77	25	20	176	80	64
86	30	24	185	85	68
95	35	28	194	90	72
104	40	32	203	95	76
113	45	36	212	100	80
122	50	40			

Rebinder Number A dimensionless group, Rb, involved in drying which relates the heat used in heating the moist body to that used in evaporation over a small increment of time:

$$Rb = \frac{d\bar{t}}{d\bar{M}} \frac{c}{L}$$

where

$\bar{t}$ = mean temperature of body
$\bar{M}$ = mean moisture content of body
c = heat capacity
L = latent heat of vaporization

Recirculation The arrangement of a dryer whereby only a part of the air that has passed over the moist material, known as exhaust air, is discharged from the system, the remainder being mixed with fresh air, reheated, and returned to the drying chamber. This type of dryer is used for products which are heat-sensitive or damaged by too rapid drying and is being used increasingly to increase energy efficiency. Also see *Recycle Ratio; Reheating Drying Air.*

Reconstitute The procedure or process followed to restore a dehydrated food to a condition resembling the original product by causing it to re-absorb water. Also known as rehydrate. Also see *Instantize.*

Recycle Ratio Defined as the ratio of recirculated solid to the product removed in a recirculating dryer or evaporator:

$$R_r = \frac{1}{x_s} - 1$$

where

R_r = recycle ratio
x_s = fraction of total flow which is removed

Redrying See *Retorrefying*.

References See *Bibliography; Books*.

Reflection The change in direction of electromagnetic radiation, which may result in regular reflection or diffuse reflection (Fig. R-2; Table R-2). Also see *Albedo; Radiation; Reflectivity; Transmissivity*.

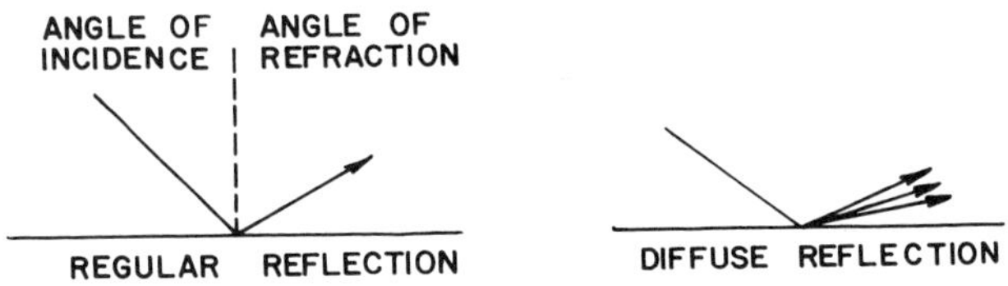

Figure R-2. Reflection.

Table R-2
Reflection of Incandescent Light

Snow	0.93
Steel	0.55
Black paper	.05- .06
Aluminum	0.77-0.81

Refraction The change in direction of electromagnetic energy as the energy passes through different materials or as exhibited by different properties of the same material, such as density of liquids (Fig. R-3). A refractometer is used to measure the refraction which can be used to determine the change in moisture/total solids in the evaporation operation. Where n is the refractive index (ratio of sine of the angle of incidence to sine of angle of refraction), the refractivity is n - 1, the specific refractivity (S.R.) is:

$$S.R. = \frac{n-1}{\alpha}$$

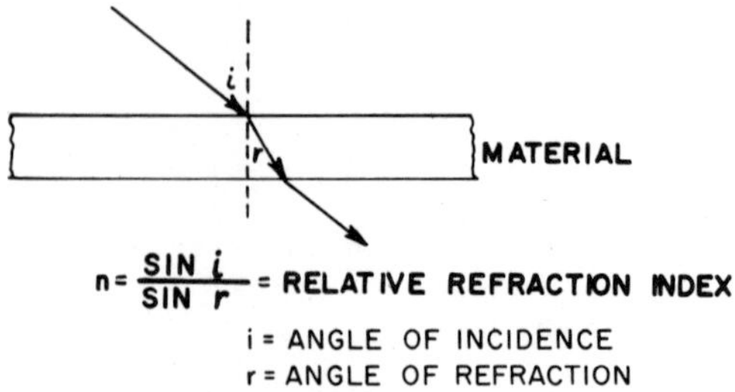

Figure R-3. Refraction.

and molecular refractivity (M.R.):

$$\text{M.R.} = \left(\frac{n-1}{\alpha}\right) \text{M.W.}$$

where

α = density of substance
M.W. = molecular weight of substance

Refractometer See *Refraction*.

Regeneration The recovery and/or re-use of heat which would otherwise be rejected by the system, such as in some dryers or pasteurizers, where heat in the exhaust air or water is used to heat the incoming product, thus increasing the total thermal efficiency.

Regression See *Standard Error*.

Reheating Drying Air In operating a drying system, some of the moist air coming from the drying bed is bled off and fresh air brought in, reheated, and sent to the dryer. The drying air is reheated to obtain greater uniformity of the product between air entering and leaving the drying bed by minimizing the temperature difference of the drying air. Characteristics resulting include: a higher thermal efficiency, damage to heat-sensitive products can be reduced, and the time of drying is normally greater, unless other parameters of drying are changed. Also see *Recirculation*.

Rehydrate See *Reconstitute; Rehydration*.

Rehydration The replacing of water in dried or dehydrated foods, also sometimes referred to as recovery, reconstitution, refreshing, regain, and restoration. Also see *Instantizing; Regain; Wetting*.

Rehydration Ratio The drained weight of product recovered by rehydrating one pound of dehydrated product, with conditions of reconstitution described. Also see *Drying Ratio*.

Relative Density The ratio of the absolute density of a substance at t°C to the absolute density of water at 4°C (39.2°F), its maximum density.

Relative Humidity A measure of the moisture content of air expressed as a percentage. It is the ratio of the weight of water vapor in a given volume of air at a given temperature to the maximum quantity of water vapor which the same volume of air could hold at the same temperature (Table R-3). The value of the relative humidity is used as an indicator of whether certain organisms will grow.

Relative Moisture Content The ratio of the mean moisture content of a body at any moment during drying to the initial mean moisture content of the body. Also see *Dimensionless Numbers*.

Relative Viscosity See *Viscosity*.

Relaxation The relieving of stresses in materials, which may be incurred as a result of heating or drying, hopefully without damage. The relieving of stresses can often result by not subjecting the product (resting) to the force causing stress. A time of relaxation is often used to avoid undesirable damage to the product.

Reliability The ability of a component, equipment, or a system to perform its function without failure or at a given level of response. Statistical procedures are used to establish the reliability of a device or system.

There are many similarities between reliability and other probability concepts. Reliability is defined on the basis of the probability of continued operation of a component or a system.

The most common and simplest reliability model is the series configuration or chain structure (Fig. R-4). This terminology is applied to a system in which the success depends on the continued operation of all the components. When applying this approach, the system configuration must be (1) a series configuration, (2) the components must be truly independent, and (3) the components must be under control of a constant failure or hazard function. The reliability of a series system is represented in the following equation by the product of the probability of success of each of the components:

$$R(t) = P(x_1) \cdot P(x_2) \ldots P(x_n) = \prod_{i=1}^{n} P(x_i)$$

If each component exhibits a constant hazard and the appropriate component model is an exponential relationship, the system reliability is:

$$R(t) = \prod_{i=1}^{n} e^{-\lambda_i t} = \exp(-\sum_{i=1}^{n} \lambda_i t)$$

Where no simple composite formula exists, it is necessary to determine the reliability of each component and then express the

Table R-3
Relative Humidities above Solutions

Salt	Temp. °F	RH, %
$BaCl_2 \cdot 2H_2O$	77	91.2
(Barium chloride)	86	90.8
	95	90.2
	104	89.7
$CaCl_2$	20	44.
(Calcium chloride)	32	41.
	50	40.
	70	35.
$Ca(NO_3)_2$	73	51.8
(Calcium nitrate)	86	46.6
	100	38.9
$Ca(NO_3)_2 \cdot 4H_2O$	68	53.6
(Calcium nitrate)	77	50.4
	86	46.6
	95	41.5
$CoCl_2$	77	65.2
(Cobaltous chloride)	86	62.6
	104	57.2
$C_{12}O_3$	68	38.5
(Chromium trioxide)	77	39.5
	86	40.0
	100	40.2
$CuCl_2 \cdot 2H_2O$	68	68.7
(Cupric chloride)	77	68.7
	86	68.3
	104	67.4
KCl	68	89.2
(Potassium chloride)	77	87.2
	86	85.3
	95	83.8
KCNS	68	47.6
(Potassium sulfocyanate)	77	45.7
	86	43.8
	100	41.1
$KC_2H_3O_2$	68	23.2
(Potassium acetate)	77	22.7
	86	22.0
	100	20.4
KNO_2	68	49.0
(Potassium nitrite)	77	48.2
	86	47.2
	100	45.9

Salt	Temp. °F	RH, %
KNO_3	32	97.6
(Potassium nitrate)	50	95.5
	68	93.2
	86	90.7
	104	87.9
K_2CO_3	68	43.9
(Potassium carbonate)	77	43.8
	86	43.6
	100	43.4
K_2CrO_4	68	86.6
(Potassium chromate)	77	86.5
	86	86.3
	100	85.6
K_2SO_4	50	97.9
(Potassium sulfate)	68	97.2
	86	96.6
	104	96.2
LiCl	68	11.2
(Lithium chloride)	77	11.2
	86	11.2
	100	11.2
$LiCl \cdot H_2O$	50	13.3
(Lithium chloride)	68	12.4
	86	11.8
	104	11.6
$MgCl_2$	73	32.9
(Magnesium chloride)	86	32.4
	100	31.9
$MgCl_2 \cdot 6H_2O$	50	34.2
(Magnesium chloride)	68	33.6
	86	32.8
	104	32.1
$Mg(NO_3)_2$	73	53.5
(Magnesium nitrate)	86	51.4
	100	49.0
$Mg(NO_3)_2 \cdot 6H_2O$	50	57.8
(Magnesium nitrate)	68	54.9
	86	52.0
	104	49.2

	Temp. °F	Acid by weight, %			
		10	20	30	40
HCl	0	83.5	56.0	27.4	8.9
(Hydrochloric)	50	83.5			
	68	83.2			
	86	84.2			

Source: Hall, C. W. 1957. Drying Farm Crops. Edwards Brothers, Ann Arbor, MI.

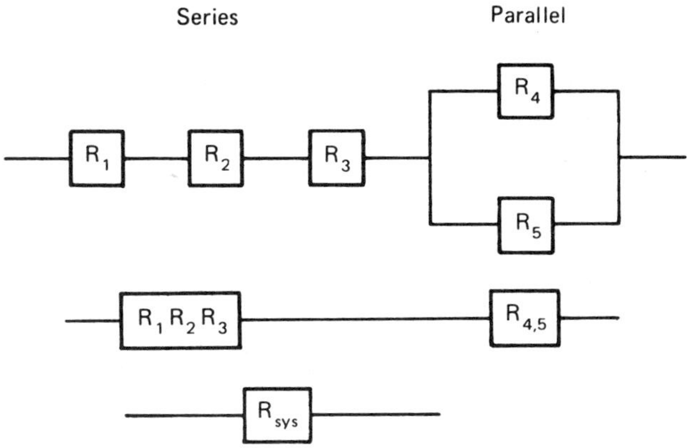

$$R_{1,2,3} = P(R_1)P(R_2)P(R_3) = \prod_{i=1}^{3} R_i$$

$$R_{4,5} = 1 - \prod_{i=4}^{5} (1 - R_i)$$

$$R_{system} = [R_{1,2,3}] \cdot [R_{4,5}]$$

$$R_{system} = [R_1 R_2 R_3] \cdot [1 - (1 - R_4)(1 - R_5)]$$

Figure R-4. Reliability, series and parallel elements.

results as a product of n terms. This procedure may be necessary when a different hazard exists for each component or where a different type of distribution affects the probability of failure.

The parallel configuration consists of a system of n elements with the system functioning when only one of the elements is operating (Fig. R-4). The reliability expression for a parallel system is expressed in terms of the probability of success of each component, x_n:

$$R(t) = P(x_1 + x_2 + \ldots + x_n) = 1 - \left[\prod_{i=1}^{n} (1 - e^{-\lambda_i t}) \right]$$

The following equation represents the general case of the system where the components are independent and governed by a constant hazard function or environment:

$$R(t) = 1 - \left[\prod_{i=1}^{n} (1 - e^{-Z_i(t)}) \right]$$

where the area under the hazard curve is:

$$Z_i(t) = \int_o^t z_i(\xi)\ d(\xi)$$

Various approaches are used in applying these basic equations. In addition, these can be combined for application to combined series and parallel configurations.

Residence Time The time of exposure of a body to a particular treatment in a piece of equipment, such as in a dryer or evaporator. Also see *Retention Time*.

Residual Moisture The moisture remaining after a normal drying process is completed, such as the moisture in air after drying, or the moisture in a product after drying. Also see *Calcium Chloride; Moisture Content*.

Resistance See *Electrical Resistance Moisture Meter*.

Resistance Thermometer An instrument for measuring temperature based on the principle that the electric resistance of a metal wire increases with an increase in temperature. Platinum is usually used as the element.

Resistance to Air Flow Ducts, floors, grain, seeds, and other products offer resistance to air flow. Aeration, cooling, and drying systems are designed on the basis of quantity and velocity of air, size of particles, depth of product, as represented by empirically obtained data (Fig. P-5). Also see *Air Flow*.

Resistance to Heat Transfer See *Conductance (Thermal)*.

Resonating Drying The process of moisture removal which involves, in addition to conventional heated air drying, vibrating the product at a resonant frequency, which for wood increases the drying rate about 15%. Also see *Acoustic Drying; Sonic Drying; Ultrasonic Drying*.

Respiration The process of gas exchange in living organisms in which oxygen is moved into the system and carbon dioxide moved out of the system. In plants the process is known as photosynthesis. Drying reduces the rate of respiration (Table R-4).

Response Time For an instrument or controller, the response time is the time required after an abrupt change in the measured variable to move to a new value shown on the indicator or pointer. The response time is also expressed in terms of the time required for an instrument or controller to indicate a specified percent of change of the measured variable after a step change in the variable (Fig. R-5).

Resting Periods See *Dryeration; Intermittent Drying; Sweating; Tempering*.

Table R-4
Respiration of Cereal Grains
(mg CO_2/24 Hr. for 100 g of Dry Matter)

Moisture Content, % (w.b.)	Temperature °C	°F	Wheat	Corn	Rice
11	38	100	0.2	0.7	0.2
13	30	85		1.5	
13	38	100	0.5	1.5	0.5
15	30	85	1.0	3.5	
15	38	100	1.2	4.7	2.4
17	30	85		12.5	
17	38	100	11.0	21.9	16.5

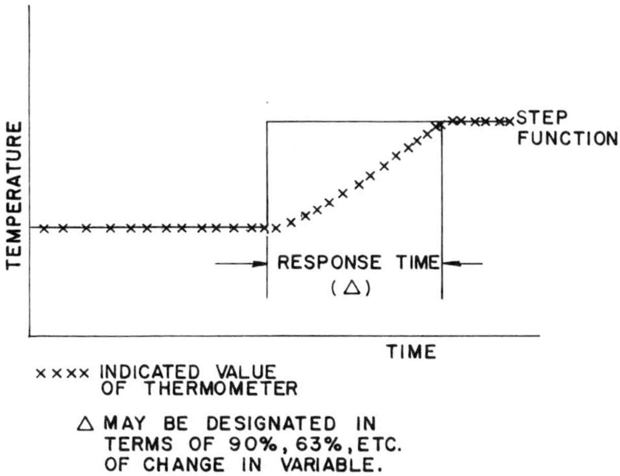

Figure R-5. Response time.

Retention Time In a batch evaporator, the time at which a certain proportion (%) of the volume is removed

$$\% \text{ removed} = (1 - e^{-t/R})\ 100\%$$

where

t = retention time, minutes
R = ratio of holding volume to discharge volume

For an evaporator (or other container) with a volume of 10 gal a feed of 2 gpm and a discharge of 1 gpm, the retention time would be 10 min.

At the end of 10 min 65% of the original volume of the vessel would be removed.

Retorrefying The redrying of barley or other grain, with hot air currents, generally restricted to the application of the hot air as in a rotary kiln.

Reverse Air Flow See *Disc Fan.*

Reversible A phenomenon or action--chemical, biological, mechanical, physical, thermal--which can take place equally readily in two directions. In contrast to irreversible reactions, such as burning.

Revolving Steel Tubular Dryer An apparatus for moisture removal made of stationary steel piping inside a rotating steel shell. The rotating shell is sloped for continuous feeding, flow, and removal of materials. Commonly used for drying granular solids, grain, spent distillery mash, and pigments. A 2 m (6 ft) diameter, 10 m (30 ft) length dryer operating at 3 kg/cm^2 (50 psi) steam demands about 980 kW (100 Bhp) of steam.

Reynolds Number A dimensionless number, Re or Rn, representing fluid flow characteristics, particular turbulence, which is a measure of the ratio of the inertia forces and viscous forces.

$$\text{Re} = \frac{DV\rho}{\mu} = \frac{\text{(diameter)(flow velocity)(fluid density)}}{\text{viscosity}}$$

where

D = diameter
V = velocity
ρ = fluid density
μ = viscosity

Rhe A unit of measure of fluidity in the cgs absolute system, which is the reciprocal of the centipoise viscosity unit. Also see *Poise; Viscosity.*

Rheology A study of the deformation and flow of materials. Measurement of rheological properties may provide an indirect measure of a variable related to wetting, drying, and moisture content of a material.

Rice Drying Combined rice normally contains about 4% excess moisture which needs to be removed before storage. Rice kernels are dense and the diffusion of moisture through the kernels is slow and limits the speed of drying. The capacity of a heated air dryer can be increased by removing the surface moisture and then storing the rice until the moisture is evenly distributed throughout the kernels and again using heated air drying. Heated air temperatures up to 54.4°C (130°F) can be used if the grain temperature does not exceed

43.3°C (110°F) and is forced through the rice at the rate of 1.34 m^3/m^3s (cfm/bu). The drying procedure for alternate drying with heated air is:

(1) First drying
  (a) Begin within 6 hr after combining.
  (b) Dry for 30 to 45 min with 54.4°C (130°F) air.

(2) Second Drying
  (a) Begin 6 to 12 hr after completion of first drying period.
  (b) Dry for 20 to 30 min at 54.4°C (130°F).

(3) Third drying and additional drying as required to reduce the moisture to 14-1/2%.
  (a) Begin 6 to 12 hr after completion of previous drying.
  (b) Dry for 20 min at 54.4°C (130°F).

(4) Last Drying
  (a) Begin within 24 hr after previous drying.
  (b) Dry without artificial heat unless the relative humidity of the atmosphere is above 75%.

Air temperatures up to 65.5°C (150°F) might be used to reduce the drying time, but there is a noticeable effect on the milling qualities and germination at these higher temperatures. Rice at 20% moisture can be dried with unheated forced air using about 0.035 m^3/m^3s (2-1/2 cfm/bu). The cost of electricity for unheated forced air drying in Texas, with electricity at 2¢/kWh, was 3.5¢/bu for drying rice from 19.6 to 12% moisture content. Multipass dryers with crossflow of air are used for rice. The flow of grain and air and the temperature of air are controlled to minimize cracking of rice.

Ring Dryer See *Pneumatic Ring Dryer*.

Rock and Sand Drying Rock and sand are used as aggregates for concrete and asphalt mixtures, the quality control of which is dependent on the moisture content of the aggregates. A rotary dryer, with direct flame heating and drying, at product temperatures of at least 149°C (300°F), with the aggregates repeatedly exposed to the flame, as flights on the inside of the rotary drum pick up and drop the aggregates. A fan supplies air for combustion, transfer of heat, removal of moisture, and transport of fines to the dust collector. Also see *Sand Drying*.

Roller Dryer Same as a drum dryer, a unit in which drying occurs on the external surface of an internally steam heated revolving drum (Fig. R-6). Also see *Drum Dryer*.

Romankov Number A dimensionless group, Ro, involved in drying relating the dry bulb absolute temperature and the product absolute temperature:

$$Ro = \frac{T_{d.b.}}{T_{Prod.}}$$

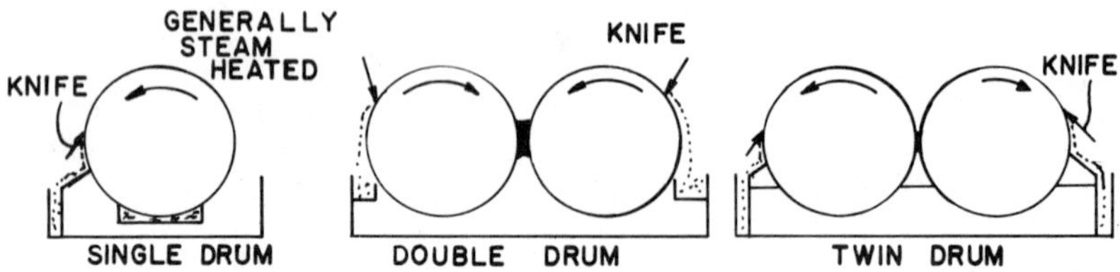

Figure R-6. Roller dryer, single drum, double drum, and twin drum.

Root Mean Square, RMS A statistical measure which is the square root of the sum of the squares of the units divided by the number of units, N:

$$RMS = \frac{\sqrt{x_1^2 + x_2^2 + x_3^2 + \ldots x_N^2}}{N}$$

and, in a frequency distribution:

$$RMS = \frac{\Sigma x^2 fx}{\Sigma fx}$$

Rotameter An instrument for measuring the rate of flow of a liquid or gas. A small float is suspended in a vertical tube carrying the flowing liquid or gas. The height of the float indicates the volume of flow. The instrument must be calibrated to insure accurate measurements. Also see *Calibration*.

Rotary Burner A device which uses centrifugal force in which oil to be burned moves from the center of a rotary cup and is thrown to the outside into the air where it ignites and burns.

Rotary Drum Dryer A device for removing moisture from products in which the product is conveyed internally through a slowly rotating drum through which heated air is forced. Direct- or indirect-heated rotary dryers may be used (Fig. R-7).

Rotary Dryer The general designation of a device for removing moisture from products in which the active drying occurs in or through a rotating chamber. Rotary dryer is the most common designation for a revolving horizontal drum, through which the product and drying air are moved, using either countercurrent or concurrent flow. The external drum dryer, with internal heating and external product drying, is more commonly designated as a drum or roller dryer. Also see *Drum Dryer; Rotary Drum Dryer.*

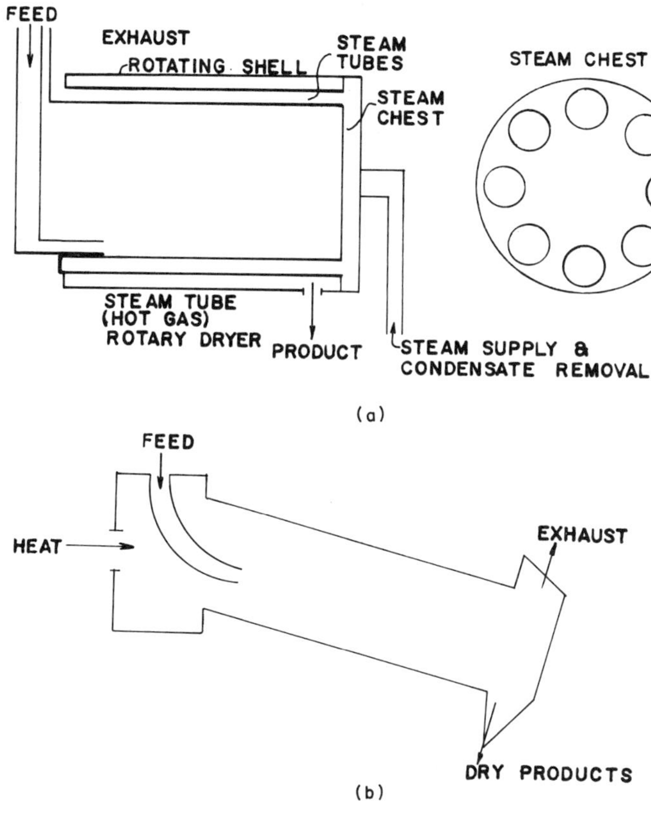

Figure R-7. Rotary drum dryer. (a) indirect. (b) direct.

S

Sack Dryer See *Bag Dryer; Platform Dryer*.

Safety (Burner) Safe operation of burners used for heated air drying systems is important to farmers, insurance companies, and equipment dealers. Important considerations of an installation designed for safe operation include the following: (1) a flame control to shut off the fuel supply in the case of ignition failure, (2) a high temperature limit switch which will stop the burner but allow the fan to continue to operate, (3) a temperature control on the bonnet of the burner to prevent overheating of heater, (4) proper electrical wiring connecting the fan and burner to the electric circuit (Fig. S-1). All units should be designed to fail "safe." Also see *Safety (Installation)*.

Safety (Installation) A properly designed heater can be used in an unsafe manner. Safety features of an installation should include the following: (1) fuel pump and piping located a safe distance from the flame of the burner, (2) the fuel feed line from the tank to the fuel pump protected from mechanical injury, (3) the fuel tank located at least 4.6 m (15 ft) from the bin and other buildings, (4) oil drums refilled a safe distance from the drying unit or the drying unit shut down when the drums are refilled, (5) separate drying and storage installations provided for safe and efficient grain drying by heated air, (6) if the crop is dried in batches, on wagons, or in a batch bin, drying equipment separated from the main building by 4.6 m (15 ft), (7) the drying unit connected to the bin by a duct of flameproofed canvas or other noncombustible material. These features are representative. Regulations and requirements change considerably over time. Many insurance companies require a special permit to use a heated air crop dryer. The insurance company representatives will determine if the installation is reasonably safe, and if approved, a permit may be purchased. A concentration of 5 g/ft^3 of organic dust

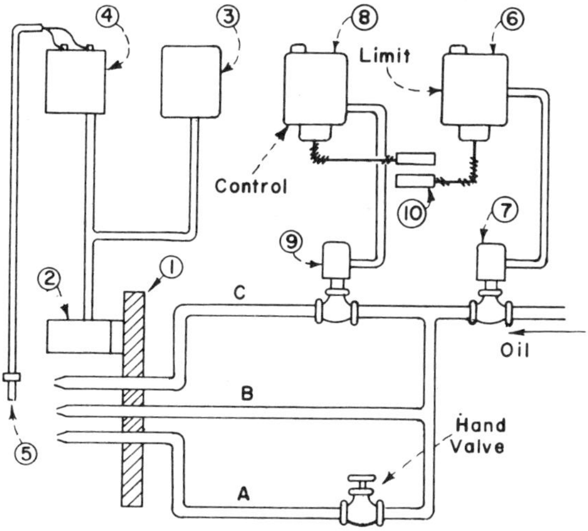

LEGEND: 1 — Oil burner; 2 — Photocell; 3 — Electronic Protectorelay; 4 — Ignition transformer; 5 — Ignition spark; 6 and 8 — Temperature controls; 7 and 9 — Solenoid valves; 10 — Bulbs in exhaust air.

Courtesy Minneapolis-Honeywell Regulator Co., Philadelphia.

Figure S-1. Safety control system for dryer.

in the air is generally considered the maximum amount from the standpoint of explosion. Also see *Electrostatics; Explosion; Safety (Burner)*.

Salmonella An organism which may cause food poisoning which when ingested causes salmonellosis. Salmonella is an ubiquitous organism carried by turtles, fowls, and their products. Salmonella organisms may enter dried food products in drying air contaminated from bird nests. Thorough cooking and refrigeration storage prevents growth in food.

Sampling A portion of a large quantity or bulk of product is used for evaluating quality and moisture content, obtained by sampling. A procedure for obtaining a representative sample must be followed. Sampling may be done manually with a probe or automatically with a catching device. The quantity of sampling ranges from 1 part in 5000 for a truckload to 1 part in 100,000 for a shipload (Fig. S-2). Procedures for sampling various containers for particular purposes are specified by different agencies and organizations. Also see *Pelican; Standard Error; Trier*.

Sand Drying Small grains resulting from the breaking down by erosion of rocks of the earth, many of which contain quartz (SiO_2), form

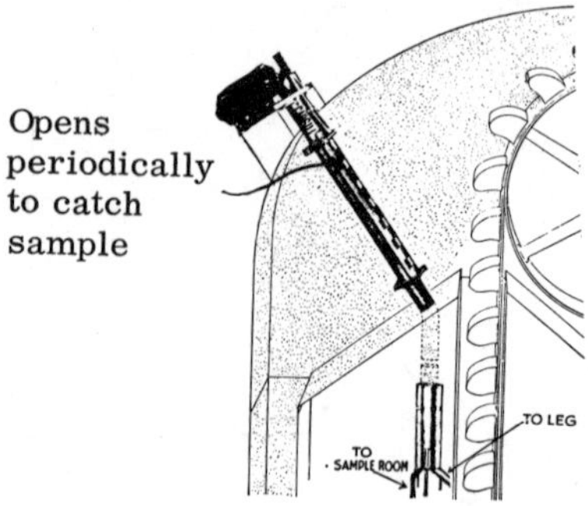

Figure S-2. Sampling device, automatic.

sand which is quarried wet and dried. Sand is often washed to remove organic matter before being used for making concrete or other structural materials. Washing may be followed by either solar drying or mechanical drying with heated air in an internal (drum) rotary dryer. Most of the moisture is removed by surface drying. Also see *Rock and Sand Drying; Rotary Dryer.*

Saponification Number or Saponification Value A measure of the chemical reaction of an alkali with a fat or fatty acid, based on milligrams of caustic potash or potassium hydroxide (KOH) necessary to react with or neutralize 1 gram of the product under test. The saponification number may be applied to products being dried, lubricants, fuels, and foods. Same as the acid number. Also see *Acid Number.*

Saturated Air See *Saturated Vapor.*

Saturated Liquid A liquid which has dissolved all of a solute, such as a salt, which the liquid can hold at a particular temperature. If by slowly cooling a saturated solution, usually without disturbance, a solution may hold more solute than is normal, the solution is known as supersaturated. The term also is used to describe a liquid which is at the boiling temperature for its pressure. It is in equilibrium with saturated vapor at that pressure and temperature. Also see *Supersaturated.*

Saturated Salt See *Salt; Saturated Solution.*

Saturated Solution A solution which contains the maximum amount of a given substance that can be dissolved in it. Also see *Equilibrium Moisture; Relative Humidity; Supersaturated Solution.*

Saturated Vapor A vapor which has a temperature corresponding to the temperature of boiling at the pressure of the vapor. As applied to water vapor and steam, a pound of liquid water at 1 atm and 100°C (212°F), to which has been added 2032 kJ/kg (970.2 Btu) were added, the quality would be 1675/2032 x 100 (or 800/970.2) or 82.4%. Heat in excess of 2032 kJ/kg (970.2 Btu/lb) would produce superheated vapor at the same pressure (Table S-1). Also see *Latent Heat; Quality; Relative Humidity; Steam.*

Table S-1
Saturated Water Vapor in a Cubic Meter, Grams

Temp., °C	0.0	2.0	4.0	6.0	8.0
-20	1.074	.909	.768	.646	.542
-10	2.358	2.026	1.736	1.483	1.264
-0	4.847	4.217	3.660	3.169	2.737
+0	4.847	5.559	6.360	7.260	8.270
+10	9.399	10.66	12.07	13.63	15.37
+20	17.30	19.43	21.78	24.38	27.24
+30	30.38	33.83	37.61	41.75	46.26

Based on data from the Smithsonian Institution. Reprinted with permission from CRC Handbook of Chemistry and Physics, 57th edition, Weast, Robert C. (ed.). Copyright 1976 The Chemical Rubber Co. CRC Press, Inc. Cleveland.

Saturation The condition in which a fluid holding another substance, such as moisture vapor in air or salt in solution, in which no additional substance can normally be held by the fluid at given environmental conditions.

Saturation Coefficient A measure of the ability of pores in brick, tile, and clay materials to absorb water. As the ratio approaches one, the ability of the pores to absorb moisture is greater.

$$\text{Saturation coefficient} = \frac{b - a}{B - a}$$

where

b = weight of brick after 24 hr immersion in cold water
a = dry weight of brick
B = weight of brick after 5 hr immersion in boiling water

Saturation Temperature The temperature of vapor or liquid at saturated conditions. At the saturation temperature liquid turns to vapor as energy is added (boiling) or vapor turns to liquid as energy is removed (condensation). In a liquid-gas two-phase system, the pressure at saturation is related to a definite temperature, called the saturation temperature at the saturation pressure. Also see *Saturation Vapor Pressure.*

Saturation Temperature Parameter   A dimensionless number, K_t, which represents the effect of the curvature of the phase boundary on the saturation temperature:

$$K_t = \frac{(L\gamma'')^2}{A\ c\ T''\ \alpha'\ [\sigma(\gamma' - \gamma'')]^{1/2}}$$

where

L = latent heat
γ'' = specific weight, condition ″
A = mechanical heat equivalent
c = specific heat
T'' = absolute temperature, condition ″
γ' = specific weight, condition ′
σ = radiation factor, $\sigma = \varepsilon\sigma_0$
= (emissivity)(Stefan-Boltzmann Number)

Saturation Vapor Pressure   The vapor pressure of saturated vapor in equilibrium with liquid, at a specified temperature and pressure (Table S-2).

Scalding   See *Blanching.*

Scale-Up Dryer   The general term applied to design of larger dryers based on experience with smaller units. Typical scale-up involves small oven sample testing, laboratory scale, pilot plant, semi-commercial unit, and full-sized plant. The oven test may be used to evaluate the effect of basic phenomena such as heat or circulation rates on drying rates. The laboratory model is used to begin work on a very small scale. The pilot plant is generally instrumented sufficiently to permit uniform results enabling market analysis of the product. The pilot plant is the first unit which uses equipment and materials similar to those expected to be suitable for the finished plant. Good pilot plant testing usually allows skipping the semi-works state for design parameters, limiting its use to cases where developing markets preclude major investment at once. Scale-up magnitudes from 2 to 1 to 5 to 1 have been replaced by values approaching 100 to 1 with increasing knowledge of the fundamental factors affecting processes. Computer simulation has been very helpful in studying the effect of various design parameters. Parameters of major significance in drying include: air circulation rate, moisture content of air in and out, average temperature of air and material being dried, state of subdivision of the material being dried, effect of depth of bed or layer, pattern of air circulation, etc. Also see *Pilot Plant.*

Schmidt Number   A dimensionless number, Sc, characterizing mass transfer by convection which relates viscosity and diffusion of a gas

Table S-2
Saturation Temperature, Pressure, and Latent Heat of Vaporization of Water

°C	cal/g	Pressure, mm Hg	°F	Btu/lb	Pressure, psia
0	595.9	4.6	32	1075.8	0.09
10	590.4	9.2	50	1065.6	0.18
20	584.9	17.5	70	1054.3	0.36
30	579.5	31.8	90	1042.9	0.70
40	574.0	55.3	110	1031.6	1.28
50	568.5	92.5	130	1020.0	2.22
60	563.2	149.4	150	1008.2	3.72
70	557.5	233.7	170	996.3	6.0
80	551.7	355.1	190	984.1	9.4
90	545.8	525.8	212	970.3	14.7
100	539.5	760.0	230	958.8	20.8
110	532.9	1074.6	250	945.5	29.8
120	525.7	1489.1	270	931.8	41.9
130	518.5	2026.2	290	917.5	57.6
140	511.1	2710.9	310	902.6	77.7
150	503.5	3570.5	330	887.0	103.1
160	495.6	4636.0	350	870.7	134.6
170	487.2	5940.9	370	853.5	173.4
180	478.6	7520.2	390	835.4	220.4

Conversion factors: *1 mm Hg = 0.133 kPa*
1 cal/g = 4.187 kJ/kg
1 Btu/lb = 2.326 kJ/kg

through itself:

$$\text{Sc number} = \frac{\mu}{\rho D}$$

where

μ = viscosity
ρ = density
D = diffusivity

Screen A device for separating materials according to size dimensions to meet grade specifications or to improve quality by removing off-size products or to improve uniformity for drying or other processing operations. Also see *Sieve; Mesh.*

Screw Conveyor Dryer A conveyor in which the trough is heated with steam, hot water, or electricity, and the screw shaft can also be heated for drying. The product may be heated with electric or gas heated infrared units located above the conveyor. The unit is used for drying pastes and granular materials. The unit is operated in a vacuum chamber for some applications. Also see *Conveyor Dryer*.

Seasoning Synonym for some drying operations, usually over a long period, for the process of removing moisture from lumber to improve its serviceability. Also see *Curing*.

Seasoning (Wood) Removing moisture from green wood in order to improve its serviceability: (1) Air-dried. Dried by exposure to air, usually in a yard, without artificial heat. (2) Kiln-dried. Dried in a kiln with use of artificial heat. Also see *Kiln; Kiln-Dried Lumber*.

Second Law of Thermodynamics The law which states that it is impossible to construct a device that operates in a cycle and produces no effect other than the transfer of heat from a cooler body to a hotter body. Stated another way, it states that it is impossible to construct a device that will operate in a cycle and produce no effect other than the raising of a weight and the exchange of heat with a single reservoir. This law sets a limit on the percentage of input heat energy which can be converted to useful work energy. Also see *Entropy; First Law*.

Seed Cotton Drying The raw cotton before ginning containing the seeds is known as seed cotton. Drying of wet seed cotton may be carried out to reduce the power for subsequent ginning. Drying to about 5% moisture is accomplished in a vertical baffled dryer using 65.6 to 71°C (150 to 160°F) air in from 15 sec to 3 min, depending on the moisture content of the cotton. The seed cotton moves through the dryer at about one-half the speed of the air in which the damp locks move at a slower rate and dries longer than the well-dried fluffy locks.

Selection of Drying System The selection of a system of drying considers the cost of operation as related to the quality of product, value of product, safety considerations, effect on the environment, and ease of installation. Factors to be considered include: installed cost, operating cost, loss of materials, safety, putting product in a desirable form if not already (such as grinding, pelleting), changes of size of load, control of drying operation as compared to effect on the product, and versatility.

Self-Contained Dryer Any dryer manufactured as a package unit consisting of the drying and cooling chamber, necessary heat source, all AMD's and duct work, along with the necessary controls and product handling equipment. These dryers may be either fixed or portable.

Semi-Indirect Fired Dryer A type of rotary dryer in which the heat from the fire box and burner (1) is used for heating the exterior of the rotating cylinder, and (2) is moved through the inside of the rotary kiln in contact with and in the opposite direction as the incoming product, such as for coal drying (Fig. S-3).

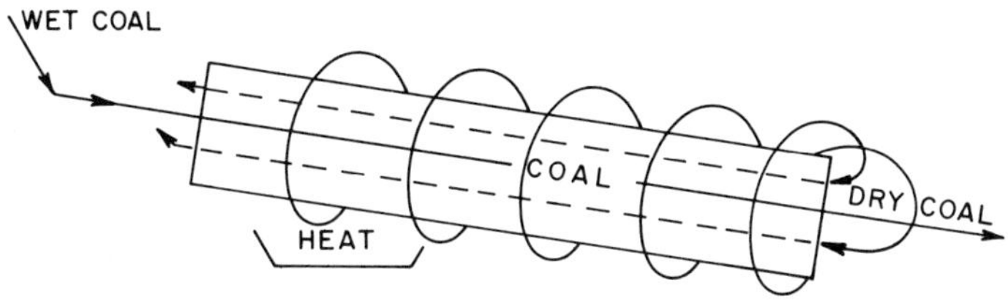

Figure S-3. Semi-indirect heated rotary dryer.

Semipermeable Membrane See *Osmosis*.

Senescence The natural life process that continues in fruits and vegetables after harvesting that leads to moisture loss, aging, and final decay of the food product.

Sensible Heat Thermal energy or heat which when added to a substance increases its temperature accordingly. Heat added to a substance can be sensed by a thermometer.

Separation Drying In ink drying, the process in which the liquid is absorbed by the paper and the solid portion of the ink is retained on the surface. Also see *Ink Drying*.

Sere Refers to a withered condition of a product, often caused by drying.

Setting See *Adhesive Hardening*.

Shade-Grown Tobacco Drying (Curing) Shade-grown tobacco is produced in the Connecticut River valley which following production must be properly cured. During the first five weeks of curing heat is required about one-fifth of the time to provide the proper environment. The storage temperature is maintained at about 8 to 11°C (15 to 20°F) above the outside temperature, with 29.4 to 37.8°C (85 to 100°F) being the most desirable. Heat is usually supplied the major part of the first week of curing, and for one or two periods of 8 to 24 hr a week until cured. LP-gas has been replacing charcoal as a fuel because of the ease of control.

Sheep (Wool) Drying Wet sheep need to be dried before shearing. As a result of weather conditions, the wool of sheep may be as wet as 125% (d.b.). The sheep can be shorn at an average moisture content of between 15 to 20% (d.b.). Approximately 8 to 10 hr are required to dry wet wool or sheep using air heated from 23.3 to 34.4°C (74 to 94°F), and an air velocity of 1 m/s (200 ft/min).

Sherwood Number A dimensionless number, Sh, representing the relationship of mass diffusivity and molecular diffusivity, which corresponds to the Nusselt number in heat transfer:

$$Sh = \frac{h_m x}{D} = f(Re, Sc)$$

where

h_m = mass transfer
x = thickness
D = diffusivity
Sc = Schmidt number

Also called the Taylor number. Also see *Nusselt Number; Schmidt Number*.

Shipping-Dry Lumber Lumber that is partially dried to prevent the development of stain and mold while the lumber is in transit.

Shrinkage Grain shrinks in both weight and volume when dried. Volume of shrinkage for shelled corn is relatively high when compared to other grain. Knowing the volume of shrinkage is important in the operation of drying systems to determine the amount of shrinkage in depth which can be expected in drying wet grains. A 2 m (6 ft) layer of shelled corn with 25% initial moisture will shrink about 0.43 m (1.4 ft) when dried to 12%. The shrinkage in volume as corn is dried is given in Table S-3. As moisture is removed from a product, the weight of the product is reduced due to the loss of water. The loss of weight is often referred to as (weight) shrinkage, and a value of the shrinkage based on original wet product is represented by:

Cost of shrinkage, cents =

$$\left[1 - \frac{(100 - \text{initial moisture content})}{(100 - \text{final moisture content})}\right][\text{price per bu, cents}]$$

In wood, volume shrinkage begins when drying proceeds below the fiber saturation point, that is, when water is removed from the cell walls. Shrinkage of wood occurs after some of the bound water is removed. Also see *Invisible Shrinkage*.

Table S-3
Shrinkage in Volume of Shelled Corn
Dried to 12% Moisture, W.B.

Original Moisture, %	Volume Shrinkage, %
30	29.1
25	22.5
20	14.5
17	9.1

Shrinkage Water Water removed from mineral grains in clays and ceramics, with collapse of the capillaries, resulting in shrinkage. In many clays and ceramics approximately one-half of the total water added to produce a plastic body is evaporated before shrinkage ceases.

SI or International System of Units The system of base units of the International Metric System. The units are the meter, m; kilogram, kg; second, s; ampere, A; kelvin, K; candela, cd; and mole, mol; and the subdivisions and multiples thereof (Table S-4). Secondary derived units are newton, N; joule, J; pascal, Pa; and watt, W. Also see *Conversion Factors.*

Siccative A substance which will absorb moisture and assist in drying of materials to which it is added, as pharmaceuticals, foods, and paints.

Sieve A material or device for separating materials according to size of openings, over which large particles are held and through which small particles pass. The high moisture materials in wet grain can often be removed thereby reducing the demand on the dryer. The size of openings is designated by mesh (Table S-5). Also see *Mesh; Screen.*

Sigmoid Isotherm The relationship between the moisture content of a particular material and its equilibrium relative humidity at a particular temperature can be expressed by means of equilibrium moisture curves. These curves are isotherms because the values plotted for each curve usually correspond to a specific temperature (Fig. S-4). According to the B-E-T (Brunauer, Emmett, Teller) classification, grain isotherms are of Type II, called S-shaped or sigmoid isotherms, and attributed to multi-molecular adsorption. Also see *Adsorption Isotherm; B-E-T; Equilibrium Moisture; Smith Equation.*

Silica Gel A drying agent produced from sodium silicate and sulfuric acid, with a very porous structure. The material has 100 to 800 m^2 per gram. Silica gel can adsorb water to 40% of its weight and as such is used as a drying agent. Also see *Drying Agent.*

Sisal Fiber Drying Sisal fibers have a moisture content of about 75% w.b. leaving the decorticator. Under mechanical pressure, the moisture content of sisal fiber can be reduced to about 50% w.b. Forced heated air at 100 to 110°C (212 to 230°F) for 25 to 30 min is used for drying to about 14% w.b. Temperatures up to 180°C (356°F) have been used for drying, but care must be taken to avoid discoloration due to excessive heating of cellulose.

Skin See *Leather Drying.*

Slumping When green clay products are first placed in a dryer, the products may be cold compared with the high relative humidity air being circulated over the product, condensation may occur, and the product may soften to the extent that the product cannot carry its own weight and slump.

Table S-4
SI (System International) Units

Basic SI Units			
Length	meter	m	1.0 m = 100 cm = 10^{-3} km = 10^3 mm = 3.2808 ft = 1.094 yd = 39.37 in = 0.62 x 10^{-3} mi = 10^{10} Å
Mass	kilogram	kg	1.0 kg = 1000 g = 2.204 lb = 0.0197 cwt = 0.06854 slug = 15,428 grains = 35.2734 ounce
Time	second	s	1.0 s = 1/60 min = 1/3600 h
Temperature	degree	K	1 K = 1.8 R
Electric current	ampere	A	
Derived Units			
Force	newton	N = mkg/s^2	1.0 N = 10^5 dyne = 0.22482 lb-force = 3.597 ounce-force = 0.10197 kg-force
Work, heat energy	joule	J = Nm	1.0 J = 10^7 erg = 0.73756 ft lb = 0.239 calorie = 0.27777 10^{-6} kWh = 0.9478 x 10^{-3} Btu
Power	watt	W = J/s	1.0 W = 0.00134 hp = 10^7 erg/s = 0.0569 Btu/min = 0.239 cal/s = 14.34 cal/min = 0.73757 ft lb/s
Pressure	pascal	Pa = N/m^2	1.0 Pa = 9.9 x 10^{-6} atm = 0.004 in. water = 0.296 x 10^{-3} in Hg = 0.145 x 10^{-3} psi = 0.0075 mm Hg

Also see *Conversion Factors.*

Slurry/Paste Dryer   A device for removing moisture from a viscous material. Conventional spray drying equipment may be used for some materials. Drum dryers may be used for drying slurries. Another

Table S-5
Sieve, Mesh and Wire Size, for Tyler Screen Sieves

Mesh No.	Wire Diameter, In.	Opening, In.
	0.148	1.050
	0.135	0.742
	0.105	0.525
	0.092	0.371
3	0.070	0.263
4	0.065	0.185
6	0.036	0.131
8	0.032	0.093
10	0.035	0.065
14	0.025	0.046
20	0.0172	0.0328
28	0.0125	0.0232
35	0.0122	0.0164
48	0.0092	0.0116
65	0.0072	0.0082
100	0.0042	0.0058
150	0.0026	0.0041
200	0.0021	0.0029

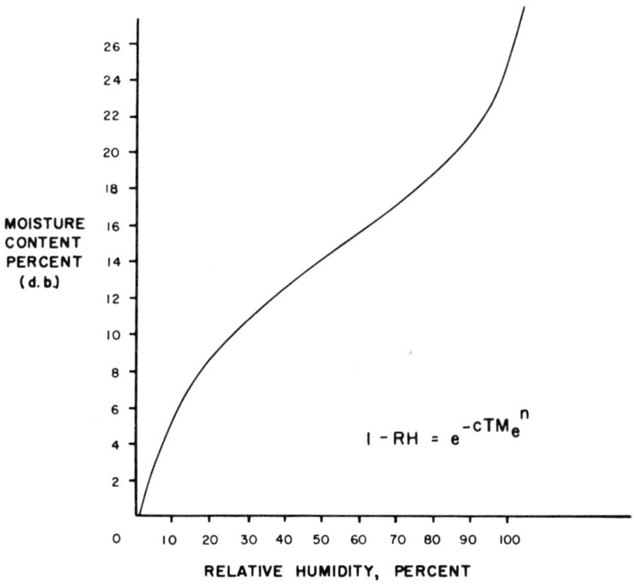

Figure S-4. Sigmoid isotherm for equilibrium moisture content.

method consists of tray drying. Another approach is to spray a thin film of slurry/paste on thousands of inert carriers which are circulated in a fluidized bed for drying. Circulating carriers may be hollow cylinders of silicone rubber, polypropylene spheres, etc. Used for dyestuffs, foods, pharmaceuticals, etc., (Fig. S-5). Also see *Drum Dryers; Fluidized Bed Dryer; Spray Dryers*.

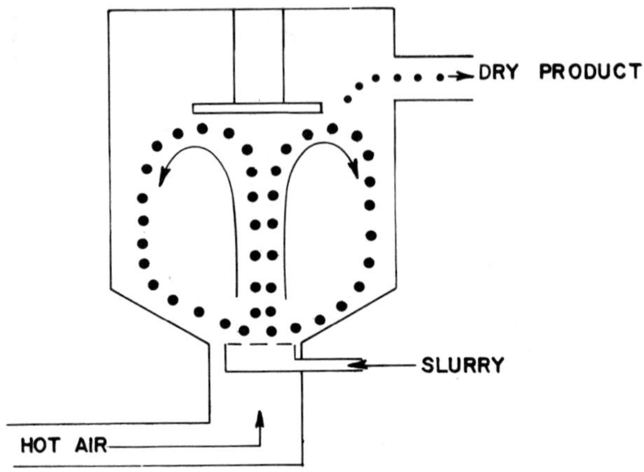

Figure S-5. Slurry dryer.

Smith Equation A relationship representing desorption isotherms for hygroscopic products, particularly applicable at higher humidities, 50 to 95%. The Smith and B-E-T equations complement each other in representing desorption isotherm data from a low to a high humidity. Also see *Adsorption Isotherm; Isotherm; B-E-T Equation; Sigmoid Isotherm*.

Smoke The product of incomplete combustion which contains particles of unburned carbon, carbon monoxide, and other gases, depending on the fuel. Smoke has a deleterious effect on health, causes discoloring of the atmosphere, buildings, and adversely affects many plants. Smoke and fog combine to make smog. Also see *Combustion; Maximum Average Contaminants; Pollution*.

Sodium Sulfate Drying An aqueous solution of 10 to 20% solids in the feed of sodium sulfate is dried in a spray dryer using heated air at 176 to 288°C (350 to 550°F).

Sol See *Gel*.

Solar Coefficient The temperature rise resulting from solar radiation, as the average 24-hr temperature rise that a solar collection will produce when receiving 4183 J/cm^2-day (1000 Langleys/ day). With 400 Langleys on a given day and a solar coefficient of 11.1°C (20°F), the temperature rise would be 11.1 x 400/1000 = 4.4°C (8°F). Also see *Absorptivity.*

Solar Constant The thermal radiation from the sun outside the atmosphere, 2.00 ± 0.04 gcal/cm^{2}min, or 2.0 Langley/min or 1395 watts/m^2, or 7.13 Btu/min st^2, or 0.168 hp/ft^2. The amount of solar energy reaching the earth varies considerably. Also see *Solar Coefficient.*

Solar Drying Open air solar drying of products--building, food, crops, aggregates, etc.--is the oldest method of drying, in which the product is placed on platforms, pavement, trays which may or may not have means of coverings (mobile roof, mobile tray, canvas) for preventing wetting from rain. The product may be stirred periodically to provide uniform heating. Downdraft of air over solar heated beds increases drying rate. Solar air collectors may be used to supply warm air, with a fan and air distribution for drying products. To dry a 5 cm (2 in) depth of grain from 20 to 15% with 190 Btu/hr ft^2 requires about 7 to 8 hr with a downdraft air flow 0.005 m^3/m^{2}s (1 cfm/ft^2). To dry the same grain without forced circulation, but with occasional stirring, requires at least twice as long, or two days of drying.

Various devices are used for collecting the solar energy from the sun called solar energy collectors. For drying the sun is used for heating air which is circulated through the product to be dried. The solar collector consists of three basic parts: the cover plate (except the bare plate collector), the absorber, and the back plate. Collectors may have one or two air channels (Fig. S-6). Solar energy collectors are (1) flat plate or (2) concentrator type. With the former the area of the absorber and the cover plate are equal; with the latter the energy is collected on a large area and absorbed on a small area, usually to get a higher temperature than that obtained with the flat plate.

During sunny days, 17,000 kJ/m^2 (1500 Btu/ft^2) day, and for cloudy but not overcast days, 10,225 kJ/m^2 (900 Btu/ft^2) day can be collected. An air flow of 0.04 to 0.06 m^3/m^{2}s (8 to 12 cfm/ft^2) over the collector surface is used for drying, which will give a temperature rise of approximately 8 to 15°C (15 to 27°F). For higher air flows over the collector, a higher efficiency is obtained, but with a lower temperature rise. A rule-of-thumb is that 1/10 ft^2 of collector area per bushel is required to dry cereal grain.

The collector efficiency is the useful energy collected by a solar energy collector divided by the amount of energy arriving at the face of the collector. The solar collector efficiencies of various types of collectors at the noon hour and for the full day are given in Fig. S-7.

Passive solar energy collection refers to the design of structures and other devices to make maximum use of the solar energy for drying without a separate collector. Also see *Solar Radiation.*

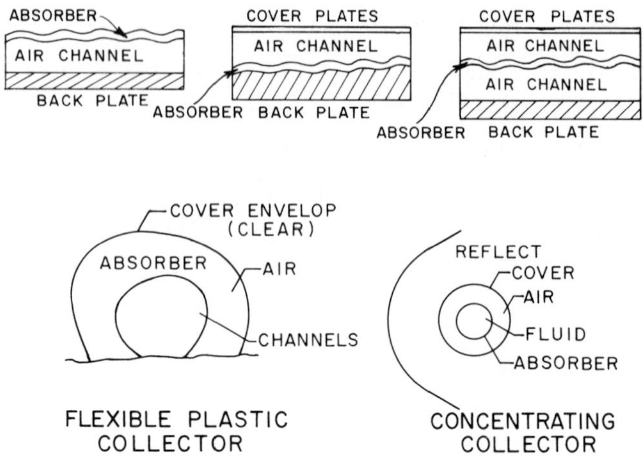

Figure S-6. Solar collectors used for drying.

Solar Radiation The radiation from the sun is made up of an extensive range of wavelengths of the spectrum. The atmosphere absorbs the very long (heat) and very short wavelengths. Dust, moisture, and bases, such as carbon dioxide, absorb certain wavelengths.

For drying or heating with solar energy, a solar constant of 1.34×10^6 erg/cm^2-sec or 424 Btu/hr-ft^2 or 1.3377 kW/m^2 is used, making adjustments for overcast conditions. Also see *Absorptivity; Insolation; Solar Coefficient; Solar Constant.*

Solar Salt Salt obtained from seawater or salt lake water by solar evaporation of trapped salt water in a pond.

Solenoid Valve A valve which is opened or closed by a solenoid (electromagnet) used on the control of the fuel system, for example, of a dryer. In the normally closed type, for example, the valve is opened by the solenoid but closed by a return spring and held closed by the fluid pressure upstream from the valve.

Solid Fuel Burner Solid fuels for supplying heat for drying provide an alternative to electricity, gas, oil, and solar sources. Of particular interest are those fuel sources obtained from the products being dried, primarily those of a cellulose base, e.g., bagasse, bark, cobs, bulbs, sawdust, shells, straw, and wood. Generally, the heating value of these fuels is about 19,000 to 19,750 kJ/kg (8000 to 8500 Btu/lb) of dry material, as listed in Table S-6. The materials may be ground, chopped, shredded, or used whole. The fuel may be pelleted to facilitate handling, transportation, and storage. Combustion may be over grates or in a fluidized bed. A fluidized bed may circulate a hot inert material, such as sand, to maintain a constant flame. Initial ignition of the bed may be facilitated with a

COLLECTOR					NOON HOUR	FULL DAY
SHAPE	COVER PLATE	ABSORBER	BACK PLATE	CHANNELS	%	%
	BARE	METAL	PLYWOOD	1	22	23
	FIBER GLASS	METAL	CHIP-BOARD-INSUL	1	68	63
	POLY	POLY	PLYWOOD	2	62	55
	GLASS	METAL	PLY-INSUL	2	75	68
	FIBER GLASS	DEEP-V METAL	PLY-INSUL	2	82	76
	FIBER GLASS MYLAR	DEEP-V METAL	PLY-INSUL W/REFLECT	2	89	83
	POLY	POLY	GROUND POLY	1/2	42	40
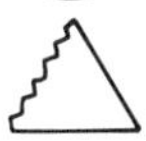	BARE	METAL	PLYWOOD	1	14	16
	FIBER GLASS	PLY-INSUL	(W/REFLECT)	1	42	41
	FIBER GLASS	CHIP-BOARD	CHIP-BOARD	2	51	51
	POLY	13" DIA PIPE	BLACK FELT	2	48	42

Figure S-7. Solar collector efficiencies, south facing, optimum angles, airflow of 8 cfm/ft^2.
From Kline, G. L. (1977). ASAE Paper 77-3007.

gas-fired unit. Coal as a solid fuel is covered elsewhere. Also see *Burner; Coal; Combustion; Fuels; Gas; Heater.*

Table S-6
Solid Fuels for Burners for Drying

Material	Heating Value (Dry Product) Btu/lb	kJ/kg
Bagasse	8,000-8,700	18,600-20,235
Bark	8,750	20,350
Charcoal	12,000-13,000	27,900-30,240
Coal	9,000-14,000	20,900-32,560
Cobs, corn	8,000	18,600
Hulls, general	7,000-9,000	16,280-22,100
Hulls, cottonseed	7,100	16,515
Peat, brown	7,800-9,200	18,140-21,400
Sawdust	8,700	20,235
Straw	5,000-6,500	11,630-15,120
Wood	7,800-9,200	18,140-21,400

Solubility   A measure of the ability of a solvent, usually a fluid, to dissolve a solid, to form a solute in a solution at a given temperature.

Sonic Drying   See *Resonating Drying; Ultrasonic Drying; Vibration Drying.*

Sorption   A general term applying to the gaining or losing of vapor or a gas by a substance which may include both adsorption and absorption. Also see *Absorption; Adsorption; Desorption.*

Soup Drying   Soup is made from a wide variety of materials from which can be made dry soup. Dry soup mix is made from either a mixture of dry ingredients or by drying a complete cooked soup. Atmospheric and vacuum drum dryers, freeze dryers, or spray dryers are commonly used for drying. Fluidized bed drying is used for pieces of soup. Continuous belt drying in a vacuum may also be used. The final moisture content of dry soup is about 3.0%.

Spalding Number   A dimensionless group, B', involved in droplet evaporation relating sensible heat and latent heat of the evaporated material:

$$B' = \frac{c_p \, \Delta t}{(L - q_r/v_m)}$$

where

c_p = specific heat
Δt = temperature difference
L = latent heat
q_r = radiation heat flux
v_m = rate of heat transfer

Specific Absorption The capacity of water holding material to absorb water after the gravitational water has been removed.

Specific Density See *Apparent Specific Density*.

Specific Gravity The ratio of the density of a substance to the density of water, at a specified temperature (Table S-7). Also see *Apparent Specific Gravity; Density; Voids*.

Table S-7
Specific Gravity of Grains and Percentage of Voids at Room Temperature

Grain	Moisture Content, %, w.b.	Voids in Bulk, %	Kernel Specific Gravity
Wheat, hard	9.8	40.1-42.6	1.29-1.30
Wheat, soft	9.8	39.6	1.32
Oats	9.4-10.3	47.6-55.5	0.95-1.06
Barley	9.7-10.4	39.5-57.6	1.13-1.33
Soybeans	6.9-7.0	33.8-36.1	1.13-1.18
Sorghum, grain	9.5-9.9	36.8-37.0	1.22-1.26
Rice	11.9-12.4	46.5-50.4	1.11-1.12
Shelled corn	9.0	40.0	1.19

Agricultural Engineers Yearbook, ASAE. 1976.

Specific Heat Quantity of heat, Btu, required to change one pound mass of a product, one degree F, or cal/°C-gram or kJ/kgK. Specific heat is also defined as the ratio of the quantity of heat to that of water for the same temperature change. The thermal capacity is often used to represent the quantity of heat required to change a unit mass of material one unit of temperature (Table S-8). Also see *Heat Capacity; Thermal Capacity*.

Specific Refractivity See *Refraction*.

Specific Surface The area per unit volume of a material, obtained by dividing the area by volume, L^2/L^3, or 1/L. Values of specific surfaces are:

Cube: 6/L
Cylinder: 2/R or 4/D
Sphere: 3/R or 6/D

The specific surface may be used to describe a bed of a product to be dried, such as sand, grain, potatoes, chips, etc.

Specific Viscosity See *Viscosity*.

Table S-8
Specific Heat, Btu/Lb °F, Approximate

Concrete	0.15
Cereal grains, 15% moisture	0.40-0.50
22% moisture	0.50-0.55
Forages, 20% moisture	0.35-0.45
60% moisture	0.65-0.75
Wood	0.40-0.70
Coal	0.3
Air	0.18-0.24
Clay	0.22
Ice	0.5
Water	1.00
Gasoline	0.5

Conversion factors: *1 Btu/lb* °F = 1 cal/g°C = 4.1868×10^3 J/kgK

Specific Volume The volume of a unit mass of a substance, such as cubic feet per pound or cubic centimeters per gram.

Specky Dried Egg An egg product which contains flecks and lumps from the material caked on the wall of a spray dryer during spray drying eggs.

Spiral Dryer An internal power driven rotor, from 0.15 to 2 m (6 to 80 in.) diameter, having spiral guide plates, inside a heating jacket, which is used for drying heat-sensitive products. The moist product is fed into the dryer in a fluidized state at the bottom of the dryer. The product moves around the spiral air guides. The revolving rotor, moving at 1 to 10 rpm, moves the product to the heated section for drying. The dry product leaves the top of the dryer and is removed in a cyclone separator. Present capacities (1973) of this type of dryer are 100 to 8000 kg/hr (220 to 17,600 lb/hr) of dry product, at 0.1 to 2.0%.

Spontaneous Combustion As a result of slow oxidation of a substance such as coal or forage, heat is produced which cannot escape, and the temperature increases until ignition occurs if sufficient oxygen is present, causing fire. Reducing the moisture content of a substance decreases the possibility of spontaneous combustion.

Spouted Bed Dryer A special fluidized bed dryer for removing moisture from particulates in which there is a convective flow of drying fluid and products being dried (Fig. S-8). As compared to stationary bed dryers, faster drying and shorter exposure time occurs while mixing is accomplished without mechanical agitators or stirrers. Also see *Fluidized Bed Dryer; Pneumatic Dryer*.

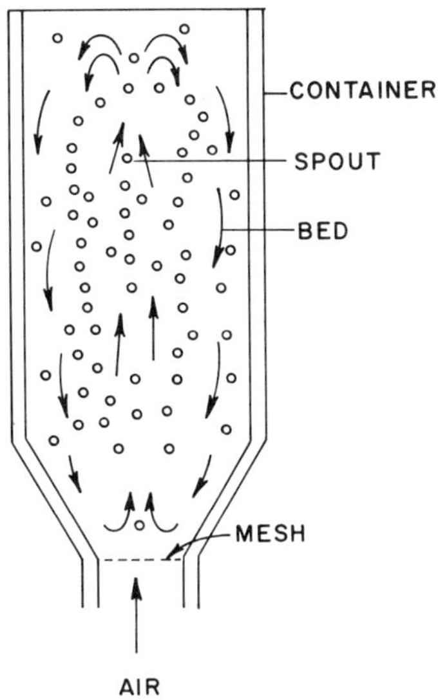

Figure S-8. Spouted bed dryer.

Spray Dryer· The spray dryer is used for removing moisture from pumpable fluids. A high moisture product, which may first be condensed in a vacuum pan or an evaporator, is pumped through an atomizing nozzle with atomization taking place inside the chamber of the dryer. The product is dried quickly by moving air, usually filtered if for food, over small droplets formed by atomization. The dried product must be removed from the dryer quickly to avoid heat damage. The inlet air temperature varies from 93.3 to 760°C (200 to 1400°F), depending upon the product, moisture content, and rate of drying. The usual inlet temperatures are from 500 to 550°C (932 to 1025°F) with an outlet temperature of 65 to 70°C (149 to 158°F). Steam heated spray dryers have an overall thermal efficiency of 35 to 40%, while direct fuel fired units have an efficiency of 80 to 85%. For spray drying, 1.2 to 1.8 kg of steam per kg of moisture evaporated are required. Units are rated in terms of production of dried product which may reach 38 kg/s (5000 lb/hr) or more or in terms of evaporation rate of 115 kg/s (15,000 lb/hr) or more. Spray dryers are characterized by continuous operation, minimal labor to operate, ability to handle many different products, short time of heat contact during drying, operational efficiency depending on large surface area produced by different atomization methods, separation of product and air after drying and a lower thermal efficiency as compared to drum dryers (Fig. S-9). Also see *Atomization; Classification; Cyclone; Foam Spray Dryer; Horizontal Chamber Spray Dryer; Milk Drying; Product Number; Spalding Number; Tower Spray Dryer; Vertical Chamber Spray Dryer.*

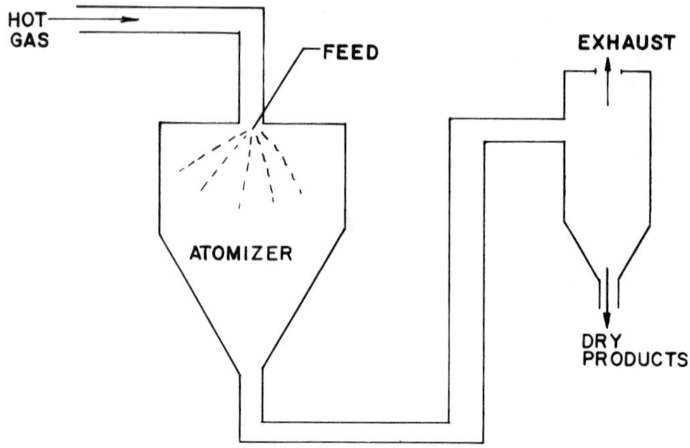

Figure S-9. Spray dryer.

Spreading Potential A function applicable to a film adsorbed on a surface of a solid which is analogous in two dimensions to the pressure of a compressed vapor in three dimensions.

Squash Drying See *Pumpkin Drying*.

S-Shaped Function A mathematical expression representing the relationship of two variables, such as relative humidity versus moisture content, of a hygroscopic material (Fig. S-10). One S-shaped general function is:

$$h(p) = \{1 - (1-p)^2\}^2 = 4p^2 - 4p^3 + p^4$$

$$\begin{aligned} \text{at } p &= 0 \; h(p) = 0 \\ p &= 1 \; h(p) = 1 \end{aligned}$$

Also see *Sigmoid Isotherm*.

Stack Effect The upward flow of hot and/or moist air due to density difference, such as in a dryer, storage, or chimney (stack).

Standard Atmosphere Used in meteorological work, an environment at 1 atm pressure, 760 mm (sea level) at 15°C (59°F). Also see *Bar; Conversion Factors; NTP; STP*.

Standard Deviation The square root of the variance of a probability distribution, which is a measure of the dispersion of values.

If ε_n denotes the deviations of various measured values of the same quantity from the mean of all measured values, the value of ε given in the equation below will be a significant measure of the observational uncertainty introduced by the errors if the quantity

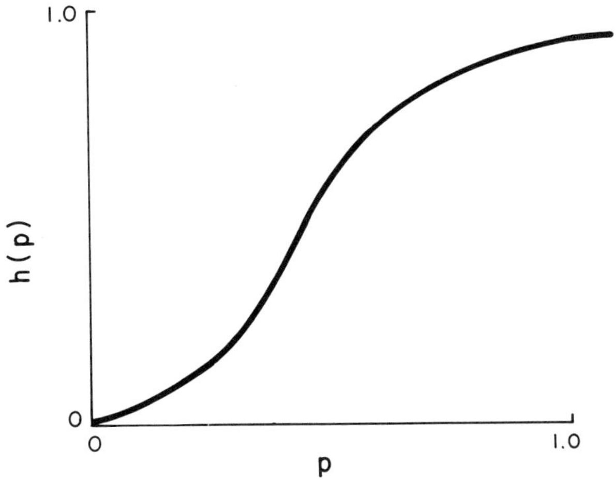

Figure S-10. S-shaped function.

under the radical is replaced by the average value of all terms under the radical, the standard deviation is represented by equation:

$$\text{Standard deviation} = \sqrt{\frac{\varepsilon_1^2 + \varepsilon_2^2 + \ldots + \varepsilon_n^2}{n}}$$

Also see *Normal Distribution; Standard Error; Variance.*

Standard Error   A designation sometimes given to the deviation of a sampling distribution. The standard error of estimate, SE, (Fig. S-11), in the y-direction is:

$$SE_y = \pm\sqrt{\frac{\Sigma y^2 - \frac{(\Sigma y)^2}{n} - n\left[\Sigma xy - \frac{\Sigma x \Sigma y}{n}\right]}{n - 2}}$$

and

$$SE_x = \pm \frac{1}{m} |SE_y|$$

$$m = \frac{\Sigma xy - \frac{\Sigma x \Sigma y}{n}}{\Sigma x^2 - \frac{(\Sigma x)^2}{n}}$$

$$b = \frac{\Sigma y}{n} - m \frac{\Sigma x}{n}$$

n = number of values

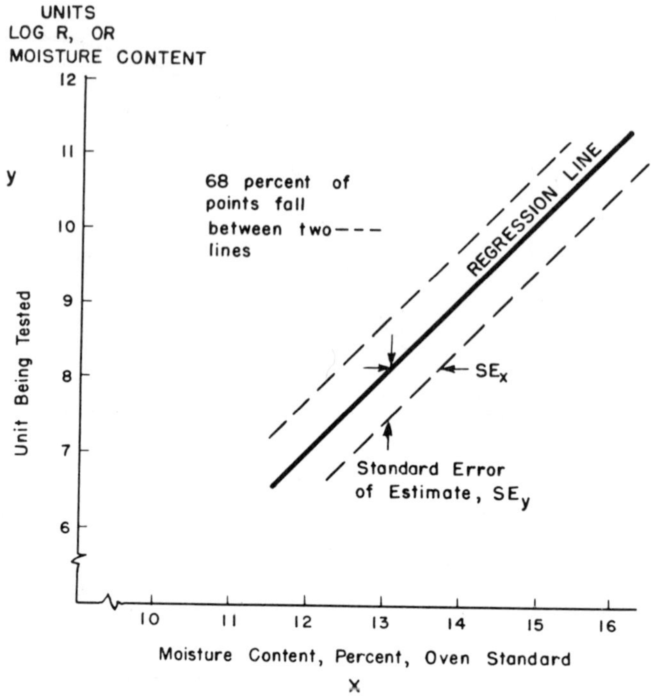

Figure S-11. Standard error of estimate.

Various representations of standard errors of measurement are given in Table S-9. Also see *Error; Sampling; Variance.*

Standard Temperature Pressure (STP) See *NTP.*

Standard Volume See *Volume.*

Stanton Number A dimensionless parameter, St, which characterizes forced convective heat transfer relating the rate of heat transfer and thermal capacity of the fluid:

$$St = \frac{h}{\rho V C_p} = \frac{Nu}{Re \cdot Pr} = \frac{Nu}{Pe}$$

where

h = surface heat transfer coefficient
c_p = specific heat
ρ = density
V = velocity

Also see *Nusselt Number; Peclet Number; Prandtl Number; Reynolds Number.*

Table S-9
Standard Error of Measurements

Standard error of measured quantities have means $m_1, m_2, \ldots, m_n$ with standard errors $\alpha_1, \alpha_2, \ldots, \alpha_n$, respectively, then the standard error of:

1. The <u>sum</u> of $m_1 + m_2$ is $\sqrt{\alpha_1^2 + \alpha_2^2}$
2. The <u>difference</u> of $m_1 - m_2$ is $\sqrt{\alpha_1^2 + \alpha_2^2}$
3. The <u>multiple</u> km_1 is $k\sigma$
4. The <u>product</u> $m_1 m_2$ is $\sqrt{m_1^2\alpha_2^2 + m_2^2\alpha_1^2}$
5. The <u>product</u> $m_1 m_2 m_3$ is α where

$$(\alpha/m_1m_2m_3)^2 = (\alpha_1/m_1)^2 + (\alpha_2/m_2)^2 + (\alpha_3/m_3)^2$$

6. The <u>power</u> m_1^p is α where

$$\alpha/m_1^p = p\alpha_1/m_1$$

$$\alpha = (pm_1^{p-1})\alpha_1$$

Note: The above relationships must be used with caution, particularly if linearity is assumed for non-linear relationships.
Based on Topping, J. 1955. Errors of Observation and Their Treatment. Reinhold Publishing Corp., New York, NY.
Source: From Hall, Carl W., Errors in Experimentation (1978). Used with permission of Matrix Publishers, Inc., Portland, OR.

Starch Drying A complex carbohydrate compound of high molecular weight, with the general formula of $(C_6H_{10}O_5)x$ with the x representing the unknown number of units in a molecule. The starch granules in vegetables determine to a large extent their texture. During dehydration of products high in starch, such as potatoes, the moisture diffuses through the starch gel, which controls the drying and heating rates.

Static Efficiency (SE) of a Fan The static efficiency of a fan is represented by:

$$SE = \frac{(cfm)(SP)}{(6356)BHP}$$

where

cfm = air flow
SP = static pressure
BHP = brake horsepower at shaft of fan

Static Pressure A measure of air pressure usually expressed in inches or centimeters of water column (W.C.) for drying beds. Also see *Conversion Factors; Manometer; Pressure Drop.*

Stationary Dryer A device which operates in a fixed position used for removing moisture from products which can be done in a batch or continuous process, in contrast to a portable unit.

Steady State A condition where the rate of exchange of mass, energy, or other substance is such that the rate of input is equal to the rate of output, so that the net change of the body is zero, a condition in which there is no dependence on time. The steady state is a condition in which the operation or process reaches equilibrium used in contrast to unsteady state.

Steam The vapor of water at or above the boiling point. Generally classified as wet steam, where all of the liquid in steam is not in a gaseous state; dry saturated steam, where all of the water is in vapor form, making 100% quality; and superheated steam, where sensible heat is added to dry steam to increase the temperature above the boiling or evaporation temperature corresponding to the pressure. The steam tables represent the pressure, temperature, and heat relationships (Table S-10). Also see *Quality; Saturated Vapor.*

Steam Dryer One of several types of dryers which uses steam, either direct or indirect, as a source of heat for drying. More commonly, the designation is used for units in which steam is used directly for drying, such as for lumber, sawdust, coal, etc.

Stefan-Boltzmann Constant A proportionality constant, usually represented by σ, in the equation representing the radiation heat loss by emission from a black body, $E = \sigma T^4$, where E is in appropriate units of heat per unit time. σ is expressed as

0.17×10^{-8} Btu/ft^{2}hr°R^4
1.36×10^{-12} cal/cm^{2}sec°R^4
5.71×10^{-5} ergs/cm^{2}sec°K^4
4.9×10^{-8} kg-cal/m^{2}hr°K^4
56.7×10^{-12} kW/m^2°K^4

Also see *Stefan-Boltzmann Law.*

Stefan-Boltzmann Law A law of thermal radiation stating that the radiation from a black body is proportional to the fourth power of the absolute temperature of the body:

Table S-10
Saturated and Superheated Steam Properties

Pressure, psia	Temp., °F	Specific Volume Ft^3/lb	Heat of Liquid, Btu/lb	Latent Heat, Btu/lb	Total Heat, Btu/lb
Saturated Steam					
14.7	212.0	26.8	180.1	970.3	1150.4
20	227.9	20.1	196.2	960.1	1156.3
30	250.3	13.7	218.8	945.3	1164.1
40	267.3	10.5	236.0	933.7	1169.7
50	281.0	8.5	250.1	924.0	1174.1
60	292.7	7.2	262.1	915.5	1177.6
70	302.9	6.2	272.6	907.9	1180.5
80	312.0	5.5	282.0	901.1	1183.1
90	320.3	4.9	290.6	894.7	1185.3
100	327.8	4.4	298.4	888.8	1187.2
125	344.3	3.6	315.7	875.4	1191.1
150	358.4	3.0	330.5	863.6	1194.1
200	381.8	2.3	355.4	843.0	1198.4
250	400.9	1.8	376.0	825.1	1201.1
300	417.3	1.5	393.8	809.0	1202.8
400	444.6	1.4	424.0	780.5	1204.5

Superheated Steam

Pressure, psia		Temperature, °F 300	400	500
	sh	88.00	188.00	288.00
14.7	v	30.52	34.65	38.75
(212°F)	th	1192.0	1238.9	1285.4
	sh	49.66	149.66	249.66
30	v	14.82	16.89	18.92
(250°F)	th	1189.2	1237.4	1284.2
	sh	25.55	125.55	225.55
45	v	9.79	11.20	12.57
(274.5°F)	th	1185.9	1235.8	128.30
	sh	7.29	107.29	207.29
60	v	7.26	8.35	9.39
(292.7°F)	th	1181.8	1234.0	1281.8
	sh		92.4	192.40
75	v		6.64	7.49
(307.6°F)	th		1232.2	1280.6
	sh		79.7	179.73
90	v		5.5	6.22
(320.3°F)	th		1230.0	1279.3
	sh		68.6	168.62
105	v		4.69	5.32
(331.4°F)	th		1227.6	1278.0

sh = superheat, °F; v = specific volume, cu ft/lb; th = total heat, Btu/lb

Conversion factors: $1 \text{ psi} = 6.8947 \text{ kPa} = 6.8947 \times 10^3 \text{ N/m}^2$
$1 \text{ ft}^3/\text{lb} = 6.243 \times 10^{-2} \text{ m}^3/\text{kg}$
$1 \text{ Btu/lb} = 2.326 \times 10^3 \text{ J/kg}$

From Van Wylen, G. J., and Sonntag, R. E., Fundamentals of Classical Thermodynamics (1976). Used with permission of John Wiley and Sons, New York.

$$E = \sigma T^4$$

where σ is the Stefan-Boltzmann constant. Also see *Boltzmann Constant; Emissivity.*

Stein Tester A mechanical device for determining the resistance of cereal grains to breakage during handling and drying. In general, the lower the temperature used for drying the better the quality and less damage during handling after drying.

Steinlite See *Dielectric Method.*

Stere A metric unit of volume equal to 1000 liters or 1 m^3.

Stickers Devices, usually made of wood, used for spacing of lumber so that air can move through and between the stacked pieces of lumber for drying. Usually placed 1 to 2 m (3 to 5 ft) apart, beginning at the ends of boards, placed one above the other over bunks or other supports to prevent warping of the lumber. Also see *Lumber Drying.*

Stoichiometry The study of material and energy balances and heat of a chemical system. The principles of stoichiometry may be used to study a drying system. Stoichiometric combustion is complete combustion of elements: $CH_4 + 2\ O_2 \rightarrow CO_2\uparrow + 2\ H_2O$ + heat. Also see *Combustion.*

Stoker A mechanical device for automatically feeding solid fuels, usually coal, into a furnace, such as a heater for a dryer.

Stomata Pores on plants, principally in leaves of plants, which have guard cells (two) around the opening, which open during the day and close at night, and which influence the ease of drying forages and plants (Fig. S-12). By maintaining a high humidity around the leaves, ease of moisture transfer through the opening is enhanced. At the same time diffusion and transfer are decreased because a lower vapor pressure difference exists with higher humidity external air. The stomates serve as the major path for gases to diffuse into and out of the leaf of a plant. The closing of the stomates greatly retard the exchange of gases, including water vapor, with the atmosphere. There are from 2000 to 100,000 stomates/cm^2 on plants. Fully open stomates approach a dimension of 10 x 40 microns in size. Stomates are generally open in light and closed in dark. Stomates have an optimum temperature for opening and even in light at optimum temperatures may remain closed in dry conditions. Water loss by vapor during transpiration occurs primarily through the stomates.

Straight-Bladed Fan See *Centrifugal Fan; Forward-Curved Fan.*

Strainer See *Filter.*

Streamline See *Laminar.*

Figure S-12. Stomata.

STP (Standard Temperature Pressure) See *NTP (Normal Temperature Pressure)*.

Stress (Drying) An internal force, exerted by two adjacent parts of the product, such as for a piece of wood, during drying. Stress is caused by uneven drying and shrinking and influence by set, the localized semipermanent deformation in wood is caused by internal tensile or compressive stresses.

Stress Cracks (Grain) See *Cracking*.

Sublimation The passage of a substance from the solid state to the gaseous state and back into the solid state without the appearance of the liquid state. Also applied to the conversion of a solid to a vapor with the return to the solid state and to conversion directly from the vapor phase to the solid state. Desublimation is sometimes used to describe the condensation from the gas to solid phase, which for condensation of water vapor to ice is called ablimation. The principles apply to freezer drying. Also see *Freeze Drying*.

Suction System Method of air movement in which the air is moved through the product in which the air duct or ducts are at a pressure lower than atmospheric. A suction system is also called an exhaust system of air movement.

Sugar A group of chemical compounds composed of carbon, hydrogen, and oxygen, which is one of the carbohydrates. Sugars are generally made of, or are combinations of, $C_6H_{12}O_6$ or $C_{12}H_{22}O_{11}$.

Sugar Beet Drying and Storage The loss of sugar in sugar beets occurs in storage when the temperature is above 18.3°C (65°F), with approximately 0.1 to 0.5 kg (1/4 to 1 lb) loss per day per ton. The use of mechanical harvesting equipment has made the potential heating losses of sugar beets significant because the beets can be harvested much faster than they can be processed. Air can be blown into the pile when the air is 2.7 to 3.3°C (5 to 6°F) colder than the hottest spot in the beets. In the initial cooling of the beets it is desirable to use from 1.0 to 1.5 x 10^{-6} m^3/kgs (20 to 30 cfm/ton), followed with approximately 0.5 m^3/kgs (10 cfm/ton) to maintain the temperature of the pile. One air distribution system for cooling sugar beet piles consists of using 200 ℓ (50 gal) drums with the ends removed and the drums welded together with strips of iron maintaining a space of about 1.2 cm (3 in.) between the drums. The fan is then placed on one end of the tunnel. The optimum relative humidity of air forced into the piles is 80%.

Sugar Beet Seed Drying To decorticate sugar beet seeds it is necessary that the moisture content be 11% or below. The seed is grown in Oregon and California and shipped to the north central section of the United States for processing and planting. As the seed is moved into the north central areas moisture is absorbed from the atmosphere. Moisture can be reduced to 11% with unheated forced air if the temperature is above 15.6°C (60°F) and relative humidity below 40%. Heated air can be used at 43.3°C (110°F) if the seed is exposed to the heat for over one hour, which would occur in deep bin drying with heated air, or 43.9°C (120°F) if a thin layer of 1.6 cm (4 in) is dried so that the seeds will not be heated for more than an hour. The static pressure at a given air flow through the whole seeds is slightly less than for shelled corn.

Sugar Drying The production of dry sugar from the juice of sugar cane by pressing or beet sugar by diffusion is similar. The juice from sugar cane is obtained by pressing between rolls with a fibrous material, bagasse, remaining. The juice is purified, concentrated in a multiple stage (two to five) evaporator, then boiled in a single stage vacuum unit and seeded with sugar crystals. The product is then centrifuged and dried. Sugar is removed from slices of beet sugar in a battery of diffusers. The beet pulp remains, which is not as fibrous as bagasse, and is often dried.

Beet pulp and bagasse are dried from 75 to 85% moisture to 5 to 15% in a direct-fired, parallel flow, forced draft rotating drum (internal with baffles) dryer. The heated air may enter the drum at about 538°C (1000°F) and leave at 93.3 to 137.8°C (200 to 280°F), the beet pulp or bagasse approaching 100°C (212°F) upon drying.

The crystallized product is first mechanically dried with a centrifugal device, reducing the moisture content to 1%. Crystal sugar has a very low equilibrium moisture content compared to many other food products which are dried and is very hygroscopic. Sugar is dried to 0.1% to get a free flowing product. Above 0.2 to 0.3% the sugar is tacky or sticky. The most common dryer is an internally heated rotary drum with internal flights. The drum is mounted at an angle to assist in moving the sugar through. Counter-current flow of sugar and air is used. The flights pick up the sugar on the bottom and drop the sugar through the air moving at about 1 m/s (3 ft/sec), the air entering at 32.2 to 43.3°C (90 to 110°F), with a residence time of the product of 5 to 10 min. The residence time in the dryer varies greatly depending on the diameter and length of drum, temperature and speed of air flow, original and final moisture content of the product, etc. Air is drawn rather than pushed through the dryer to minimize escape of dust into the room in which the dryer is located.

Vertical dryers, with discs mounted on a rotating shaft, are also used. The wet sugar enters on top, is thrown out centrifugally, and moves down to the next tray, as air moves up through the dryer removing moisture. Fixed trays may also be used in which the sugar is scraped from one tray and falls to the next in the heated air, which removes moisture from the product.

Upon leaving the dryer the sugar is passed through a screen to remove lumps. Also see *Granulator*.

Sugared Whole Egg or Yolk Solids Eggs to which a carbohydrate such as sucrose has been added before drying in order to improve the stability of properties.

Sulfiting See *Sulfuring*.

Sulfuring The treatment of a fruit or vegetable with sulfur dioxide gas or of a sulfite to control browning. Also known as *Sulfiting*. Also see *Apple Drying; Apricot Drying; Ascorbic Acid; Banana Drying; Celery Drying*.

Sumac Leaf Drying The leaves of sumac contain about 25% tannin which is used for processing leather and preparing ink. The seeds of the sumac are undesirable for processing purposes. Sumac has about 60% moisture when cut. It is important that the leaves do not get wet during drying because the tannin will be lost through leaching. Unheated forced air can be used for drying in much the same procedure as for hay. Air heated about 5 to 7°C (10 to 12°F) above atmospheric is better than natural air, even though more expensive, as long as the leaves are not heated above 60°C (140°F).

Sun Curing Tobacco Developed as a cottage industry in many areas of the world in which leaves of the tobacco are strung on racks, placed in open fields, along roads and streets, so that the tobacco is exposed to the sun to heat. The exposure is provided for about 1 month until the tobacco leaves turn a golden yellow.

Superficial Air Velocity The calculated velocity of air through a cross section containing a product as if no product is present. For a product with 40% void space where the average velocity is 100 ft/min the superficial velocity is 250 ft/min. Also see *Air Velocity; Mean Moisture Ratio*.

Superheat The heat added to saturated vapor to increase the temperature of the vapor above the temperature of vaporization corresponding to the pressure of the vapor. Also see *Superheated Steam*.

Superheated Steam Steam to which heat has been added to increase the temperature above that temperature corresponding to the temperature of boiling for the pressure of the vapor. By adding sensible heat to dry saturated steam, while maintaining the same pressure, superheated steam is obtained. Each degree F increase is identified as a degree superheat. Heaters for vapor heat transfer, known as superheaters, are used. The greater the degree of superheat, the greater the amount of heat that can be removed without condensation. Also see *Quality; Saturated Vapor.*

Supersaturated Solution   A solution in which more solute is held by the solvent than in a saturated solution. The normal maximum amount of solute held in solution is a saturated solution. A supersaturated solution is relatively unstable and will often drop part of the solute from solution when thermally or physically disturbed or shocked. Also see *Saturated Liquid; Saturated Solution.*

Supplemental Heat   Heat added to the heat already in the atmosphere to obtain a limited temperature rise, usually less than 6°C (10°F), to accomplish drying within the maximum permissible drying time to prevent spoilage. Also see *Low Temperature Drying.*

Supplemental Heater   A unit designed for a limited temperature rise, usually less than 6°C (10°F), is designated as a supplemental heater. A heat output of about 29,300 to 73,250 J/s (100,000 to 250,000 Btu/hr) is provided. Supplemental heaters are particularly valuable for drying crops in deep bins during periods of high humidity. The operation of the heater can be controlled by a time clock, thermostat, or humidistat.

Surface Active Agent   A substance, basically an emulsifier, added to another material which changes the surface properties of their solutions, generally lowering the surface tension of the solvents. Generally used for obtaining uniform wetting in a manufacturing process operation. Examples are monoglyceride diglycerides, pectins, gums, lecithin, and gum. Also called surfactants. Also see *External Moisture; Surfactants; Wetting Agent.*

Surface Area   See *Atomization.*

Surface Moisture   The free moisture on the surface of material.

Surface Tension   A property of liquids resulting from intermolecular forces, and appearing as a force drawing the liquid together into a body of minimum surface area, dynes/cm, lb-force/ft, or newtons/m.

Surfactants   Surface active agents which when added to a liquid increase spreading and wetting of a surface. Materials used as surfactants placed in a solvent in the dry cleaning industry, as oil-soluble surfactants, are of three types: anionic, non-ionic, and cationic. Also see *Surface Active Agents; Wetting Agent.*

Suspension Drying   See *Fluidized Bed Drying.*

Sweating   The procedure of holding products in large bins for a time after drying to equalize the moisture content throughout the product. Also see *Equalization; Resting; Tempering.*

Sweet Corn Drying Kernels which are somewhat soft are removed from the cob of sweet corn and blanched before drying. The corn kernels can be frozen and stored before drying. Sweet corn is dried on trays in a tunnel dryer or continuously on a belt dryer. A two-stage tunnel dryer, with air entering the first stage at 82.2°C (180°F) and the second stage at 73.9°C (165°F). The tunnel dryer may be followed by a bin dryer to dry to 5% moisture or less.

Single stage drying can be carried out in trays placed in a tunnel or cabinet dryer, using air at 82.2°C (180°F). The temperature may be reduced as drying occurs until the sweet corn is 5 to 6%.

Sweet Potato Curing Low temperatures, especially those below 4.4°C (40°F), will injure sweet potatoes. To heal after harvest sweet potatoes should be cured using air at a temperature of 29.4°C (85°F) and a relative humidity of 85%, requiring about 10 days. If the relative humidity is too low healing will not take place even though the temperature is high. The curing period, for which heat is required in most areas of the United States, will take from 10 to 20 days. If the curing period is too long excessive sprouting will occur. The optimum temperature for storage is 10 to 12.8°C (50 to 55°F) and the relative humidity should be maintained at about 80 to 85%. Ventilation is required during storage to provide the proper uniform temperature and to prevent condensation. A heater with an output of 105 to 180 kJ/m^3 (3 to 5 Btu/ft^3) storage capacity is adequate for most installations.

Sweet Potato Drying Sweet potatoes are peeled, diced, blanched, and sulfited before drying. Sweet potatoes are dried on a conveyor with heated air at 121°C (250°F) forced through the product or tunnel dryer, first stage at 79.4 to 100°C (175 to 212°F) and second stage at 60 to 71°C (140 to 160°F), to provide 10 to 12% moisture content product. Bin drying using 60 to 71°C (140 to 160°F) can be used for final drying to 4% moisture content after conveyor or tunnel drying. Also see *Sweet Potato Curing; Sweet Potato Flake Drying*.

Sweet Potato Flake Drying Sweet potatoes for flaking are peeled, cut, and blanched. The pieces, which are about 2.5 cm (1 in.) thick, are puréed and adjusted to 20% solids with hot water. The slurry is dried to produce instant flakes at 2% moisture content on a double drum dryer, heated with steam at 516 to 585 kN/m^2 (kPa) (75 to 85 psi). A sheet of dried product is stripped from the drums which is ground to produce flakes.

Symbols and Units Commonly used symbols, definitions, and SI units are given in Table S-3. Also see *Number Prefixes*.

Syneresis The consolidation by reduction in volume or contraction of a gel so that the interstitial water is forced out, making drying easier.

System Internationale See *SI Units*.

T

Tablet See *Block; Compaction.*

Tag-Heppenstall See *Electrical Resistance Method* of moisture determination.

Tanning The process of making leather from dehaired hides by treatment with chemicals to dry and to convert certain biological components providing a pliable, flexible material. Also see *Leather Drying.*

Tapahan Drying See *Copra Drying.*

Tapered Duct A specially designed duct to carry and/or distribute air uniformly to a product being ventilated. The large area end of the duct carries a greater quantity of air than the smaller area end. Also see *Duct.*

Tau, Time Constant See *Time Constant.*

Taylor Number See *Sherwood Number.*

Tea Drying Leaves from the tea plant are at 88% moisture content when harvested which is followed by drying. Tea leaves are fermented from which an extract is made. The liquid extract may be concentrated to 25 to 30% solids with a falling film evaporator from which the volatiles may be recovered. Drying of the concentrate may also be done with a vacuum (8 mm pressure absolute) continuous belt dryer or vertical spray dryer tower using air at 190°C (375°F).

Temperature A measure of the intensity of heat which is proportional to the kinetic energy of the molecules of a body. Temperature is measured with thermometers of which there are several. Various temperature scales are used: Fahrenheit, Celsius, Rankine, Kelvin, and Réaumur. Also see *Critical Temperature; Drying Temperature; Temperature Scales*.

Temperature Gradient The difference in temperature per unit distance perpendicular to the isothermal surfaces. The temperature gradient provides a measure of the rate of heat transfer:

$$|\Delta t| = \lim_{\Delta s \to 0} \frac{t_2 - t_1}{s} = \lim_{\Delta s \to 0} \frac{\Delta t}{\Delta s}$$

where

t_2, t_1 = temperatures
Δs = distance

Temperature Quotient The ratio of the rate of a particular reaction or process to the rate of a specified higher temperature, represented as q_{10} for 10°F or Q_{10} for 10°C. Often applied to respiration processes.

Temperature Rise As applied to crop drying, temperature rise refers to the difference between ambient temperature and the temperature of the drying air resulting from the addition of heat by the dryer burner. Also see *Low Temperature Drying*.

Temperature Scales The scales for representing temperature (hotness or coldness) are developed with a value of zero at a zero level of molecular activity, with zero degree at zero activity for absolute zero temperature. For practical use the freezing point of water is taken as zero and the boiling point is taken as 100° of the Celsius (C) scale; 0° and 80° on the Réaumur scale; and 32° and 212° on the Fahrenheit (F) scale. Relationships of temperatures are:

$t_F = 9/5\ t_C + 32°$
$t_C = 5/9\ (t_F - 32°)$
Temperature Kelvin, $T_K = t_C + 273.16°$
Temperature Rankine, $T_R = t_F + 459.69$
Temperature Réaumur, $T_{Re} = 4/5\ t_C$

Tempering The process of equalization of moisture or temperature throughout a product or the bringing of a product to a desired moisture content or temperature for processing. The product is held in a container with the desired environmental conditions maintained. Also see *Dryeration; Intermittent Drying; Resting Period*.

Tension Dryer See *Tenter Dryer*.

Tensionless Dryer See *Festoon Dryer*.

Tenter Dryer A dryer designed to remove moisture from pieces of cloth under tension or in which the cloth is stretched while being dried. The two ends of the cloth or product are clipped to conveyor chains and moved through forced air. High velocity jets may be directed on the material to accelerate drying.

Textile Drying Several drying and moisture control processes are involved in the manufacture of fabrics and in the laundering of textiles. During manufacture cloth is dried under tension on a tenter frame on a steam heated can or without tension on a loop dryer. Wet raw silk or skeins (a quantity of fibers removed from a reel) is dried in a centrifugal extractor in the presence of circulating warm air. During manufacture, following washing, wool and other fabrics are dried in a dryer which holds the cloth at a desired width in the hot air (Table T-1).

One process uses a series of perforated rotating drums, with heated air pulled into the drums through the material which is placed over the drum perforations, for drying loose fibers in the textile industry. Yard (cotton and wool), knitted fabrics, rayon fiber, cellulose acetate, and sisal are examples of products which can be dried with this unit. Also see *Festoon Dryer; Hydrophilic Fiber; Loop Dryer; Tenter (Frame) Dryer.*

Table T-1
Textile Fiber Characteristics*

Fiber	Density, g/cm^2	Moisture Regain %, d.b.**	Effect of Heat, °F		
			Melting	Decomposes	Safe Ironing
Acetate	1.32	6-6.5	500	-	420
Acrylic	1.17	1.0-2.5	-	-	325
Cotton	1.54	6.0-8.5	-	464	425
Flax, linen	1.50	6-10.0	-	500	450
Nylon	1.14	8	420-480	-	300
Polyester	1.38	0.4-0.6	455-554	-	300
Rayon	1.52	12.0	300	464	250
Silk	1.30	10-11	-	-	300
Wool	1.32	15-17	-	400	275

**Partially based on Linton, George E. 1966. Natural and Manmade Textile Fibers. Duell, Sloan and Pearce, N. Y., NY. 420 p.*
***At 70°F and 65% RH.*

Therm In English units, may refer to the heat equivalent of 100,000 Btu. In older literature, therm was another name for the British thermal unit or calorie. Also, used to represent the amount of heat equivalent to 1000 kilo-calories or 1,000,000 (gram) calories. Also see *Q; Quadrillion; Quintillion.*

Thermal Capacity The heat capacity of a body on the basis of amount of heat required to raise the temperature 1°C. Also see *Specific Heat; Thermal Conductivity.*

Thermal Conduction A process in which heat is propagated as a result of the thermal motion of the elementary particles of the material (molecules, atoms, free electrons). Essentially, conduction occurs in solid bodies and in very thin, stationary layers of liquid or gas. Also see *Thermal Conductivity*.

Thermal Conductivity A measure of the movement of heat through substances represented by k, by kinetic molecular activity. k has units of cal/cm-sec-°C in the cgs system and Btu-in/hr ft^2°F or Btu/hr ft°F in the English system and J/ms°K or W/mK in SI units (Table T-2). A conversion factor is 1 cal/cm sec°F = 0.80620 Btu-in/ft^2sec°F = 418.7 J/ms°K = 418.7 W/mK. The rate of heat transfer, q, Btu/unit time, by conduction,

$$a = \frac{k}{x}(t_2 - t_1)a$$

where

k = thermal conductivity
x = distance perpendicular to the face
A = the cross sectional area
$t_2 - t_1$ = the temperature difference

These relationships are used for steady-state heat transfer. Also see *Conduction*.

Thermal Convection The transfer of heat due to movement of the fluid because of difference in density or temperature, in natural convection, or with the additional movement caused by forced agitation, known as forced convection. The heat transfer properties of a film or layer of the fluid adjacent to the heat transfer surface controls the rate of heat transfer. Heat moves through the film by conduction, which in turn controls the rate of heat transfer by convection. The film coefficient or surface thermal conductance, h (represented with f by some authors) (Table C-8), in W/m^2K or Btu/hr ft^2°F is used to calculate the heat transfer by convection.

$$q = h\,A\,(t_2 - t_1)$$

Also see *Convection Heat Transfer*.

Thermal Diffusivity The rate at which a substance with a non-uniform temperature approaches equilibrium based on thermal conductivity, k, specific heat, c, and the density, rho, ρ:

$$\alpha = k/c\rho$$

The thermal diffusivity is used in calculations of unsteady state heat conduction problems. Calculations of unsteady-state heat transfer (where the difference in temperature varies with time), are usually solved from prepared graphs or charts, using thermal diffusivity as one of the parameters. Also see *Diffusion; Thermal Conductivity*.

Table T-2
Thermal Conductivity, k, fph, and Thermal Diffusivity, α

	c_p Btu/lb °F	p Lb/ft³	Temp., °C	°F	k, Btu/hr. Ft °F	α, Ft²/hr.
Coal, anthracite	0.30	75-94	20	68	0.15	.005-006
Coal, powdered	0.31	46	30	86	0.067	.005
Cotton	0.31	5	20	68	0.034	0.075
Ice	0.46	57	0	32	1.28	0.048
Sawdust		12	24	75	0.04	
Wood, Ash, Pine	0.43	40	30	85	0.10	0.0057
Wood, Oak	0.41	44	30	85	0.11	0.0062
Air	0.24	0.08	0	32	0.013	0.7
Beef	0.77		50	122	0.29	.005
Snow			50	122	0.07	
Water	1.0	62.4	0	32	0.32	.0051
Iron	0.10	490	0	35.0		
Water vapor	0.47	.05	16	60	0.01	
Steam, saturated vapor	0.48	.4	149	300	0.2	
		.66	205	400	0.3	
Cereal grains, 12-20%	0.3-0.4	40-50	32	90	0.8-1.1	

Conversion factors: *1 Btu/lb °F = 4.187 kJ/kgK*
1 lb/ft³ = 16.018 kg/m³
1 Btu/hr ft°F = 1.731 W/mK
1 ft²/hr = 25.8 x 10^{-6} m²/s

Sources: *Ingersoll, L. R., Zobel, O. J., and Ingersoll, A. C. 1954. Heat Conduction. University of Wisconsin Press, Madison, WI. Marks, L. S. 1958. Mechanical Engineers Handbook. McGraw-Hill Book Co., N. Y., NY.*

Thermal Efficiency The heat output divided by the heat input of a thermal device, such as a furnace, heater, heat exchanger, steam engine, and dryer, expressed as a decimal or percent. Typical values of thermal efficiencies are given in Table O-4. Also see *Efficiency; Overall Thermal Efficiency*.

Thermal Expansion Most substances increase in length or volume with increasing temperature. Liquid water is the notable exception, which decreases in volume above 0°C (32°F) until a temperature of approximately 4°C (39.2°F) is reached and then increases in volume with a temperature increase. The ratio of the change in length per °C to the length at 0°C (32°F) is the coefficient of linear expansion, α, 1/°C; the coefficient of volumetric expansion is β, or the cubical coefficient of expansion, 1/°C (Table T-3), is approximately three times α.

Table T-3
Coefficient of Thermal Expansion, 0-100°C, 1/°C

Solids	Linear, λ	Cubical, β
Iron	.000012	.000035
Aluminum	.000025	
Copper	.0000165	.00005
Ice (before freezing)	.000051	.000113
Glass	.0000085	.000026
Wood, across fiber	.00004	
Monel metal	.000016	
Stainless steel	.000012	
Rubber		.00048
Grain	.000034	.0001008

Gases, cubical, β		Liquids, cubical, β	
Air	.00367	Water	.000207
Ammonia	.00379	Mercury	.000181
Carbon dioxide	.00372	Ethyl alcohol	.00112
Chlorine	.00383	Acetic acid	.00107
Ethylene	.00373	Petroleum	.000955
Hydrogen	.00366		
Oxygen	.00367		
Water vapor	.00420		
Nitrogen	.00367		
Most gases	.00367		

Source: Smithsonian Institute Tables (Table 269).

$$\ell_t = \ell_o \ (\ell + \alpha t)$$
$$V_t = V_o \ (\ell + \beta t)$$

Thermal Hysteresis A physical or electric property, which shows a different path or curve in a body when heated through a particular temperature range than when cooled through the same range.

Thermal Properties of Grain The thermal properties of some grains, such as conductivity and diffusivity, related to temperature, as shown in Table T-4.

Thermal Radiation The energy emitted by all bodies (at a temperature above absolute zero) due to excitation by molecular vibration. Energy is transferred from one body to another without heating the space between, except when the space has substances which absorb energy. Thus, heat transfer by radiation is determined on the basis of net gain or net loss of energy and temperature change. The Stefan-Boltzmann constant, σ, is used in thermal radiation calculations. The heat loss from a black body, emissivity, $\varepsilon = 1$, is

$$q = \sigma AT^4$$

Table T-4
Thermal Properties of Selected Grains

Grain	Moisture Content, % w.b.	Temp. Range, °F	Specific Heat, Btu/lb°F	Conductivity, Btu/hr ft°F	Diffusivity, ft^2/hr
Corn	0.9	47-84	0.366	0.0812	0.00395
	5.1	47-84	0.404	0.0847	0.00381
	9.8	47-84	0.438	0.0878	0.00364
	14.7	47-84	0.484	0.0919	0.00351
	20.1	47-84	0.531	0.0945	0.00336
	30.2	47-84	0.588	0.0996	0.00358
Oats	9.1	80-88		0.0370	
	11.7-17.8	80-88	0.305		
	27.7	80-88		0.0537	
Rice,					
finished	10.8-17.4			0.282	
rough	10.2-17.0			0.265	
shelled	9.8-17.6			0.287	
Soybeans	17.7	75-130	0.47		
	21.7	73-190	0.49		
Wheat	dry		0.334	0.0676	0.00359
	9.2		0.370	0.081	0.00446
	12.5	80-88		0.074	
	17.8	80-88		0.094	
	23.0	79.4	0.51	0.0867	

Source: ASAE Data 243.1.
Conversion factors: 1 Btu/lb°F = 4.187 kJ/kgK
1 Btu/hr ft°F = 1.7308 W/mK
1 $ft^2/hr = 2.5806 \times 10^{-5}\ m^2/s$

and the heat exchange

$$h = \sigma A \varepsilon \, (T_1^4 - T_2^4)$$

Also see *Emissivity; Stefan-Boltzmann.*

Thermal Similarity in Physico-Chemical Transitions A parameter of thermal similarity, K, which is a measure of the relation between the heat flow required for a phase transition in a material and the (latent) heat of super-heating (or super-cooling) of one of the phases:

$$K = \frac{L}{c\Delta T}$$

where

L = latent heat; c = specific heat; Δt = temperature difference

Thermization See *Heat Treatment.*

Thermocouple The sensing part of an instrument for measuring temperature in which a pair of two dissimilar metal wires are joined. One pair of wires becomes a reference junction (RJ). The other pair of wires becomes the sensing junction (Fig. T-1). If a temperature difference exists, there is an EMF difference, which is measured and a current flows, from which the temperature difference is determined.

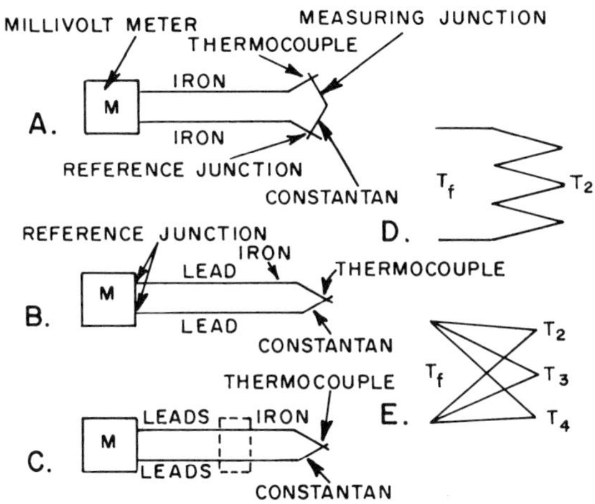

Figure T-1. Thermocouple.

Thermodynamics A field of study dealing with conversion of energy from one form to another and with the relationships between thermodynamic properties. Also see *First Law of Thermodynamics; Irreversible Thermodynamics; Second Law of Thermodynamics; Third Law of Thermodynamics.*

Thermograph A recording thermometer which provides a graphical representation of the temperature versus time.

Thermometer An instrument for measuring temperature. The measurement is based on expansion of a gas (gas-filled) or of a liquid (liquid-filled) in a glass or metal container with an increase of temperature. A common thermometer uses mercury as a liquid. Other principles used include: expansion of metal, bi-metal units, change in resistance to flow of electricity, and semi-conductor (thermistor).

Thermophilic A term meaning "heat loving" applied to bacteria which grow in the range 0 to 71°C (100 to 160°F) at an appropriate moisture content. Growth can be reduced by drying.

Thermopile A grouping of several thermocouples connected to a galvanometer which is calibrated to give the value of temperature measured by the thermopile. It is widely used for solar and other radiation measurements.

Thermostat A device for automatically controlling the temperature within a certain range of values of a system or space or location. The device will activate units calling for heat if the temperature goes below the desired control temperature or will inactivate heating units when the maximum allowable temperature is reached.

Third Law of Thermodynamics The law which states that the entropy of a perfect crystal is zero at the absolute zero of temperature. Absolute values of entropy are of particular importance in problems of chemical equilibrium.

Through-Circulating Dryer See *Band Through-Circulation Dryer.*

Through-Draft Dryer An arrangement in which the drying air flows through a relatively thick bed of wet material. Also called *Through-Flow Dryer.*

Through-Flow Dryer See *Through-Draft Dryer; Updraft.*

Timber See *Lumber.*

Time Constant A measure of the time of response of a system, circuit, or device, often represented by tau, τ (Fig. T-2). For an instrument, the time constant, τ, is the time to reach 1/e of the final value. For drying, see *Drying Constant.*

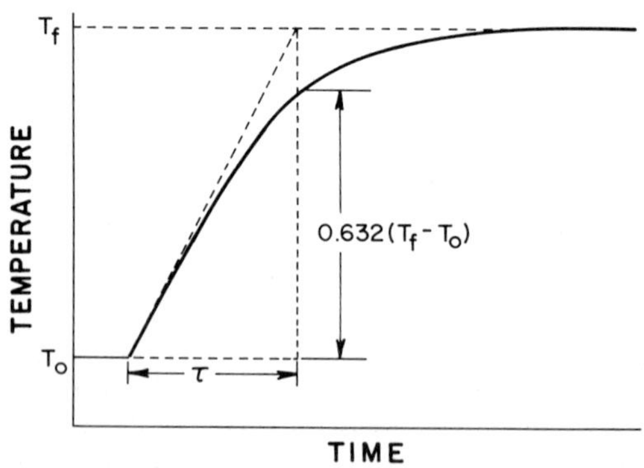

Figure T-2. Time constant, τ.

Time of Drying The elapsed time from the start of the drying process to the time the drying front arrives at a point in the product.

Time of One-Half Response Time required to dry fully exposed products halfway to equilibrium.

Time of Response, Drying The time of response for drying is similar to the time of radioactive decay. At a time when one-half of the original material remains, the time elapsed from the beginning is known as the time of one-half life or time of one-half response, in minutes or hours. A more descriptive term is "time of retention." The same terminology may be applied to heat transfer or moisture movement. In these cases the difference between the initial and final equilibrium values represent the original material available for change. Thus, the time of one-half response in a drying process would be the number of hours necessary to obtain a moisture content ratio of one-half:

$$\theta_{1/2} = \frac{\log_e 2}{k}; \quad e^{-k\theta} = \frac{1}{2}$$

Also see *Response Time*.

Time of Retention Similar to time of response for drying of products, but a more descriptive term. Also see *Time of Response*.

Time Unit The time of one-half response of drying of a particular product, sec, min, hr. The time of one-half response depends on the product, its moisture content, and the vapor pressure and temperature of the air. The time unit may be used as one axis, usually the abscissa, in plotting drying information. Also see *Time of One-Half Response; Depth Factor*.

Tobacco Curing The process of using heat to reduce the moisture content of the leaves from 80 to 20%. Tobacco is harvested when some leaves begin to yellow. The heat used in drying is also the first step in fermenting. Without fermentation, tobacco smoking would be like smoking cabbage. Four general methods are used for curing, depending on the weather, region, type of tobacco, and use--fire, flue, air and sun curing. Also see *Air Curing; Bright-Leaf; Burley; Fire Curing; Flue Curing; Shade Grown Tobacco; Sun Curing Tobacco*.

Toluene Also called toluol, which is a colorless hydrocarbon with a boiling point of 110.5°C (230°F) and insoluble in water. Toluene is methyl benzene, $C_6H_5CH_3$. Toluene is used for the official toluene distillation method of moisture determination of grain. Finely ground grain (about 25 grams) is heated in a container with toluene (about 75 ml) as a heating fluid with the moisture evaporated and collected in a graduated tube, from which the moisture content can be determined.

Toluol See *Toluene*.

Tomato Drying Sliced tomatoes may be dried on trays to make

flakes. Two-stage tunnel dryers are used to dry to 4% moisture content. The slices are made into flakes or powder. A more popular procedure is to make a tomato purée, adjusted to 20% moisture, and dried on a double drum dryer. Sulfur dioxide is used as a means of reducing scorching of particles and flavor deterioration. The product is dried to 2% moisture content to minimize caking. Also see *Vegetable Drying; Vegetable Juice Drying.*

Ton A measure of weight, which in the U. S. is 20 hundredweight (20 cwt) where 1 cwt = 100 lb, is 2000 lb, also known as the short ton; in England the ton of 20 British hundredweight, where 1 cwt = 112 lb, is 2240 lb, also known as the long ton or gross ton; and the metric ton which is 1000 kg or 2204.6 lb, represented by tonne in the SI terminology.

Toned Milk A product prepared by adding dry nonfat milk to a high fat milk, such as buffalo milk, primarily in southeast and central Asia. The purpose is to reduce the percentage of fat but to maintain the total solids of the product. The dilution of the milk with water would decrease the percent of fat and of solids.

Tonne See *Ton.*

Tons Refrigeration A unit for designating the capacity of a refrigeration or air conditioning system in which 1 ton will freeze 1 ton of water at 32°F in 24 hr. A one ton unit will remove 12,000 Btu/hr. A one ton unit is equivalent to a 3.52 kW unit in SI terminology.

Torr A unit of pressure employed in vacuum work, such as vacuum in freeze-drying, in which 1 atm = 760 torr and 1 torr = 1 mm mercury = 133 Pa.

Torrefier A heating apparatus used to dry barley and other small grains, using hot air currents.

Total Heat The sum of the sensible heat and the latent heat in a substance; in air-vapor mixtures the total heat is also called enthalpy. Also see *Enthalpy.*

Total Pressure The sum of the partial pressures of all of the constituents of a mixture of gas and vapor and the measured pressure of the mixture.

Total Solids The colloidal, soluble, and suspended solids which remain after the evaporation of water. Expressed as percentage and/or decimal form of initial solution weight.

Tower Spray Dryer A device for removing moisture having a chamber in the shape of a vertical tower in which the product is atomized, usually at the top of the tower, with the dry product removed at the bottom. Also see *Spray Dryer.*

Transducer A device which changes the energy, signal, or sensed variable to another form or mode. Thus, an ultrasonic transducer converts the electrical energy to mechanical vibrations at a frequency above the audible range. A sonic or ultrasonic transducer may be used to produce vibration for drying.

Transient Heat Transfer Steady-state conditions do not normally occur in product storages surrounded by the environment. The temperature will change with time giving a transient condition. For unsteady or transient heat flow in one direction, an instantaneous heat balance gives the relationship as shown below with the heat removed from the surroundings on the right side of the equation and the heat gained on the left side of the equation:

$$c\rho\ dx\ dy\ dx\ \frac{\partial t}{\partial \theta} = k\ dy\ dz\ \frac{\partial t/\partial x}{\partial x}\ dx$$

In most heat transfer calculations it is assumed that the thermal properties, specific heat and conductivity, remain constant. The heat transfer relationship is:

$$\frac{\partial t}{\partial \theta} = \left(\frac{k}{c\rho}\right)\left(\frac{\partial^2 t}{\partial x^2}\right)$$

The general transient heat transfer equation is represented by:

$$\frac{t - t_s}{t_o - t_s} = \Phi\ \left(\frac{k\theta}{c\rho r_m^2}\right)\left(\frac{hr_m}{k}\right)\left(\frac{r}{r_m}\right)$$

where

t = temperature at any time, °F or °C (K)
t_s = temperature of surroundings, °F or °C (K)
θ = time, hr or sec
t_o = initial temperature, °F or °C (K)
r_m = cylinder radius, ft or m
k = thermal conductivity, Btu/hr ft^2°F or W/mK
h = surface conductance, Btu/hr ft^2°F (usually 1.65 for still air, and 6.0 for 15 mph wind) or W/mK (9.35 and 34, respectively).

A graphical solution of the transient heat transfer equation is presented in Fig. T-3. The chart can be used for determining the temperature at the center of a bin of grain with the other conditions given. Or, if the various temperatures are known, including the temperature changes in the bin at a given time, the value of k can be determined.

Transmissivity; Transmittance; Transmission Coefficient; Transmissibility The portion of radiation energy falling on a surface which is transmitted through the body. Radiation energy which falls on a body may be absorbed, (α), (absorptivity); reflected, (ρ),

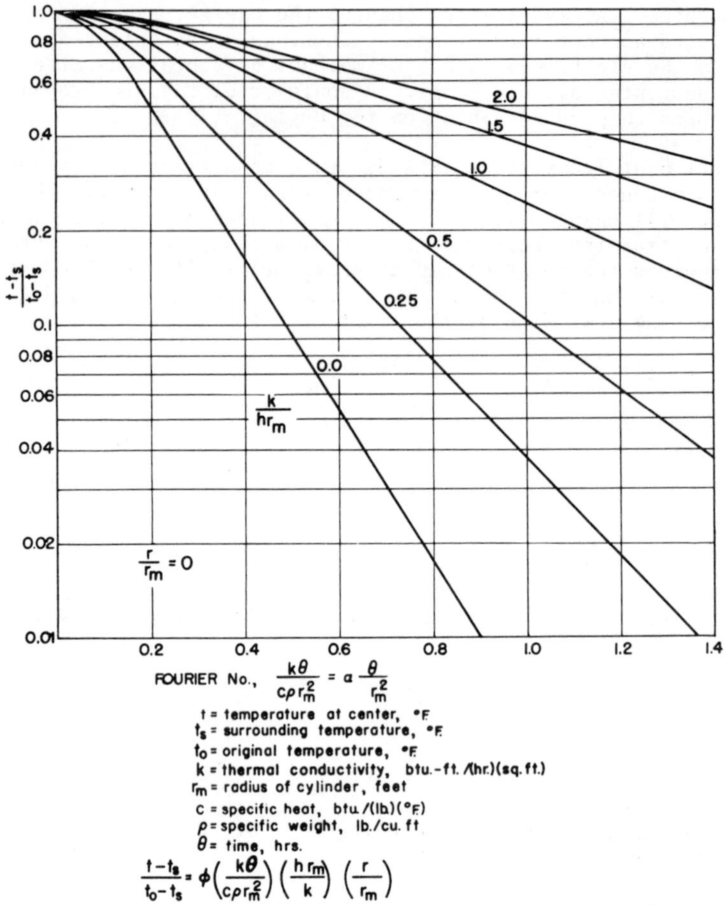

Figure T-3. Transient heat transfer in center of cylinder.

(reflectivity); and transmitted (τ), (transmissivity), expressed in percent of each involved, decimal form. Thus, $\alpha + \rho + \tau = 1$. Most gases have a high value of τ.

Traveling Screen Dryer A device for moisture removal from products in which a controlled quantity of product, usually granular or flake in nature, is placed and moves through a drying medium, usually heated air.

Traverse Time The in-product travel time of air from entrance to any point or to the exit, usually expressed in seconds.

Tray Dryer A device for removing moisture in which the product is placed on perforated trays or tray through which air can move, and through which heated air is forced. The units may be designed and operated for continuous flow of product or as a batch system. Also see *Atmospheric Tray Dryer; Continuous Agitated Tray; Continuous Band Dryer; Continuous Tray Dryer; Vacuum Tray Dryer.*

Tree Nut See *Nut Drying.*

Tree Seed Drying Seeds of trees fall into three groups as related to extraction involving drying:

(1) Tree seeds readily extracted from dry fruits, such as cones (baldcypresses, cypresses, firs, larches, pines, spruces, white-cedars); conelike clusters (yellow-poplar); pods (Kentucky coffee-tree, honeylocust, locust); or capsules (aspens, cottonwoods, poplars, willows).
(2) Dry fruits with seeds surrounded by a tightly adhering fruit wall, such as the nuts (chestnuts, oaks), and omaras (ashes, elms, maples, yellow-poplar).
(3) Seeds of fleshy fruits, such as dupes (cherries, dogwoods, plums, walnuts), and multiple or collective fruits (mulberries, Osage-orange), and berrylike conelets (junipers).

Seeds of the second group are seldom extracted from the fruits because that is either unnecessary or very difficult. Those of the first and third groups are separated from the fruits by drying, threshing, tumbling, depulping, fanning, or sieving.

The simplest method of drying is to spread the fruits in shallow layers so that there is free circulation of air around each fruit. Where the climate is dry, drying may be done in the open. Where the climate is damp or the amount of fruit is great, drying is usually done under a roof. Protection from rodents and birds often is necessary to prevent serious seed losses during drying.

Some cones do not open readily and must be heated artificially in special kilns. These kilns provide the highest dry heat (usually between 37.8 and 65.6°C (100 and 150°F)) that the seeds can stand without injury, and these predetermined safe limits must not be exceeded.

Two general types of kilns are used in extracting seeds from cones--simple convection and forced-air kilns. The first is the oldest, least expensive and simplest to operate. The second is more complicated and expensive but more efficient.

Recommended temperatures and schedules in convection kilns for several pines are: Jack pine, 2 to 4 hr at 62.8 to 65.6°C (145 to 150°F); loblolly and slash pines, 6 to 48 hr (usually 8 to 10 hr) at 48.9°C (120°F); longleaf pine, 12 to 72 hr at 48.9°C (120°F); ponderosa pine, 3 hr at 48.9°C (120°F) or less; red pine, 24 to 72 hr at 54.4 to 60°C (130 to 140°F); and Scotch pine, 5 to 24 hr at 54.4°C (130°F).

In forced-air kilns, comparable schedules are 8 to 16 hr at 46.1°C (115°F) for longleaf pine, 5 hr at 76.6°C (170°F) for red pine, and 4 to 8 hr at 54.4°C (130°F) for Scotch pine.

Seeds of the following genera and species usually are extracted by air or kiln drying: Aspens, baldcypresses, chestnuts, cottonwoods, cypresses, Douglas-firs, elms, hemlocks, incense-cedar, larches, the pines, poplars, sequoias, spruces, sweetgum, sycamores, thujas, white-cedars, and yellow-poplar. Normally kilns are necessary for the hard-to-open cones of these pines: Bishop, jack, knobcone, lodgepole, Monterey, pond, and sand.

After drying, the cones are tumbled in revolving screened cages or drums to shake out and separate the seeds.

The separation of the seeds of many dry fruits from the bunches, pods, or capsules in which they grow requires flailing, treading under foot, or treatment in agricultural threshing machinery or special apparatus, such as a macerator, hammer mill, or mixer.

Threshing or screening commonly is required to extract seeds of the alders, American beech, Kentucky coffee-tree, firs, hickories, honeylocust, black locust, Siberian pea-tree, eastern redbud, and walnut.

Some small fleshy fruits are dried whole, but the seeds of most fleshy or pulpy fruits must be extracted promptly to improve germination and to prevent spoilage. Small lots can be cleaned by hand methods, but larger lots should be processed mechanically. Reference: Seeds, Yearbook of Agriculture, U.S.D.A., 1961, p. 223.

Trier   A device used for sampling stored products, usually bulk. The sample is used to make quality evaluations such as moisture content, mold, and color (Fig. T-4).

Triple Point   The temperature and pressure at which gas, liquid, and solid phases (or states) of a substance are in equilibrium (Fig. T-5). For the vapor, water, and ice phases, the triple point is at 0.01°C (32°F) and 0.613 kPa (4.7 mm of mercury). In freeze-drying it is usually desired that the liquid phase is avoided, with ice passing to the vapor phase, which is done at pressures below 4.7 mm. Practically, most freeze-drying is done at 4 mm.

Trough Dryer   A container or containers to hold the product for drying, shaped in a U or V, in which the product is moved by a spiral conveyor, spiral ribbon, or chain-driven paddles which may be operated under vacuum or atmospheric pressure. The heat is transferred primarily by conduction for steam or electric heated walls to the product and air is moved over or through the product to carry away moisture.

Truck Dehydration   In food dehydration, a low-wheeled dolly which supports a stack of drying trays which move as a unit through a drying tunnel. Also see *Tunnel Dryer*.

Trunking   See *Air Trunking*.

T-s Diagram   A graphical representation, usually with T (absolute temperature) on the ordinate and s (entropy) on the abscissa. The area under the curve of a reversible process represents the heat transfer during the process. Also see *i-x Diagram*.

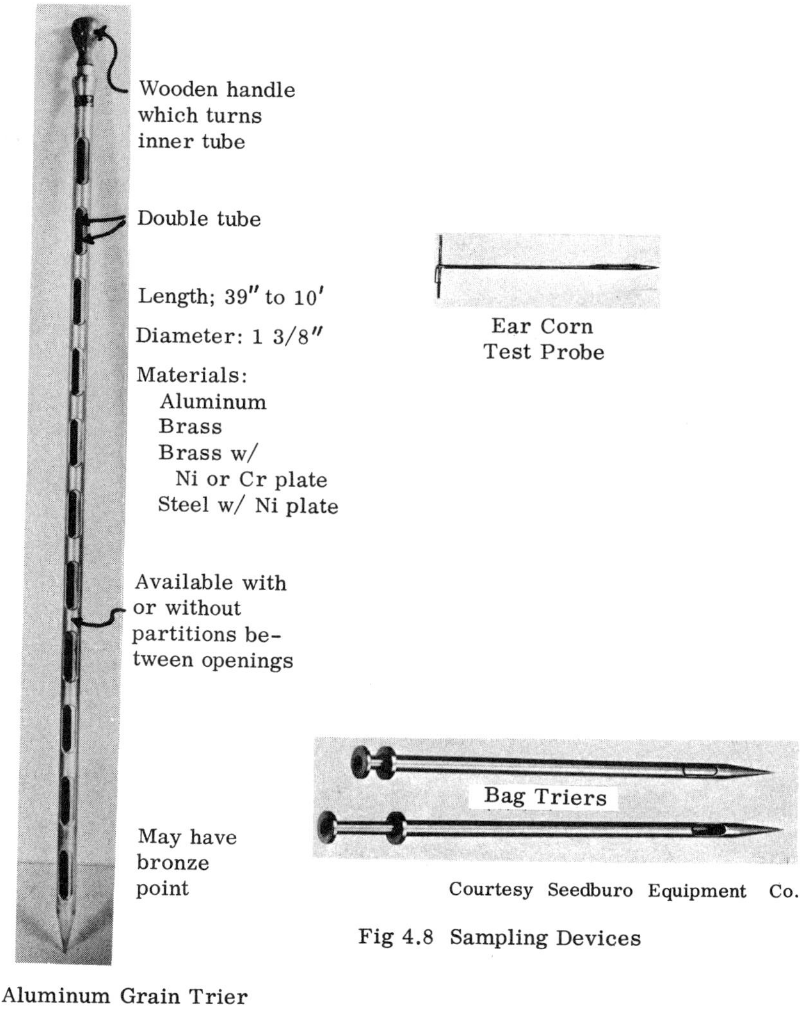

Figure T-4. Triers.

Tube-Axial Fan An air moving device (AMD) used for drying of the axial flow type in which the air is moved parallel to the turning shaft of the fan using short blades and a large hub. The tube-axial fan is similar to the vane-axial fan except that there are no guide vanes. Also see *AMD; Axial Flow Fan; Vane-Axial Flow Fan*.

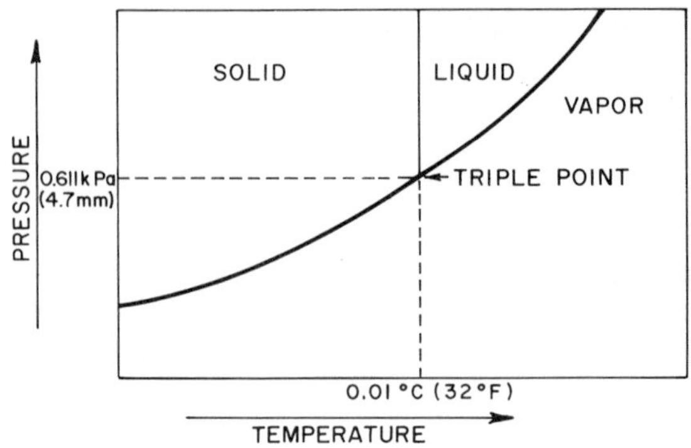

Figure T-5. Triple point for water. *From Chhinnan, M. S., and Miller, M. W. (1977). ASAE Paper 77-6605.*

Tubular Dryer See *Revolving Steel Tubular Dryer.*

Tumble Dryer An apparatus for removing moisture from clothes, consisting of a perforated cylinder which is rotated while heated air is moved through the cylinder. Heating is done by steam, gas, or electricity.

Tunnel Dryer A type of dryer wherein the product being dried is conveyed through a tunnel-like chamber which may be of the continuous or batch-type (Fig. T-6). Also see *Compartment Dryer; Progressive Dryer; Truck Dehydration.*

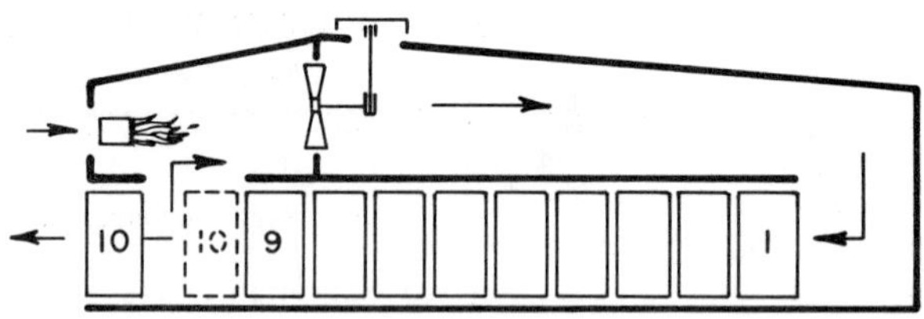

Figure T-6. Tunnel dryer/dehydrator.

Turbulent Flow A type of flow of fluids, gases or liquids, in which there is considerable mixing as applies to a combustion chamber, burner, and heater, and the flow of air through ducts and products, such as for drying; and in heat transfer for gases and liquids, etc. Turbulent flow in a duct or pipe is characterized by a rather uniform velocity at a cross-section with the variation 3% or less from the mean velocity; and with Reynolds Number criteria, in which R_N is greater than 4000. Turbulent flow is contrasted to laminar or streamline flow in which the product moves in a straight pattern with the average velocity of flow at a cross-section being one-half of the maximum velocity of flow (Fig. T-7). Turbulent flow and laminar flow are separated by a transition zone. Also see *Critical Velocity; Darcy's Law; Laminar Flow.*

Figure T-7. Turbulent flow.

Turning The process of moving a product through the air within a bin or storage structure, or from one bin or storage structure to another, for the purpose of equalizing the product condition. Also see *Aeration; Equalization.*

Two Fluid Atomizer The method of atomizing the liquid being dried in a spray dryer by using a high-velocity air jet. Also see *Atomization.*

Two Stage As applied to drying and dehydration, drying a product under two different sets of drying conditions. A two-stage moisture tester uses two different time-temperature conditions so as to get the maximum rate of moisture removal without excessive damage to the product.

Two Stage Moisture Testing See *Oven Method.*

Types of Dryers See *Classification of Dryers.*

Types or Classifications of Dryers Categorization of dryers according to the method of handling the product: tray or compartment dryers which may be atmospheric or vacuum; through-circulation which may be batch or continuous; agitated which may be batch or rotary, atmospheric or vacuum spray; pneumatic conveying; rotary which may be direct or indirect; drum; fluidized bed. Also see *Classification of Dryers.*

U

u   A symbol for internal energy. See *Internal Energy*.

UHF, Ultrahigh Frequency   Radio frequency of 300,000 to 3,000,000 kilocycles/sec. Also see *HFC; Radio Frequency; VHF*.

Ultrasonic   Waves which have a frequency above, or larger than, audible frequencies, generally defined as above 20,000 cycles/sec. In general, frequencies from 20,000 cycles/sec to 200,000 cycles/sec are produced with quartz crystal transducers. Sonic and ultrasonic energy have been used to assist in the drying process. Also see *Transducer*.

Ultraviolet Curing   The use of intense ultraviolet energy to treat certain solvent-free resin formulations applied like paint, to make a hard surface coating, which will dry in a couple of minutes. A high pressure mercury vapor lamp may be used as an energy source. Whereas conventional paints may release 50 to 60% of their weight during curing and drying, about 2% of the weight of the coating is released with ultraviolet-cured resin. The curing process consists of converting the monomers of an ultraviolet-sensitive coating to a polymer.

Ultraviolet Ray   A band in the electromagnetic spectrum, just beyond the visible end of short wavelength. The wavelength is less than 4×10^{-5} cm and longer than 1.8×10^{-5} cm. Ultraviolet is needed by the body to make vitamin D. Ultraviolet will produce fluorescence of some materials and is used to identify impurities in grain or potential food products, such as animal or rodent urine.

Unaccomplished Moisture Content Change   The ratio of the free water remaining in a product at a particular time to the total free water originally present. Also see *Moisture Ratio*.

Unavailable Energy   The amount of energy which is dissipated in an irreversible process and is thus unavailable for work in the process.

Unbound Moisture   See *Bound Moisture; Moisture Content.*

Unit   Fundamental units are length, weight, force, and time. Derived units are based on measurements in fundamental units. Derived units are area, velocity, density. Also see *Absolute Unit; Conversion Factors; Fundamental Units.*

Unit Operation   The study of physical changes which occur during processing of materials and the design of equipment and systems based on these operations, such as drying, dehydration, heat transfer, and distillation.

Unit Process   The study of chemical changes which take place in materials subjected to processing and handling such as in combustion, oxidation, and polymerization. Used as the basis to find the best condition for the reaction to provide for design of processes and associated equipment to improve the suitability of product and equipment.

Units   See *Conversion Factors; Force Units; Fundamental Units; Symbols and Units; SI Units.*

Universal Constants   See *Constants.*

Universal Meter   See *Electrical Resistance Method.*

Unsaturated Solution   A solution in which the solvent can hold additional solute as contrasted to a saturated solution. Also see *Saturated Solution.*

Unwatering   The removing of surface water and the free water from solid materials of an open porous nature.

Updraft   A type of ventilation, cooling, or drying system in which the air is moved upward in the structure or through the product. The opposite direction of air flow is called downdraft. Often used to designate the direction of air flow in a dryer (Fig. U-1). Also see *Through-Draft Dryer.*

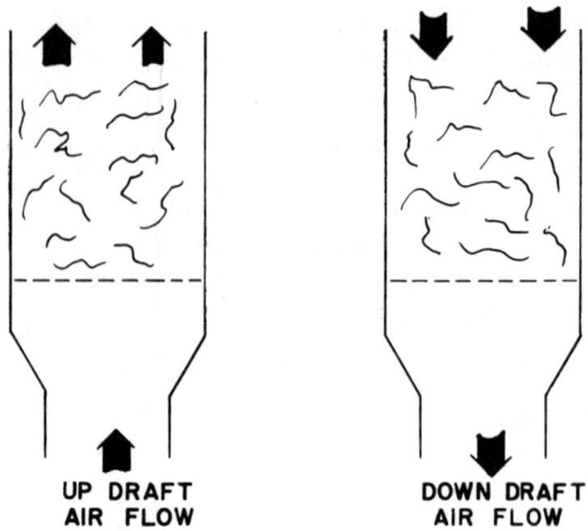

Figure U-1. Updraft/downdraft flow of air.

V

V,v A letter used to represent velocity, volume, and volts.

Vacuole A part of the structure of a plant cell which contains most of the water and water-soluble constituents. The vacuole is the cavity in the cytoplasm of a cell which contains a fluid.

Vacuum A space devoid of matter which occurs as a space is evacuated and the pressure reduced with a vacuum pump.

Vacuum Belt Dryer See *Belt Dryer*.

Vacuum Contact Dehydration (V.C.D.) Drying is carried out in a vacuum (absolute pressure of 10 to 15 mm) in which heating plates are moved in contact with opposite sides of the product. Also referred to as a vacuum contact dryer.

Vacuum Desiccator An apparatus for removing moisture from products under a vacuum. Often refers specifically to a laboratory unit of approximately one-fourth liter capacity, 23 cm (10 in.) diameter with a tray for holding the product to be dried. The unit can also be used as a lyophilizing chamber for freeze-drying small samples.

Vacuum Dryer A device for removing moisture from products in which evaporation takes place in an environment at a pressure below atmospheric pressure (Fig. V-1). Also see *Freeze Dryer; Vacuum Tray Dryer*.

Vacuum Evaporation The process of changing from a liquid or solid to a vapor in a vacuum environment. As the vacuum is increased the temperature of vaporization is decreased.

Vacuum Freeze Dryer See *Freeze Drying*.

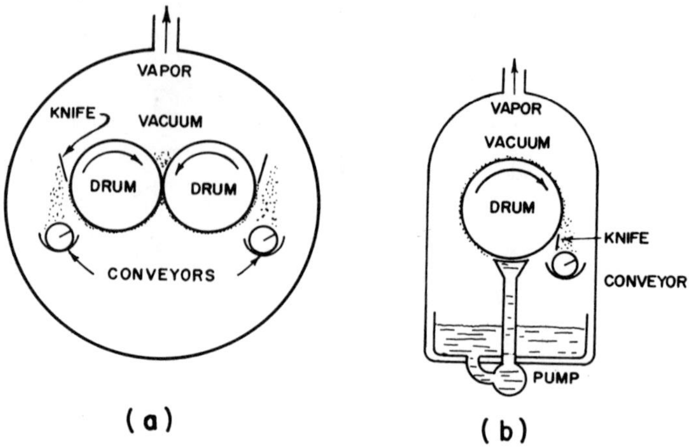

Figure V-1. Vacuum drum dryer. (a) double drum (top feed). (b) single drum (pan feed). *From Hall, C. W., and Hedrick, T. I., Drying of Milk and Milk Products, 2nd edition (1971). Used with permission of Avi Publishing Co., Westport, CT.*

Vacuum Oven Method The vacuum oven method is used for determining the moisture content of products based on original and final weight of a sample. For grain, the product is ground and placed in the oven at approximately 100°C (212°F) and the oven maintained at 40 mm of vacuum for about 5 hr. By using a lower temperature or a shorter time, there is less possibility of a loss of weight due to deterioration of the dry matter, which is particularly important for fruits and vegetables. For meat and meat products, a sample representing 2 grams of materials dried in a vacuum oven to consider weight at 95 to 100°C under an absolute pressure not greater than 100 mm Hg for about 5 hr. This method should not be used for high fat products for which an air drying method should be used.

For accurate moisture content determinations it is necessary to prevent the sample from gaining moisture after the moisture has been removed from the sample. Moisture adsorption can be prevented by placing the sample in an alumina desiccator and weighing after the sample and container have cooled. Also see *Oven Method of Moisture Determination*.

Vacuum Pan Condensed or evaporated milk is produced by evaporating part of the moisture from skim or whole milk, usually 2 to 1 or 3 to 1 volume ratio before and after evaporation in a vacuum pan. Evaporation takes place at approximately 88 kPa (26 in Hg) of vacuum. The product is preheated to about 93.3°C (200°F) before entering the evaporator. For a 1 m (3 ft) diameter pan, 180 to 400 kW (18 to 40 Bhp)of steam is needed to provide a pan output of 56.7 to 453 kg/hr (125 to 1000 lb/hr) (Fig. V-2).

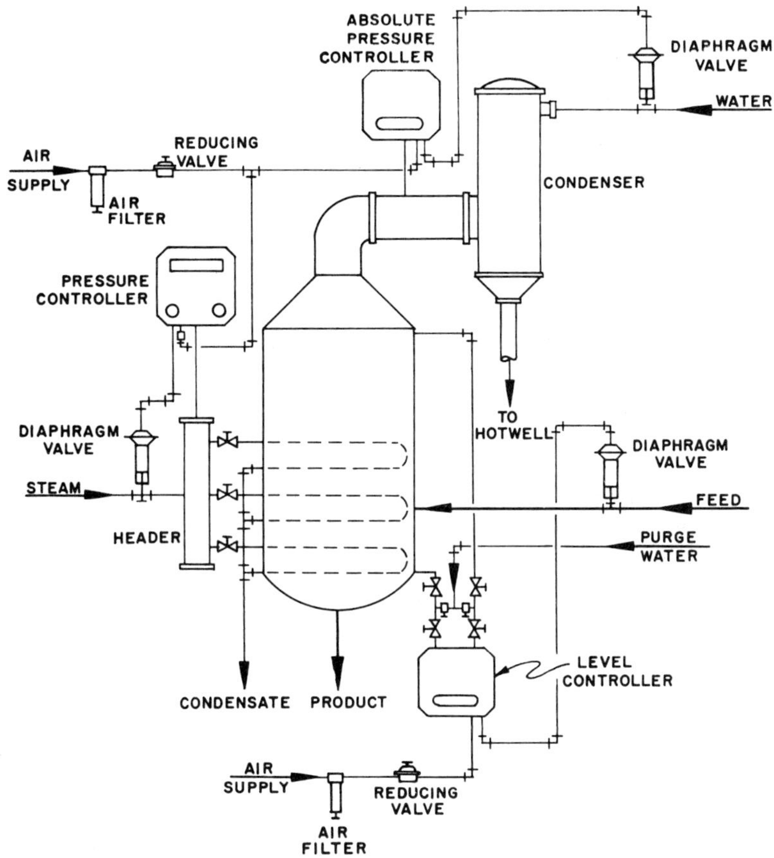

Figure V-2. Vacuum pan control system. *From Hall, C. W., and Hedrick, T. I., Drying of Milk and Milk Products, 2nd edition (1971). Used with permission of Avi Publishing Co., Westport, CT.*

Vacuum Tray Dryer   A device for moisture removal in which trays of product are placed on heated shelves in a vacuum chamber (Fig. V-3). The vacuum tray dryer is used for drying heat-sensitive materials, such as pharmaceuticals, food, dyestuff, etc., usually at not less than 40°C (104°F). Usually operated as a batch unit.

Valve   See *Excess Flow Valve; Pressure Relief Valve; Quick-Acting Valve; Solenoid Valve.*

Van der Waals Equation   An equation which relates the pressure,

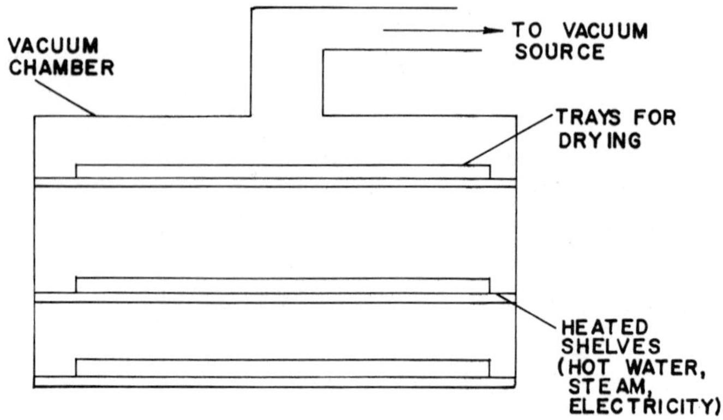

Figure V-3. Vacuum tray batch dryer.

volume, and temperature of a gas involving the gas constant, R. The equation includes a factor for the reduction of the total pressure by the attraction of molecules and for reduction of the total pressure by the volume occupied by the molecules.

$$(P + a/V^2)\cdot(V - b) = RT$$

where

P = gas pressure
V = volume
T = absolute temperature
R = gas constant
a and b = constants, available for several gases

Also see *Gas Constants*.

Vane Anemometer An instrument using a rotating wheel, blade, or turbine for measuring the velocity of air flow. The number of revolutions per unit of time represents the air velocity.

Vane-Axial Fan A device for mechanically moving air with the air moving through nearly parallel to the drive shaft with the fan characterized by a large hub and short blades, with straight vanes on the discharge side to prevent rotation or swirling of the air (Fig. V-4). Also see *Axial Flow Fan; Centrifugal Fan; Tubeaxial Fan*.

Vapor A gaseous state of matter, which can be changed to a liquid by lowering the temperature (such as condensation of water vapor) or raising the pressure. A vapor is a gas at a temperature below its critical temperature. Also see *Critical Temperature; Gas; Saturated; Superheated; Water Vapor*.

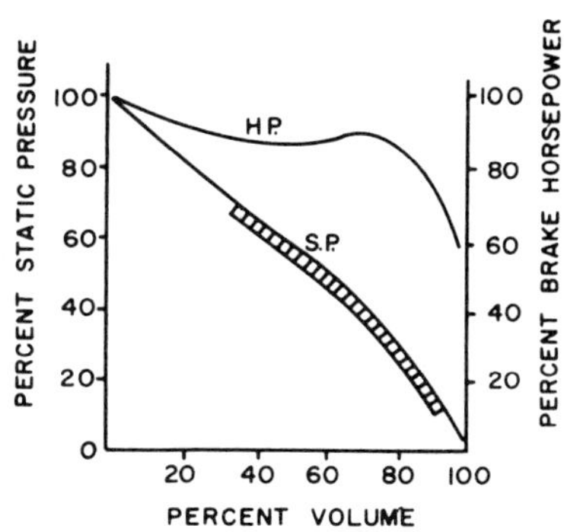

Figure V-4. Vane-axial fan characteristics.

Vapor Barrier See *Insulation*.

Vapor Density The mass of a unit volume of vapor or gas which may also be expressed relative to the density values for hydrogen or air.

Vapor Diffusion Flow of vapor through the pores of a solid provided that a vapor pressure gradient exists as caused by a temperature gradient. Vaporization may occur on one surface of a solid where heating occurs, with diffusion through the solid and drying occurring at the opposite face of the solid.

Vapor Drying A procedure for drying by subjecting wood to hot vapors produced by boiling an organic chemical such as xylene.

Vaporization See *Evaporation; Latent Heat*.

Vaporizer In an LP-gas system, the vaporizer is a heat exchanger wherein heat is supplied to change the liquid fuel to vapor, ready for combustion. The vaporizer may be integral with the burner so that part of the heat of combustion is used for vaporization.

Vaporizer-Burner An integral vaporizer-burner unit, dependent upon the heat generated by the burner as the source of heat to vaporize the liquid fuel.

Vapor Pressure The pressure exerted by the vapor of a solid or a liquid when in equilibrium with the material. The vapor pressure of a solvent is lowered as additional solute is added. This principle is used in adjusting the vapor pressure of environments to determine the equilibrium moisture content of materials exposed to different vapor pressures. The lowering of the vapor pressure by adding solute is related to the decrease of freezing point and increase of boiling point. Also see *Boiling; Freezing; Moisture Distribution; Partial Pressure; Raoult's Law.*

Vapor Pressure, Fuel Commonly taken to mean the saturated vapor pressure which is the vapor pressure of a vapor in contact with its liquid form. An example is the pressure in an LP-gas storage tank. The term is also used for the pressure of the vaporized fuel being fed to the burner orifice.

Variance In statistical representation, the square of the standard deviation, represented by σ^2. Also see *Standard Deviation.*

VCD See *Vacuum Contact Dehydration.*

Vegetable Drying Generally, vegetables are prepared for drying by washing, chopping or dicing, blanching, and often sulfiting. A common method of drying is to place the prepared wet product on trays at 6 to 7.3 kg/m^2 (1.2 to 1.5 lb/ft^2) through which heated air is forced. For a two-stage tunnel dryer, temperatures of 71 to 93.3°C (160 to 200°F) entering the first stage and 54.4 to 73.9°C (130 to 165°F) entering the second stage. The moisture content of the product is reduced from 80 to 90% to 15 to 35% in the first stage and 6 to 10% in the second stage. From 5 to 15 hr are required for drying, except for leafy products. Drying is completed in a batch bin using 37.8 to 60°C (100 to 140°F) air for 7 to 48 hr, drying to 3 to 5%. Some products are dried in a tunnel on a tray in a single stage dryer or in a cabinet dryer. Continuous flow perforated belt dryers, with higher temperatures than for tunnel dryers, with carefully controlled air flow may be used. Granules, particles, or powder can be produced. Some vegetable products are puréed and dried on a drum dryer (steam heated at 240 kPa (75 psi)) to produce flakes or powder, or in a spray dryer (air heated to about 176.6°C (350°F) to produce powder (Table V-1). Also see *Tomato Drying.*

Vegetable Juice Drying Tomato is the principal vegetable juice dried. For beverage quality tomato powder, the juice is first concentrated in an evaporator, then dried in a spray dryer. Puff drying, foam-mat drying, and foam-spray drying offers other approaches with minimum effect on the product.

Drum drying is less expensive but gives a product with evidence of heat effect on flavor. Thus, drum drying is used for vegetable juices to be used for food manufacture or animal food. A product is puréed ahead of the dryer. Flaked product is produced which can be forced through a screen or put through a hammer mill to get a more uniform particle size.

Velocity See *Air Velocity; Superficial Air Velocity.*

Velocity Pressure The force per unit area exerted by a fluid as a result of its velocity, with a higher velocity exerting a greater pressure. The velocity head, in terms of flowing fluid (water, air, etc.), $h = v^2/2g$, can be converted to dynes/cm^2, psi, newtons/m^2, Pascal, etc.

Table V-1
Moisture Content of Fresh Vegetables, Percent, W.B.

Commodity	Range	Mean
Beans, lima	58.9-71.8	66.5
Beans, snap	78.8-94.0	88.9
Beets	82.3-94.1	87.6
Cabbage	88.4-94.8	92.4
Carrots	83.1-91.1	88.2
Celery	89.9-95.2	93.7
Corn, sweet	61.3-86.1	73.9
Onions	70.2-95.2	87.5
Peas, green	56.7-84.1	74.3
Potatoes	66.0-85.2	77.8
Potatoes, sweet	58.5-82.7	68.5
Tomatoes	90.6-96.7	94.1

Source: Vegetable and fruit dehydration. Misc. Pub. 540, USDA, 1944.

Vena Contracta The property of a free jet of fluid upon passing through an orifice under pressure to contract to a diameter less than the orifice (Fig. V-5).

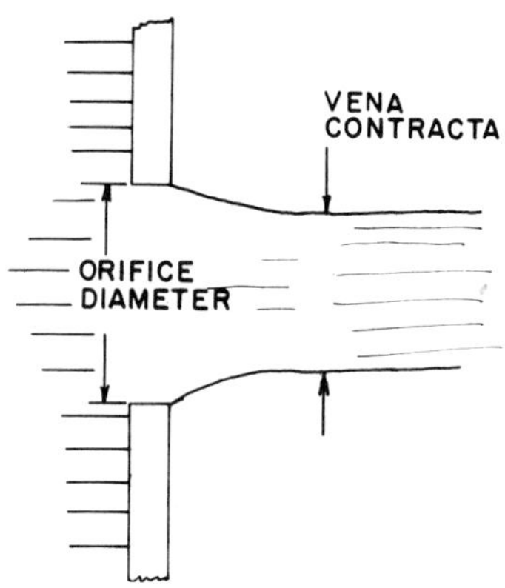

Figure V-5. Vena contracta.

Veneer Drying Thin sheets of wood, 0.12 to 0.25 cm (1/20 to 1/10 in.) thick, may be dried for manufacture. Heated air or steam drying, jet drying, and fluidized bed drying may be used. For temperatures above 315.6°C (600°F) steam is more effective than air. Prefreezing of wet wood before drying reduces shrinkage. In veneer drying, the major cause of fire is from ignition of pitch in the wood.

Ventilated Silo Dryer Grain storage bins or silos with ventilating equipment for drying of grains.

Ventilation Air movement through space.

Ventilation Front The locus of all points of equal traverse time in the product being conditioned or dried. Also see *Drying Front*.

Venturi Meter An instrument for determining the velocity of air or gas based on a pipe or tube with a constriction or throat, in which the differential in pressure as a result of the constriction is related to the velocity of flow (Fig. V-6).

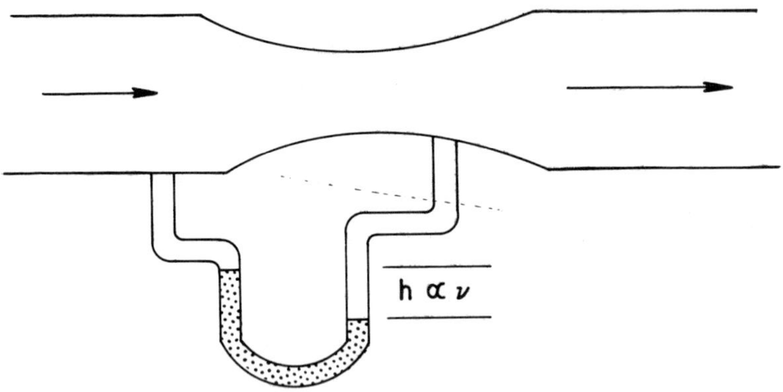

Figure V-6. Venturi meter.

Vertical Agitated Batch Vacuum Dryer A vertical or upright heated vessel in which the material to be dried is in contact with the heated metal wall and is stirred or agitated (Fig. V-7). Units can be operated under vacuum or at atmospheric pressure. Also called pan type dryer with a vertical paddle agitator.

Vertical Chamber Dryer The name applied to dryers in which the product is dried in a vertical or upright container. Most commonly applied to a classification of spray dryers (Fig. V-8), in which the drying air and product may enter together at either end or may enter at opposite ends.

Vertical Column Dryer A device for drying granular material with heated air in which the wet product enters at the top and moves vertically downward by gravity. Heated air is directed horizontally through the product known as crossflow (Fig. V-9). Coal may be dried using a vertical column dryer preceding pulverizing. Also see *Column Dryer*.

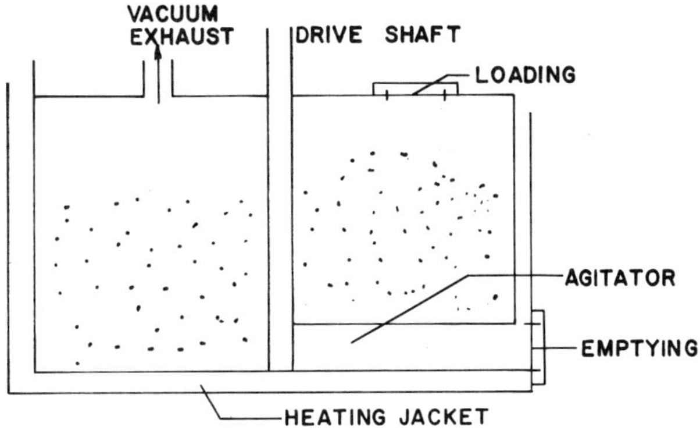

Figure V-7. Vertical agitated batch vacuum dryer.

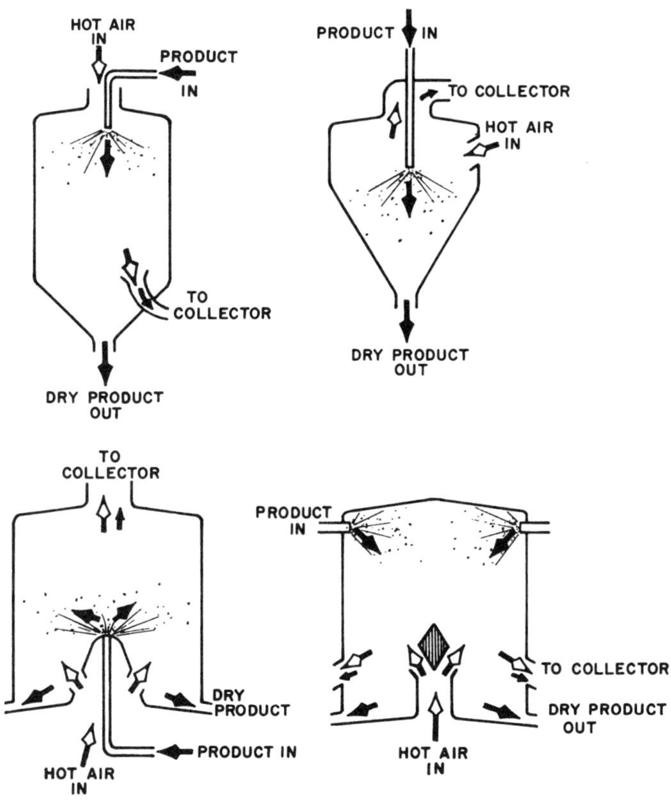

Figure V-8. Vertical chamber spray drying arrangements. *From Hall, C. W., and Hedrick, T. I., Drying of Milk and Milk Products, 2nd edition (1971). Used with permission of Avi Publishing Co., Westport, CT.*

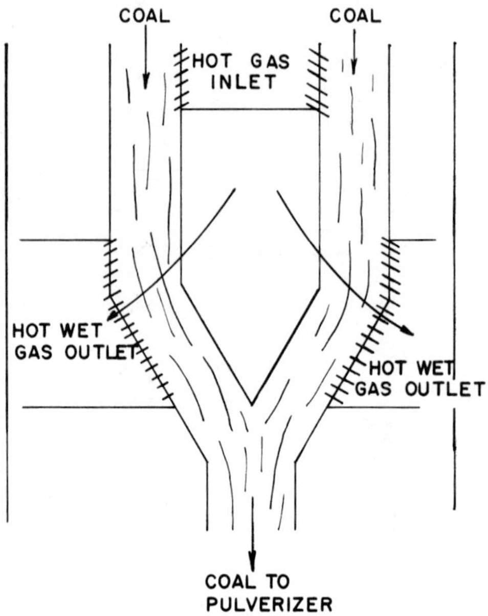

Figure V-9. Vertical column dryer.

Vertical Tube Evaporator A device having tubes for heating in a vertical position for removing water from liquids. In a short tube batch unit, the heating medium, usually steam, is contained in tubes which heats the product in a vessel, usually closed and under vacuum (Fig. V-10). In a continuous operating vertical tube evaporator, long tubes are used, with the product inside the tubes and the heating medium external to the tubes. Also see *Long Tube Vertical Evaporator*.

VHF, Very High Frequency Radio frequency of 30,000 to 300,000 kilocycles/sec. Also see *HFC; Radio Frequency; UHF*.

Vibration Drying Also see *Acoustic Drying; Resonating Drying; Transducer; Ultrasonic*.

Virtual Temperature The temperature required to maintain a constant density of an air-water vapor mixture when water vapor is removed. The virtual temperature of air, t_V, with water is greater than the existing temperature, t:

$$t_V = \frac{t}{1 - 0.38\ (vp)/P}$$

where

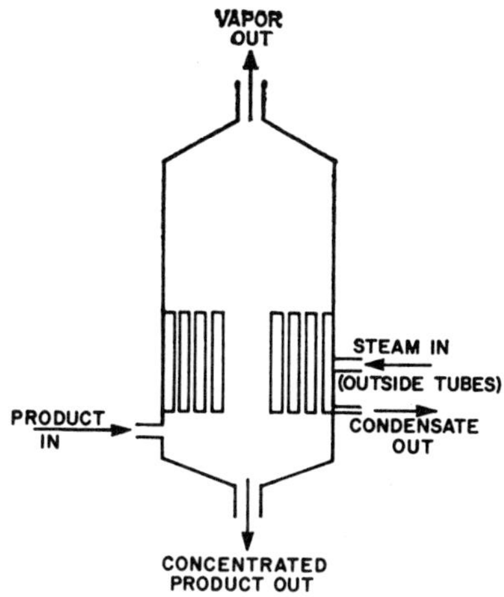

Figure V-10. Vertical tube (short) evaporator. *From Hall, C. W., and Hedrick, T. I., Drying of Milk and Milk Products, 2nd edition (1971). Used with permission of Avi Publishing Co., Westport, CT.*

t_V = virtual temperature
t = existing temperature
vp = vapor pressure of water vapor
P = total pressure of the air

Viscosity A measure of the resistance of fluids (gases or liquids) to flow. The coefficient of viscosity, η, is the shearing stress divided by the rate of shear. The coefficient of viscosity decreases with an increase in temperature for liquids and increases with an increase in temperature for gases (Table V-2). The common unit of viscosity is the poise in cgs units, of 1 dyne sec/cm^2, with centipoise equal to .01 poise, which is absolute viscosity. To convert centipoises to lb-mass per ft sec or poundal sec/ft^2, also called Reyns, multiply by 1/1488; to lb-mass/ft hr, multiply by 2.42; to lb-force sec/ft^2, multiply by 0.209×10^{-4}. The relative viscosity is the ratio of the viscosity of a fluid to that of water at 20°C (68°F). The specific viscosity is the ratio of the viscosity of a fluid to the viscosity of a standard fluid, usually water, at the same temperature. The kinematic viscosity is the viscosity μ in lb-mass/ft sec divided by the density of the fluid, ρ, lb-mass/ft^3. The cgs unit is the stoke, cm^2/sec of kinematic viscosity. Also see *Absolute Viscosity; Conversion Factors; Kinematic Viscosity; Poise*.

Vitamins Complex organic substances considered necessary for life in small quantities. There is considerable difference in source, availability, physical and chemical properties, effect of heat, and

Table V-2
Viscosity of Common Fluids, Centipoise

Material	Temperature, °C	Centipoise
Air	0	173×10^{-4}
	65	201×10^{-4}
	100	220×10^{-4}
Creamed soup		100 to 800
Evaporated milk		20 to 80
Gasoline	0	0.50
	65	0.25
Milk	30	1.64
Molasses	20	100,000
Oil, SAE 10	20	100
	65	10
Skim milk	30	1.33
	80	0.57
Water vapor	0	1.79
	20	1.00
	100	0.284

Conversion factor: Centipoise x 10^{-3} = Pa•s

physiological function served by various vitamins, as summarized in Table V-3, as related to drying and dehydration.

Void Space The space between particles in a bulk of stored product, usually expressed as percent of total volume (Table V-4). Also see *Porosity; Specific Gravity*.

Volume The space occupied by the body; the space enclosed by the surfaces of a body; in dimensional terms of length, L, the volume is L^3.

A standard volume is occupied by one gram molecular weight of a gas at 0°C (32°F) at 1 atmosphere pressure. The critical volume is occupied by one gram of liquid or gas at its critical temperature and critical pressure.

Volumetric Efficiency The ratio of volume of a fluid discharged by a propelling or pumping device, such as a fan, blower, compressor, propeller, and piston in cylinder, divided by the volume displaced by the moving part during the same time.

Volumetric Shrinkage As drying occurs the volume of a product decreases. From clay the volume decreases uniformly with moisture removal until a moisture content of about 10% is reached. The volumetric shrinkage of clay, based on percent of the final dry volume, is:

$$= \frac{S + M - V_d}{V_d}$$

where

S = apparent solid volume of clay as a plastic mixture
M = original moisture content, % d.b.
V_d = dry volume of 100 g of clay

Also see *Clay Drying; Shrinkage; Shrinkage Ratio.*

von Karman Number A dimensionless parameter, Ka, which is a measure of stream turbulence:

$$Ka = \sqrt{\frac{v_x^2 + v_y^2 + v_z^2}{3\, w_o^2}}$$

where

v_x = stream velocity in x-direction
v_y = stream velocity in y-direction
v_z = stream velocity in z-direction
w_o = flow velocity

Table V-3
Vitamins: Some Characteristics Related to Drying

Vitamin	Major Source	Temperature, °C	Solubility and Stability
Vit. A (Carotene)	green, leafy and yellow vegetables, whole milk, egg yolk, liver, cod liver oil	mp = 62-64; bp = 120 at 0.65 atm	fat soluble; not destroyed by cooking; vitamin made in body from carotene
Vit. B_1 (Thiamine)	whole cereal grains, yeast, legumes, peas, beans, pork	mp = 250	water soluble; destroyed by heat or in water; when dry, stable at 100°C; very hygroscopic
Vit. B_2 (Riboflavin) or Vit. G	whole milk, yeast, liver	dec at 280	water soluble; not easily destroyed by heating; solutions are heat stable
Vit. B_6 (Pyridoxine)	meats, vegetables, whole grains	mp = 205	water soluble; may be heated to 120° for 30 min
Vit. B_{12}	muscle meats, eggs, dairy products	darkens at 210 did not melt at 300	water soluble; stable in heat and light; stable to air when dry
Vit. C	citrus and tomato juices, cantaloupe, paprika	mp = 192 some dec	water soluble; readily destroyed by heat
Vit. D	milk, eggs, cod liver oil, salmon, tuna fish	mp = 115	fat soluble
Vit. E	seeds, liver, eggs, whole grain, lettuce	bp = 200	fat soluble; stable in heat; slowly oxidized in atmosphere
Vit. G	see Vit. B_2		
Vit. K	green vegetables, sunflower seeds, fatty substances in rice and corn	dec above 120	fat soluble; stable to air and moisture

Table V-3--Continued

Vitamin	Major Source	Temperature, °C	Solubility and Stability
Choline	egg yolks, liver, grains, legumes		water soluble; heat stable in dilute water solution; melts on heating
Folic acid Vit. M	widely distributed in plants and animals	chars at 250	water soluble; very unstable to heat; deteriorates in sunlight
Niacin nicotinic acid	liver, yeasts, salmon, poultry, grains, legumes	mp = 236	water soluble; stable in air and light; required by man
Vit. H biotin	legumes, meats, vegetables	mp = 232; some dec	water soluble

mp = melting point, °C; bp = boiling point, °C; dec = decomposition, °C.
Temperature effects based on Altman, P. L., and Dittmer, D. S. 1964. Biological Data Book. Federation of American Societies for Experimental Biology, Washington, DC. 633 p.

Table V-4
Voids in Selected Bulk Materials

Product	Moisture Content, %, w.b.	Voids in Bulk, %
Barley	9.7-10.7	39.5-57.6
Corn, shelled	9.0	40
Flaxseed	5.8	34.6
Oats	9.4-10.3	47.6-55.5
Rice	11.9-12.4	46.5-50.4
Rye	9.7	41.2
Sand, coarse	dry	28
Sand, screened	dry	40-45
Sorghum	9.5-9.9	26.8-37.0
Soybeans	6.9-7.0	33.8-36.1
Stone aggregates	dry	25-55
Wheat	9.8	40.1-42.6
Wheat	9.8	39.6

W

Wagon Dryer   A wagon normally used for transporting products designed and equipped for drying with heated air. Wagon dryers are used for forages and nuts. Product handling is minimized with a wagon dryer. It may be necessary to place a cover over the top of the wagon, if an upflow system is used, to prevent excessive condensation on the product at the outlet. A cover may be placed over the top of the wagon to form a plenum to force the heated air downward (Fig. W-1).

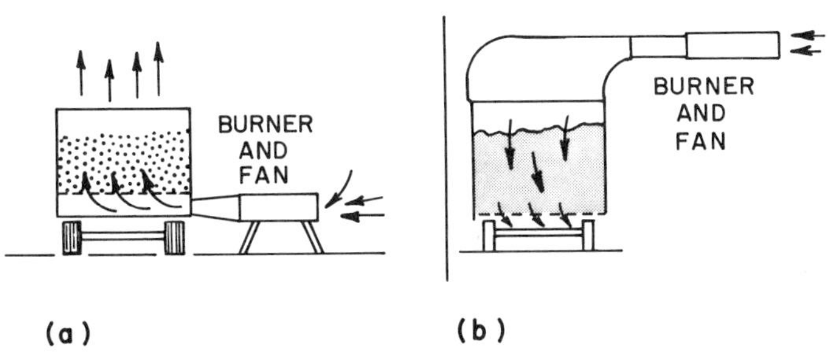

Figure W-1. Wagon dryer. (a) upward flow. (b) downward flow.

Wash and Wear   See *Drip Dry*.

Waste Materials Drying   A wide variety of waste material can be made into useful by-products by drying. Garbage, waste from meat and fish processing, animal and plant wastes can be dried in a wide variety of dryers. The direct-fired horizontal rotary dryer is commonly used where incoming temperatures of 537.8°C (1000°F) are acceptable. Size reduction often follows drying. Also see *Garbage; Paunch Drying; Manure Drying*.

Water A colorless, odorless, tasteless liquid, formula of H_2O, a freezing point of 0°C (32°F), and a boiling point of 100°C (212°F) at 1 atmosphere pressure. At 20°C (68°F) water has a viscosity of 1 centipoise; critical temperature of 374°C (705.2°F); critical pressure of 217.7 atmospheres; critical density, 0.4 gram/cm^3; specific heat, 1.0 at 15°C (59°F); dielectric constant, 81 at 18°C (64.4°F); thermal conductivity of 0.32 Btu/ft^2°F-hr/ft at 0°C (32°F), 0.36 at 37.8°C (100°F), and 0.39 at 93.3°C (200°F). A water molecule has a diameter of about 2.9 Å. Also see *Adsorptionally Bound Water; Bound Water; Capillary Bound Water; Moisture; Moisture Content; Osmotically-Bound Water*.

Water Activity or Water Availability A means of relating the quantity of water present in a food product related to its preservation by drying. The level of water below which the food is preserved, in which the fresh tissues or products have a water activity equal to 1 (Table W-1). Generally, dry foods have a water activity of approximately 0.2 or less. Also, a measure of the activity, a, of water in a dehydrated product. The water activity decreases as the amount of dehydration increases. Growth of microorganisms, activity of enzymes, and browning generally decrease as the water activity decreases. A representation of water activity, a, is:

$$\frac{a}{m\ (1 - a)} = \frac{1}{m_1\ c} + \frac{c - 1}{m_1\ c}\ a$$

where

m = moisture content, dry basis
m_1 = monolayer value
c = constant

Also see *Intermediate Moisture Foods*.

Water Column The vertical distance which a water column is moved, above or below atmospheric pressure, as a result of change in air pressure, such as with a fan in a drying system. The pressure in centimeters or inches of water column is often used for small values in contrast to inches of centimeters of mercury for large values. One inch water column is equivalent to 0.036 psi and 0.2488 kPa. Also see *Conversion Factors*.

Water of Compaction Water furnished by destruction of pore space owing to compression of sediments.

Water of Dilation Water in excess of water of saturation held by sedimentary material in an expanded state.

Table W-1
Water Activity and Moisture Content

Products	Moisture Content, %, w.b.	Water Activity
Pastes of tropical fruits		
banana, cashew, guava, quinces		
hard	18	0.70
soft (jelly)	30	0.80
Dehydrated figs		
hard	15	0.62
soft	25	0.75
Dehydrated banana	35	0.80
Honey	17	0.54
Corn syrup	17	0.62
Molasses	26	0.76
Orange juice concentrate	32	0.76

Source: Quast, D. G., and Teixeiro Neto, R. O. 1976. Food Technology 39(5):98-105.

Water of Hydration   Water which combines with certain compounds other than oxides, particularly salts, to form hydrates. The hydrate of calcium chloride is $CaCl_2 \cdot 6H_2O$. When heated the water is driven off. Also see *Hydrated*.

Water of Saturation   Water which completely fills the interstices of rock formation or of earth whose particles are in contact (i.e., not expanded).

Water Properties   The properties of water in different states (solid, liquid, and gas), at different temperatures and pressures, and in various materials provide a basis of developing and designing drying systems (Table W-2).

Water Vapor   Gaseous state of water in the atmosphere with the amount present designated by absolute humidity, relative humidity, or dew point. Also see *Vapor*.

Watt   A unit of power, W, in the cgs system and SI system which is a rate of energy output or consumption of 1 joule/sec or 1/746 horsepower. The power consumed over a period of time is measured by a watt-hour meter, which integrates the watts x hours.

Watt-Hour   See *Kilowatt-Hour*.

Wave   A means of representing certain phenomena in which there is a characteristic shape, amplitude, period, etc., such as a sine wave,

Table W-2
Water Properties

Density		lb/ft^3	kg/m^3
4°C	40°F	62.4	1000
20	68	62.3	998
Specific heat (heat capacity)		Btu/lb °F	kJ/kgK
20°C	68°F	1.0	4.19
Thermal conductivity		Btu/ft^2°F hr/ft	W/mK
0°C	32°F	0.32	0.561
20	68	0.34	0.596
100	212	0.36	0.631
200	392	0.39	0.684
Vaporization, heat of		Btu/lb	kJ/kg
100°C	212°F at 1 atm.	970	2257.0
Viscosity, dynamic		centipoise	kg/ms
20°C	68°F	1.0	0.00101

Conversion factors: *1.0 Btu/ft hr °F = 1.7296 W/mK*
1.0 Btu/lb °F = 4184 J/kgK = 4.184 kJ/kgK
1.0 lb/ft^3 *= 16.0185 kg/m*^{3}

representing an alternating current. Also see *Periodic; Range*.

W.B. An abbreviation for wet basis moisture content and wet-bulb temperature. Usually written as w.b. Also see *D.B. (d.b.); Moisture Content*.

W.C. See *Water Column*.

Web Dryer See *Continuous Web Dryer*.

Weber Number A dimensionless parameter, We, of surface tension which is a measure of the relation between surface tension and gravity forces:

$$We = \frac{\sigma}{(\gamma' - \gamma'')\ell^2}$$

where

σ = coefficient of surface tension
$\gamma' - \gamma''$ = difference in specific weight
ℓ = length

Weight The property of a body to exert a force as a result of the earth's gravity which depends on the quantity of matter and its distance from the center of the earth or other gravitational body.

Weight is used in contrast to mass which is invariant for a particular quantity of matter.

Weighting A procedure of empirically adjusting the values of certain measurements so proper representation of these values will occur in relation to other values or in mathematical or statistical calculations. The weighting factor, wf, for any individual determination is inversely proportional to the square of the uncertainty of that determination:

$$wf = \frac{1}{\text{uncertainty}^2}$$

then

$$q = \frac{wf_1 q_1 + wf_2 q_2 + \dots}{wf_1 + wf_2 + \dots}$$

for values obtained q_1, q_2.

Well-Dried A term applied to nut meat, means "that portion of the kernel is firm and crisp, not pliable or leathery."

Wet Basis See *Dry Basis; Moisture Content.*

Wet-Bulb Depression The difference between the dry-bulb and the wet-bulb temperature of a body of air. Also see *Wet-Bulb Temperature.*

Wet-Bulb Drying Period The surface of the drying product assumes the wet-bulb temperature corresponding to the temperature, humidity, and quantity of drying air. If the quantity and condition of the drying gas remains constant the wet-bulb temperature will also be constant. In practice, the wet-bulb drying period is very similar to the constant rate drying period.

Wet-Bulb Temperature The temperature indicated by a thermometer of which the bulb is kept wet by a thin film of water and which is exposed to a current of air of at least 4.56 m/s (15 ft/sec) is known as the psychrometric wet-bulb temperature. The knowledge of weather data (Fig. W-2) wet-bulb temperature is particularly useful for drying with unheated air. The temperature reached by moist air and water if the air is adiabatically saturated by the evaporated water is the thermodynamic wet-bulb, nearly equal to the psychrometric wet-bulb for water.

Wetting Agent An organic substance that causes a liquid such as water to reduce its surface tension so that it will spread evenly on a surface of the solid which in turn will usually increase the absorption, or will aid in removing dirt or impurities from a surface. Also see *Surface Active Agent; Surfactants.*

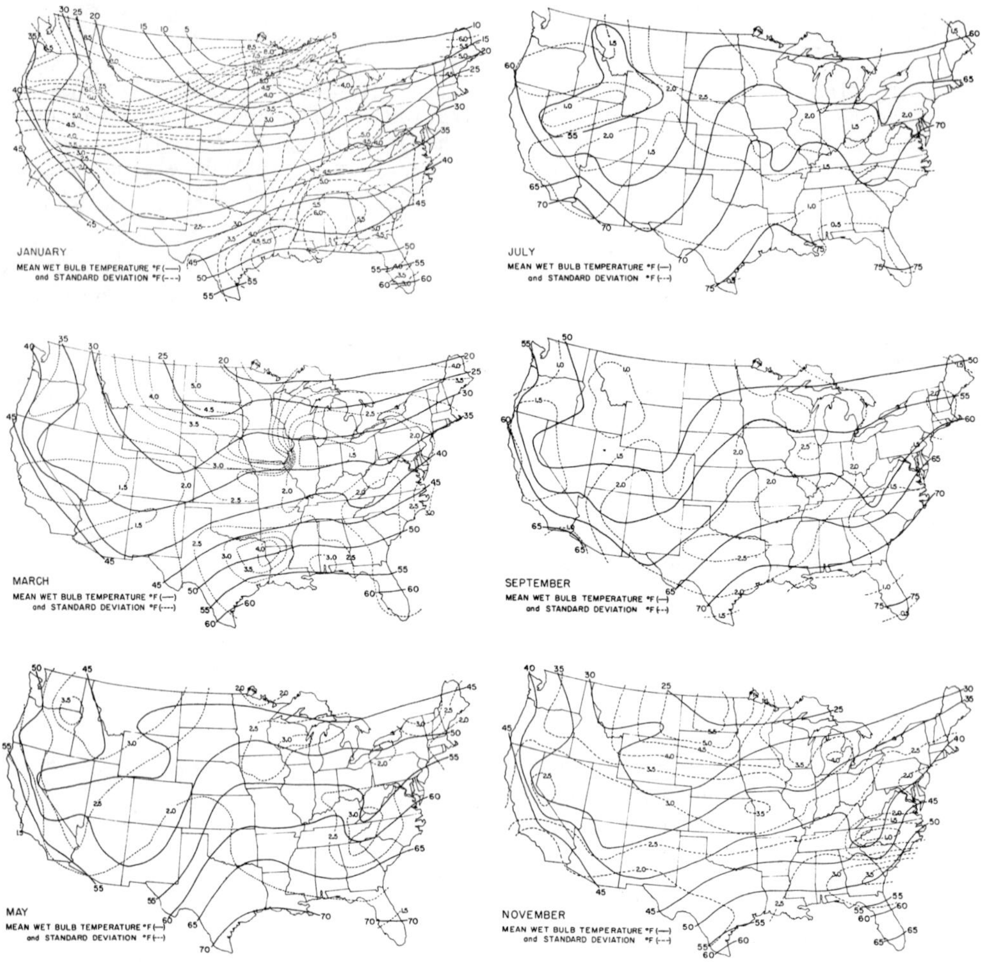

Figure W-2. Wet bulb temperature map. *From ASAE (1976). Agricultural Engineers Yearbook, St. Joseph, MI.*

Wheat Drying See *Cereal Grain Drying.*

Whey The liquid portion of milk separated from the curd formed in making cheese. Whey is often dried to provide a source of soluble minerals and to reduce the pollution that would otherwise result from sewer or dilution disposal of the whey.

Whirl Dryer A name sometimes given to a centrifugal dryer, particularly in the wood industry.

Whizzer A term applied to centrifugal devices for drying (hydro-extractors), especially those used for grains.

Wilting The decrease in strength and rigidity of plants, particularly leaves, due to lack of moisture as a result of droughty conditions or to cutting of the stem or other part of the plant. The moisture load on the dryer for grasses and hays and forages is often reduced by wilting the plants and field drying before drying on the dryer.

Wind Chill Factor An index used in meteorology which represents the cooling power of the wind expressed in terms of the equivalent temperature without the wind. Thus, with an ambient air temperature of -6.6°C (20°F), a wind speed of 4.47 m/s (10 mph) gives an equivalent temperature of -15.6°C (4°F), from the standpoint of the normal skin or body sensitivity (Fig. W-3).

Wind Pressure and Air Velocity The quantity of air which moves through a granular product in storage by natural ventilation depends upon the wind velocity, direction of the wind, resistance to air flow offered by the material, and difference in temperature of air in storage. The resistance of air flow depends upon the material in the storage and the length of the path which the air must pass in going through the bin.

The relationship between the wind speed and the pressure exerted is given in the equation:

$$p = 0.00256\ \omega^2$$

where

p = pressure normal to surface, lb/ft^2
ω = wind speed, mph

Also see *Cowls; Natural Ventilation.*

Wood Combustion The burning of the solid components of wood or bark which is preceded first by drying and second by distillation of other volatiles, the latter beginning at a temperature of about 275°C (527°F). The high heating (calorific) values of hardwoods are 18,850 to 20,446 kJ/kg (8100-8900 Btu/lb) and of softwoods 20,469 to 22,795 kJ/kg (8800-9800 Btu/lb).

Wood Chip Drying See *Particleboard Furnish.*

Wood Drying See *Bark Drying; Lumber; Particleboard Furnish; Vapor Drying; Veneer Drying.*

Wool A portion of the processing of wool consists of drying. In the cleaning process, wool first passes through a duster which opens up the fleeces to remove some of the loose dirt. Then the wool is handled gently through a series of tubs, called a scouring train,

WIND SPEED MPH	AIR TEMPERATURE, °F																	
	45	40	35	30	25	20	15	10	5	0	-5	-10	-15	-20	-25	-30	-35	-40
	CHILL INDEX - equivalent temp. in cooling effect on exposed flesh																	
5	43	37	32	27	22	16	11	6	0	-5	-10	-15	-21	-26	-31	-36	-42	-47
10	34	28	22	16	10	3	-3	-9	-15	-22	-27	-34	-40	-46	-52	-56	-64	-71
15	29	23	16	9	2	-5	-11	-18	-25	-31	-38	-45	-51	-58	-65	-72	-78	-85
20	26	19	12	4	-3	-10	-17	-24	-31	-39	-46	-53	-60	-67	-74	-81	-88	-95
	COLD		VERY		BITTER						EXTREME							
25	23	16	8	1	-7	-15	-22	-29	-36	-44	-51	-59	-66	-74	-81	-88	-96	
			COLD		COLD						COLD							
30	21	13	6	-2	-10	-18	-25	-33	-41	-49	-56	-64	-71	-79	-86	-93		
35	20	12	4	-4	-12	-20	-27	-35	-43	-52	-58	-67	-74	-82	-89	-97		
40	19	11	3	-5	-13	-21	-29	-37	-45	-54	-60	-69	-76	-84	-92			

Wind speeds above 40 mph have little additional chilling effect.

Enter top of table at outside air temperature and left side of table at wind speed. The number at the intersection is the WIND CHILL INDEX.

Figure W-3. Wind chill factor.

containing a detergent solution, followed by rinsing. The washed wool passes through squeeze rolls and conveyed to a dryer where it is dried to 12% moisture content. The dried wool is put into 227 kg (500 lb) bales for later use.

Wool Drying See *Sheep (Wool) Drying*.

WVTR An abbreviation for water vapor transmission rate, the movement of moisture through products during freeze-drying or through a structural wall.

XYZ

Yard Brown Stain A chemical stain that may develop during air seasoning or storage of lumber.

Yield The percent yield is the weight of output of a process divided by the weight of input, times 100%. Yield may be expressed in weight produced from a given process. The term yield may be applied to a step, series of steps, or the entire process. Also see *Overall Yield*.

Zero, Absolute See *Absolute Zero*.

Zero Order Reaction A reaction which is not dependent on the pressure but only on the surface area.

Zeroth Law of Thermodynamics In an isolated system, if each of two bodies is in thermal equilibrium with a third body, the two bodies are also in thermal equilibrium.

REFERENCES

References to books on drying are listed under Books. Other specific references are included with the definition. The following general references were of considerable value in assembling the definitions and descriptions.

Altman, P. L., and Dittmer, D. S. 1964. *Biological Data Book*. Federation of American Societies for Experimental Biology, Washington, DC. 633 p.

American Fabrics Magazine. 1960. *Encyclopedia of Textiles*. Prentice-Hall, Englewood, NJ. 702 p.

American Heritage Dictionary of the English Language. 1969. Houghton Mifflin Company, Boston, MA. 1550 p.

Bender, Arnold E. 1965 (2nd Ed.). *Dictionary of Nutrition and Food Technology*. Butterworths, London.

Encyclopedia Britannica. 1961. *Encyclopedia Britannica*. Chicago, IL. 24 volumes.

Gaynor, Frank. 1959. *Concise Dictionary of Science*. Philosophical Library, New York, NY. 546 p.

Gray, Peter. 1970. *Encyclopedia of Biological Sciences*. The Van Nostrand Reinhold Co., New York, NY. 1027 p.

Harper Encyclopedia of Science. 1967. Edited by James R. Newman. Harper and Row, New York, NY. 1379 p.

James, Glenn, and James, Robert C. 1968. *Mathematics Dictionary*. D. Van Nostrand Company, New York, NY. 517 p.

Kirk, Raymond E., Othmer, D., et al. 1963-1970. *Encyclopedia of Chemical Technology*. Interscience Publishers, New York, NY. 22 volumes.

Linton, George E. 1963. *The Modern Textile Dictionary*. Duell, Sloan, and Pearce, New York, NY. 1077 p.

McGraw-Hill Encyclopedia of Science and Technology. 1971. McGraw-Hill Book Company, New York, NY. 15 volumes.

Mandel, Siegfried. 1969. *Dictionary of Science*. Dell Publishing Company, New York, NY. 407 p.

Speck, G. E., and Jaffe, B. 1956. *A Compact Science Dictionary*. Fawcett Publications, Greenwich, CT. 272 p.

Thewlis, James. 1961-1964. *Encyclopedic Dictionary of Physics*. Macmillan, New York, NY. 9 volumes.

Thorpe, J. F., and Whitely, M. A. 1947-1954. *Thorpe's Dictionary of Applied Chemistry*. Longman, Green & Company, London and New York. 12 volumes.

Van Oss, J. F. 1972. *Chemical Technology: An Encyclopedic Treatment*. Barnes and Noble, New York, NY. 6 volumes.

Webster's New Collegiate Dictionary. 1975. G. & C. Merrian Co., Springfield, MA. 1536 p.